Spectroscopic studies can reveal a wealth of information about the rotational and vibrational behaviour of the constituent molecules of gases and liquids. This book reviews the fundamental concepts and important models which underpin such studies, dealing in particular with the phenomenon of spectral collapse, which accompanies the transition from rare gas to dense liquid.

Following a description of the relevant basic physics, including a discussion of collisions between rotating molecules, the theory of spectral collapse is covered in terms of both quantum mechanical and classical models. A critical comparison of gas-like and solid-like models, combined with theoretical descriptions of rotational and vibrational relaxation, is interwoven with analysis of experimental results. These include data from optical, NMR, ESR, and acoustic investigations. *Ab initio* semiclassical calculations and phenomenological fitting laws are used to estimate the rotational transition rates, and the cross-section of rotational energy relaxation. The book concludes with a discussion of the latest theories describing the mechanism of rotational diffusion in liquid solutions.

This comprehensive review of theoretical models and techniques will be invaluable to theorists and experimentalists in the fields of infrared and Raman spectroscopy, nuclear magnetic resonance, electron spin resonance and flame thermometry. It will also be useful to graduate students of molecular dynamics and spectroscopy.

Spectroscopy of Molecular Rotation in Gases and Liquids

SPECTROSCOPY OF MOLECULAR ROTATION IN GASES AND LIQUIDS

A. I. BURSHTEIN
Weizmann Institute of Science, Israel

S. I. TEMKIN
Institute of Chemical Kinetics and Combustion, Russian Academy of Sciences

CAMBRIDGE
UNIVERSITY PRESS

Published by the Press Syndicate of the University of Cambridge
The Pitt Building, Trumpington Street, Cambridge CB2 1RP
40 West 20th Street, New York, NY 10011-4211, USA
10 Stamford Road, Oakleigh, Melbourne 3166, Australia

First published 1994

Printed in Great Britain at the University Press, Cambridge

A catalogue record of this book is available from the British Library

Library of Congress cataloguing in publication data

Burshtein, A. I. (Anatolii Izrailevich)
Spectroscopy of molecular rotation in gases and liquids/
A. I. Burshtein and S. I. Temkin.
p. cm.
Includes bibliographical references and index.
ISBN 0 521 45465 4
1. Molecular spectroscopy. 2. Molecular rotation. 3. Gases.
4. Liquids. I. Temkin, S. I. (Semen Isaakovich) II. Title.
QD96.M65B86 1994
539′.6–dc20 93–36766 CIP

ISBN 0 521 45465 4 hardback

TAG

Contents

0
Introduction

As the density of a gas increases, free rotation of the molecules is gradually transformed into rotational diffusion of the molecular orientation. After 'unfreezing', rotational motion in molecular crystals also transforms into rotational diffusion. Although a phenomenological description of rotational diffusion with the Debye theory [1] is universal, the gas-like and solid-like mechanisms are different in essence. In a dense gas the change of molecular orientation results from a sequence of short free rotations interrupted by collisions [2]. In contrast, reorientation in solids results from jumps between various directions defined by a crystal structure, and in these orientational 'sites' libration occurs during intervals between jumps. We consider these mechanisms to be competing models of molecular rotation in liquids. The only way to discriminate between them is to compare the theory with experiment, which is mainly spectroscopic.

Line-shape analysis of the absorption or scattering spectra supplies us with normalized contours $G_\ell(\omega)$ which are the spectra of orientational correlation functions $K_\ell = \langle P_\ell\ [\boldsymbol{u}(t) \cdot \boldsymbol{u}(0)]\rangle$. The full set of averaged Legendre polynomials unambiguously defines the orientational relaxation of a linear or spherical rotator whose molecular axis is directed along the unit vector $\boldsymbol{u}(t)$. Unfortunately, only the lowest few K_ℓ are available from spectroscopic investigation. The infrared (IR) rotovibrational spectroscopy of polar molecules gives us $G_1(\omega - \omega_v)$ which is composed of some rotational branches around vibrational frequency ω_v. In the case of a linear molecule such as CO the P- and R-branches in the rare gas spectrum transform into a single Q-branch in the liquid phase which is forbidden for the free rotator (Fig. 0.1). This effect, known as a rotational structure collapse [3, 4], is the spectral manifestation of the transition from free rotation to rotational diffusion. The collapsed line is progressively narrowed as the density increases. The anisotropic

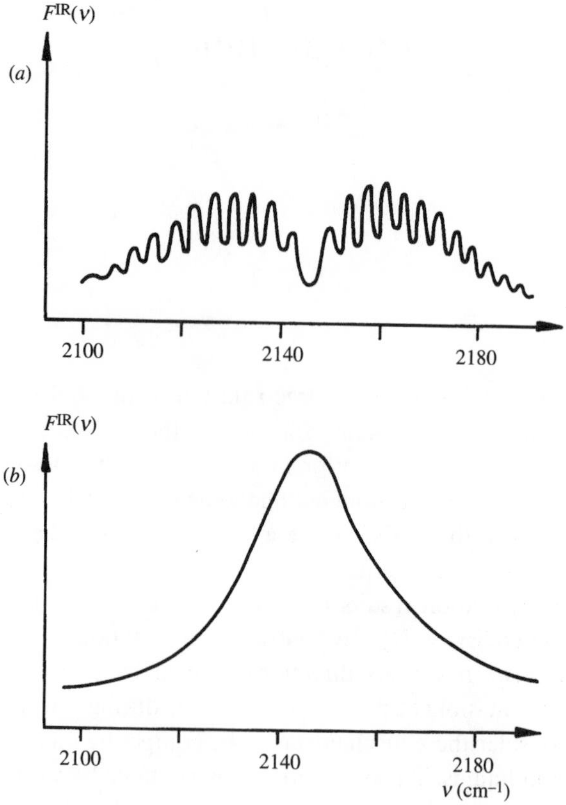

Fig. 0.1. Infrared absorption spectrum of CO in gas (*a*) and liquid (*b*) at 80 K [6].

Raman scattering spectrum $G_2(\omega - \omega_v)$ undergoes a spectral collapse of the same kind. The spectrum of the gaseous linear rotator is composed of three branches whose rotational structure becomes unresolved at relatively low pressure (Fig. 0.2). With further increase of pressure it is reduced to a single quasi-Lorentzian Q-branch which, in the liquid phase, is much narrower than the spectrum before collapse (Fig. 0.3). Similar information is available from the pure rotational spectra found from the Rayleigh scattering contour $G_2(\omega)$ or from the light absorption coefficient in the far infrared (FIR) region which is proportional to $\omega G_1(\omega)$. The dielectric data commonly used to determine Debye's relaxation time $\tau_D = \int_0^\infty K_1(t)\,dt = \pi G_1(0)$ supplement the spectroscopic information.

Since the mechanics of a rotator is set by two canonical variables $\boldsymbol{u}$

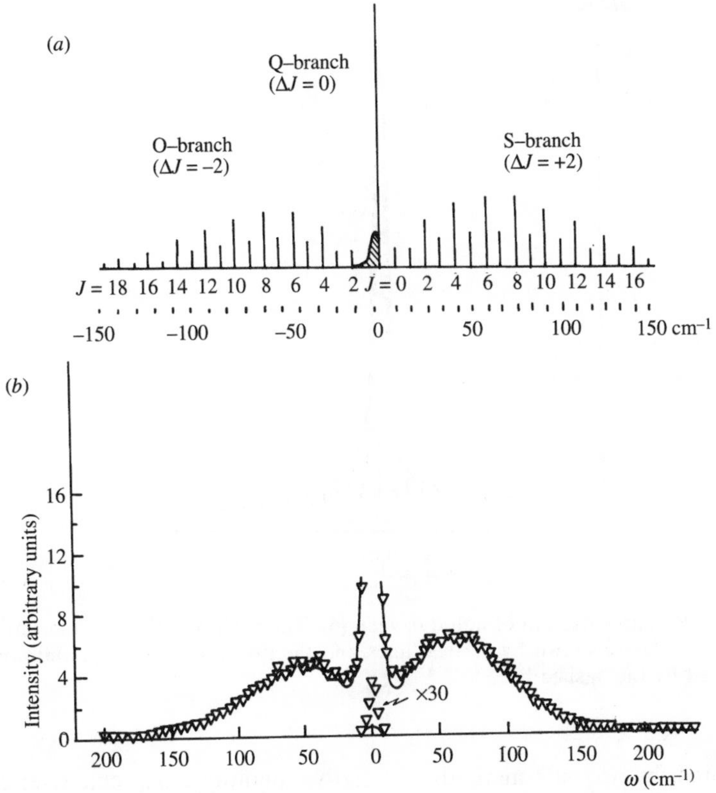

Fig. 0.2. (*a*) The 'comb' spectrum of N_2 considered as a quantum rotator. The envelope of the rotational structure of the Q-branch slightly split by the rotovibrational interaction is shaded. (*b*) The depolarized rotovibrational spectrum of N_2 at corpuscular density $n = 92$ amagat, $T = 296$ K and pressure $p = 100$ atm. The central peak, reported in a reduced ($\times 30$) scale is due to a polarized component [5]: (∇) experimental; (—) best fit.

(axis) and $\boldsymbol{J}$ (angular momentum), the rotational relaxation presented by $K_J(t) = \langle \boldsymbol{J}(t)\boldsymbol{J}(0)\rangle$ and its spectrum $g(\omega)$ must be studied together with $K_\ell(t)$ and $G_\ell(\omega)$. The periphery of $G_\ell(\omega)$ gives direct but difficult to obtain information about the far wings of $g(\omega)$. More frequently NMR data are used to find $\tau_J = \int_0^\infty K_J(t)\,\mathrm{d}t = \pi g(0)$ from both longitudinal and transverse relaxation times determined by spin-rotation interaction [7]. The relaxation of the rotational energy $E \propto J^2$ is also of great importance. Its kinetics is characterized by the correlation function $K_E(t) = \langle J^2(t)J^2(0)\rangle$ but only $\tau_E = \int_0^\infty K_E(t)\,\mathrm{d}t = \pi g_E(0)$ has been

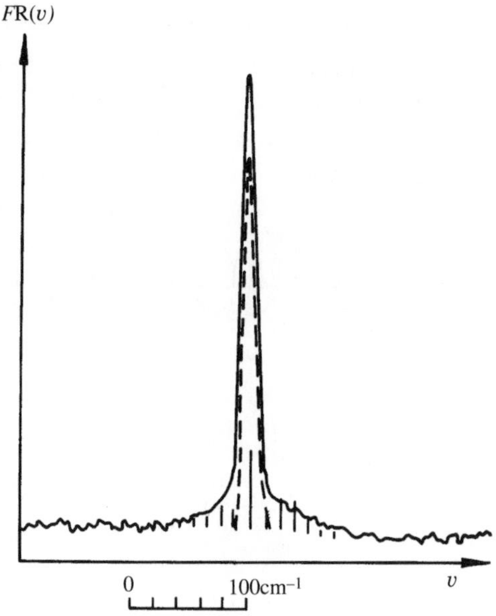

Fig. 0.3. Raman spectrum of liquid oxygen [6]. The positions of the free rotator's j-components are shown by vertical lines and the isotropic scattering contour is presented by the dashed line.

measured, by acoustic methods. A relative quantity τ_E/τ_J characterizes the strength and adiabaticity of collisions. In principle the far wings of $g_E(\omega)$ may be seen in the periphery of the isotropic Raman spectrum $G_0(\omega) = (1/\pi)\int_0^\infty K_0(t)\, dt$, when broadened mainly by rotational energy relaxation. Unfortunately such work has not yet been reported.

Since the information about all $G_\ell(\omega)$ and especially about $g(\omega)$ and $g_E(\omega)$ is never complete, the importance of models capable of predicting them is greatly enhanced. The more developed gas-like models consider J to be a classical variable. They are appropriate at relatively large pressures when the rotational structure of the branches is smoothed. One of the best models is Gordon's impact theory of rotational and orientational relaxation [2, 8]. The impact theory implies that collisions are instantaneous and retard the molecular rotation by changing both the molecular orientation and the angular momentum. If collisions are adiabatic they do not change J^2 but simply turn $\boldsymbol{J}$ and $\boldsymbol{n}$ (Gordon's m-diffusion model). However, the overwhelming majority of molecules (excluding hydrogen) undergo non-adiabatic collisions which change both the direction and magnitude of $\boldsymbol{J}$ but do not change a molecule's orientation (Gordon's 'J-diffusion model'). The change in angular velocity, proportional to J, results in frequency exchange in each spectral branch, while the change in direction of rotation causes an exchange between the branches. To the extent that the collisions conserve the phase of rotation, the related radiation is only frequency modulated. Since the modulation is not accompanied by rotational dephasing, it must result in spectral narrowing after collapse. The averaging and narrowing of absorption and scattering spectra in the J-diffusion model amount to a direct analogue of the motional narrowing in NMR discovered earlier [9]. The residual width of the absorption spectra is obviously the rate of the Debye 'rotational diffusion', which decreases as the collision frequency increases.

The chapters in this book are arranged in accordance with both logical expediency and the authors' intention to complicate the problem gradually. We start from rotational (angular momentum) relaxation (Chapter 1), which does not change as radically with pressure as orientational relaxation (Chapter 2). Our main achievement is the utilization (through integral equations of impact theory) of the Keilson–Storer kernel originally proposed to describe the translational velocity distribution after collision [10]. It implies that collisions are purely non-adiabatic but makes it possible to solve exactly the Feller integral equation describing J-diffusion. The 'extended diffusion' model and the 'Langevin model' of rotation widely used in the literature were found to be opposite particular cases for strong and weak collisions respectively. Although in this book we consider primarily impact theories, non-model methods such as memory function formalism (the Mori chain) and perturbation theory (cumulant expansion) are also used to account for finite collision times. These methods as well as very fresh and disputable ideas (like non-Poissonian collisional statistics) are exploited to calculate $K_J(t)$ outside the binary collision approximation, where it is expected to be of alternating sign, as MD simulations of the liquid show.

The perturbation theory presented in Chapter 2 implies that orientational relaxation is slower than rotational relaxation and considers the angular displacement during a free rotation to be a small parameter. Considering $\boldsymbol{J}(t)$ as a random time-dependent perturbation, it describes the orientational relaxation as a molecular response to it. Frequent and small chaotic turns constitute the rotational diffusion which is shown to be an equivalent representation of the process. The turns may proceed via free paths or via sudden jumps from one orientation to another. The phenomenological picture of rotational diffusion is compatible with both models. However, they are not equivalent because the original jumping theory completely ignores rotational relaxation and the very existence of angular momentum [11]. This is a significant demerit since the orientational relaxation time $\tau_{\theta,\ell} = \int_0^\infty K_\ell(t)\,dt = \pi G_\ell(0)$ is usually plotted against τ_J. This plot is a way to check the famous Hubbard relation [12] derived in the lowest order of perturbation theory

$$\tau_{\theta,\ell} \cdot \tau_J = I/\ell\,(\ell+1)\,kT$$

(I is the moment of inertia, T is the temperature). To introduce τ_J into a jump-wise model (at least semiquantitatively) it is necessary to consider the dynamics of above-barrier rotation. As soon as this is done the model becomes self-consistent and also leads to the Hubbard relation in Debye's limit.

Although long-time Debye relaxation proceeds exponentially, short-time deviations are detectable which represent inertial effects (free rotation between collisions) as well as interparticle interaction during collisions. In Debye's limit the spectra have already collapsed and their Lorentzian centre has a width proportional to the rotational diffusion coefficient. In fact this result is model-independent. Only shape analysis of the far wings can discriminate between different models of molecular reorientation and explain the high-frequency pecularities of IR and FIR spectra (like Poley absorption). In the conclusion of Chapter 2 we attract the readers' attention to the solution of the inverse problem which is the extraction of the angular momentum correlation function from optical spectra of liquids.

Chapter 3 is devoted to pressure transformation of the unresolved isotropic Raman scattering spectrum which consists of a single Q-branch much narrower than other branches (shaded in Fig. 0.2(*a*)). Therefore rotational collapse of the Q-branch is accomplished much earlier than that of the IR spectrum as a whole (e.g. in the gas phase). Attention is concentrated on the isotropic Q-branch of N_2, which is significantly narrowed before the broadening produced by weak vibrational dephasing becomes dominant. It is remarkable that isotropic Q-branch collapse is indifferent to orientational relaxation. It is affected solely by rotational energy relaxation. This is an exceptional case of pure frequency modulation similar to the Dicke effect in atomic spectroscopy [13]. The only difference is that the frequency in the Q-branch is quadratic in J whereas in the Doppler contour it is linear in translational velocity v. Consequently the rotational frequency modulation is not Gaussian but is still Markovian and therefore subject to the impact theory. The Keilson–

Storer model used in this theory enables us to describe classically the spectral collapse of the Q-branch for any strength of collisions. The theory generates the canonical relation between the width of the Raman spectrum and the rate of rotational relaxation measured by NMR or acoustic methods. At medium pressures the impact theory overlaps with the non-model perturbation theory which extends the relation to the region where the binary approximation is invalid. The employment of this relation has become a routine procedure which puts in order numerous experimental data from different methods. At low densities it permits us to estimate, roughly, the strength of collisions.

The quantum theory of spectral collapse presented in Chapter 4 aims at even lower gas densities where the Stark or Zeeman multiplets of atomic spectra as well as the rotational structure of all the branches of absorption or Raman spectra are well resolved. The evolution of basic ideas of line broadening and interference (spectral exchange) is reviewed. Adiabatic and non-adiabatic spectral broadening are described in the frame of binary non-Markovian theory and compared with the impact approximation. The conditions for spectral collapse and subsequent narrowing of the spectra are analysed for the simplest examples, which model typical situations in atomic and molecular spectroscopy. Special attention is paid to collapse of the isotropic Raman spectrum. Quantum theory, based on first principles, attempts to predict the j-dependence of the widths of the rotational component as well as the envelope of the unresolved and then collapsed spectrum (Fig. 0.4).

This theory presented in Chapter 4 is applied to the nitrogen–argon mixture in Chapter 5. It is chosen as an example since the inter-molecular interaction in this system is well known and the cross-section of rotational energy relaxation is measured in the bulk and in molecular beam experiments. The semiclassical versions of infinite order sudden and centrifugal sudden approximations are used to calculate the impact operator which is bilinear in the scattering matrix and responsible for both isotropic Q-branch collapse and rotational energy relaxation. The theory fits well all experimental data over a wide range of temperatures, especially when the small adiabatic corrections to the quantum J-diffusion model are taken into account. In the conclusion of Chapter 5 the phenomenological approach to Q-branch collapse, based on the widely used semi-empirical 'fitting laws', is discussed and applied to pure nitrogen.

The orientational relaxation, considered in Chapters 6 and 7, is a more complex problem. The impact theory is the only model capable of tracing the transition from quasi-free rotation in the rare gas to

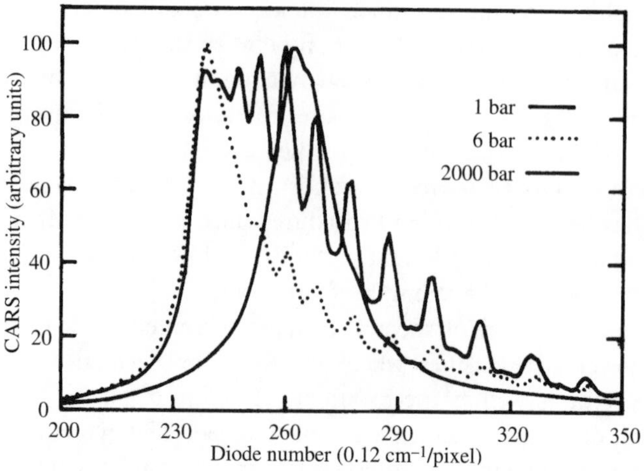

Fig. 0.4. Experimental nitrogen Q-branch of coherent anti-Stokes Raman scattering spectrum (CARS) measured at 700 K and different pressures [14].

rotational diffusion in dense media. In principle it may describe the gradual transformation of multi-branched spectra shown in Fig. 0.1(*a*) and Fig. 0.2(*b*) to a single Q-branch which arises from collapse and turns into a narrow line similar to that shown in Fig. 0.1(*b*) and Fig. 0.3. Classical impact theory presented in Chapter 6 ignores the quantum structure of rotational branches but is the simplest model that one can use to link the opposite cases of gases and liquids. Even in this approximation the problem is so complicated that it may be solved analytically solely in the limits of strong and weak collisions. The difference between them is clearly seen when $\tau_{\theta,\ell}$ is plotted against τ_J. In dense media one obtains the universal Hubbard dependence which is linear in a log plot. At lower pressures it declines from linearity and splits into two branches corresponding to strong and weak collisions which set upper and lower bounds for collisions of arbitrary strength. Rotation is quasi-free when the experimental points are located in this corridor and hindered if they fit the Hubbard relation. The only disadvantage is that the impact theory, being binary, is unlikely to be valid in very dense gas and liquid regions.

The fluctuating cage model presented in Chapter 7 is an alternative. The idea came from comparison of the different kinds of absorption spectra of HCl found in liquid solutions (Fig. 0.5). In SF_6 as a solvent the rotational structure of the infrared absorption spectrum of HCl is well resolved [15, 16], while in liquid He it is not resolved but has

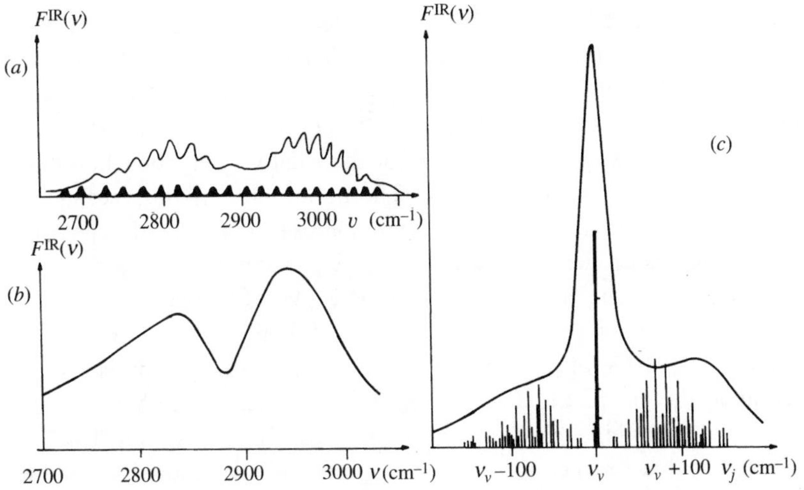

Fig. 0.5. IR absorption spectra of HCl in different liquid solvents: (*a*) in SF_6 [16] (the triangles mark the positions of the rotational components in the resolved spectrum of the rarefied gas); (*b*) in He [15]; (*c*) in CCl_4 (the vertical lines mark the frequencies ν_j and the intensities of the Stark components of the linear rotator spectrum split by the electrical field of the cage)[17].

pronounced P- and R-branches (Fig. 0.5(*a*),(*b*)). Clearly the rotation may be either quantized or classical, but is practically free in these media. This observation might be a physical reason to explore the gaseous impact theory for describing the double-branch spectrum and to expect the appearance of a single Q-branch after (and as a result of) its collapse. However, in CCl_4 and many other solvents the spectrum consists not of two nor of one but of three branches simultaneously (Fig. 0.5(*c*)). Since the existence of such a triplet is absolutely excluded by the impact theory, it indicates the dubious impact origin of a doublet also. It is more likely that a molecule trapped in the cage rotates in the permanent random field created by surrounding solvent particles.

The envelope of the Stark structure of the rotator in a constant orienting field, calculated quantum-mechanically in [17], roughly reproduces the shape of the triplet (Fig. 0.5(*c*)). The appearance of the Q-branch in the linear rotator spectrum indicates that the axis is partially fixed, i.e. some molecules perform librations of small amplitude around the field. Only molecules with high enough rotational energy overcome the barrier created by the field. They rotate with the frequencies observed in the far wings of the P- and R-branches. It is well known that the orienting field in the centre of a spherical cell is zero but increases towards its border [6]. Thermal motion of the molecule between the borders as well as random temporal deformations of the cell modulate the value and direction of the field. In other words, the orienting field appears at any distortion of the spherical symmetry of the rotator's neighbourhood and changes randomly with time [18]. The importance of this mechanism depends on dielectric properties of the solution. One would expect the highly polar hydrogen halides to induce a particularly strong orienting field, which must be larger for solvent molecules of higher polarizability. From this point of view it is even more surprising that in He the axis of HCl is not fixed at all (Fig. 0.5(*b*)) and in SF_6 the rotation is even as free as in the gas phase (Fig. 0.5(*a*)). To explain this paradox, we have developed an original theory of rotation in the fluctuating cell [19]. The oriented field is assumed to change randomly and instantaneously and its successive values form the Markovian chain. The theory is first reviewed and is formalized, as is its impact alternative. This shows that free rotation in a liquid does not consist of free paths as in gas, but occurs in a permanent random field when it is small or changes so rapidly that it practically averages itself. This conclusion is confirmed by evidence that the rotational structure in SF_6 is better resolved at higher temperatures, when thermal motion is faster.

This brief discussion of the physical meaning and mutual correspondence of different models and theories of rotational motion is intended as a guide for those who do not intend to examine the book systematically. Setting forth the material consistently, one cannot avoid certain formalisms peculiar to angular momentum theories. We hope, however, that a detailed commentary will enable readers to form a clear notion of the most important assumptions and results without referring to proofs.

The authors are happy to acknowledge the help and support of those people and institutions who assisted in fulfilling the task of writing this book. The present English edition includes numerous original results obtained by the authors in cooperation with their graduate students and coworkers after publication of the Russian version of the book in 1982. We appreciate very much their collaboration and the important role of the seminar of the Theoretical Chemistry Laboratory in the Institute of Chemical Kinetics and Combustion of the Russian Academy, where most of them work.

Our special thanks go to Professor James R. McConnell (Dublin Institute for Advanced Studies, Ireland), who encouraged us to publish this book in English. We are greatly indebted to the first reader of the manuscript, Professor David H. Waldeck (University of Pittsburgh, U.S.A.) for his creative criticism and outstanding help in shaping the language of the book.

The authors are also thankful to the Weizmann Institute of Science (Israel) for their support, in particular, for the Einstein Centre sponsorship of Dr Temkin's stay in Rehovot in Spring 1992.

1
Rotational relaxation

In a rarefied gas the time-dependent intermolecular interaction is partitioned into a sequence of weakly-correlated separate splashes. In the impact theory which ignores completely their correlation and duration, they are considered to be instantaneous binary collisions. In non-adiabatic theories they result in change of the molecular angular momentum $\boldsymbol{J}$ either in direction or in magnitude. In the frame of such an idealization the intermolecular interaction results in time alteration of $\boldsymbol{J}(t)$ as a purely discontinuous Markovian process: $\boldsymbol{J}$ is preserved during a free path and is instantaneously changed when molecules collide. The new $\boldsymbol{J}$ value depends solely on the previous one (Fig. 1.1). Furthermore the original impact theory is for binary collisions which are distributed according to Poisson's law with average free path time τ_0.

However, as the gas density increases, ternary, etc. collisions gain in significance, and the binary theory should not be valid. In the case of a hard sphere fluid, collisions are always instantaneous but their distribution at high densities may deviate from the Poissonian law. Moreover, in a real gas and liquid interaction with the neighbourhood is no longer divided into isolated collisions, since the collision time τ_c becomes comparable to the mean free path time τ_0. This manifests itself in a non-exponential relaxation of the rotational moment which has to be considered as non-Markovian even in gases. The latter can be described with the memory function formalism which takes into consideration the finite duration of collisions and their correlation. The demerit of this approach is an *a priori* choice of memory functions which makes the theory semi-empirical. However, it may be used to determine the limits of validity for the impact phenomenology and to show that rotational relaxation in a dense medium qualitatively differs from that in a gas. The applicability of the

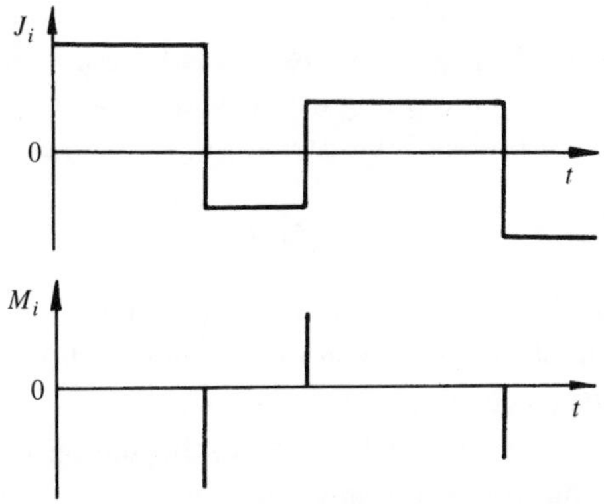

Fig. 1.1. Time-dependence of the components of angular momentum J_i (Markovian process) and the torque M_i (white noise) in the impact approximation.

memory function formalism to the high-density region has to be checked separately.

To make evaluations more definite, we use optical and microwave experimental data, as well as calculations of molecular dynamics of certain simple liquids which usually fit the experiment. Rotation is everywhere considered as classical, and the objects are two-atomic and spherical molecules, as well as hard ellipsoids.

1.1 The Feller equation

In impact theory the result of a collision is described by the probability $f(\boldsymbol{J}', \boldsymbol{J})\mathrm{d}J$ of finding angular momentum $\boldsymbol{J}$ after the collision, if it was equal to $\boldsymbol{J}'$ before. The probability is normalized to 1, i.e. $\int f(\boldsymbol{J}', \boldsymbol{J})\, \mathrm{d}\boldsymbol{J}=1$. The equilibrium Boltzmann distribution over $\boldsymbol{J}$ is

$$\varphi_B(\boldsymbol{J}) = \prod_{i=1}^{r} (2\pi d)^{-1/2} \exp\left(-J_i^2/2d\right), \tag{1.1}$$

where $d = IkT$ and $r = 2$ (linear rotators) or 3 (spherical rotators). It has to remain unchanged in any time cross-section of the process. This becomes possible, if the detailed balance principle

$$\varphi_B\left(\boldsymbol{J}'\right) f\left(\boldsymbol{J}', \boldsymbol{J}\right) = \varphi_B(\boldsymbol{J}) f\left(\boldsymbol{J}, \boldsymbol{J}'\right) \tag{1.2}$$

is satisfied, which makes the choice of $f(\boldsymbol{J}', \boldsymbol{J})$ less arbitrary.

The above information is enough to define $\boldsymbol{J}(t)$ as a random process, and allows one to calculate the probability density for a multi-dimensional distribution $\mathbf{P}(\boldsymbol{J}_0, t_o; \boldsymbol{J}_1, t_1; ...\boldsymbol{J}_n, t_n)$. The latter determines the probability of the given sequence $\boldsymbol{J}_0, \boldsymbol{J}_1, ...\boldsymbol{J}_n$ realized at the corresponding moments of time $t_0, t_1, ..., t_n$. As we are dealing with a Markovian stationary process, the multi-dimensional probability distribution is given by

$$\mathbf{P}(\boldsymbol{J}_0, t_0; ...; \boldsymbol{J}_n, t_n) = \varphi_B(\boldsymbol{J}_0) P(\boldsymbol{J}_0, t_0; \boldsymbol{J}_1, t_1)...P(\boldsymbol{J}_{n-1}, t_{n-1}; \boldsymbol{J}_n, t_n),$$

where $\varphi_B(\boldsymbol{J}_0)P(\boldsymbol{J}_0, 0; \boldsymbol{J}, t - t_0)$ is a two-dimensional density and $P(\boldsymbol{J}', 0; \boldsymbol{J}, \tau) \equiv P(\boldsymbol{J}, \boldsymbol{J}'; \tau)$ is a conditional probability of finding the rotational momentum equal to $\boldsymbol{J}$ at the end of the interval (t_0, τ), if it is known to have been equal to $\boldsymbol{J}'$ at its beginning (no matter how many collisions occur within the interval). Since we are concerned with a purely discontinuous process, this quantity is defined by the Feller equation [20, 21]

$$\frac{\partial}{\partial \tau} P(\boldsymbol{J}, \boldsymbol{J}_0; \tau) = -\frac{1}{\tau_0}\left[P(\boldsymbol{J}, \boldsymbol{J}_0; \tau) - \int P(\boldsymbol{J}', \boldsymbol{J}_0; \tau) f(\boldsymbol{J}', \boldsymbol{J})\, \mathrm{d}\boldsymbol{J}'\right], \quad (1.3)$$

where τ_0 is a mean free path time and the initial condition is $P(\boldsymbol{J}, \boldsymbol{J}_0; 0) = \delta(\boldsymbol{J} - \boldsymbol{J}_0)$. Thus the problem is reduced to solving this equation. If the solution is found, then calculations of the correlation function $\boldsymbol{J}(t)$ of any order is reduced to quadratures. In particular,

$$\begin{aligned} K_J(\tau) = \langle \boldsymbol{J}(t)\boldsymbol{J}(t')\rangle &= \int \boldsymbol{J}\boldsymbol{J}'\mathbf{P}(\boldsymbol{J}', t'; \boldsymbol{J}, t)\, \mathrm{d}\boldsymbol{J}'\mathrm{d}\boldsymbol{J} \\ &= \int \boldsymbol{J}'\boldsymbol{J}\varphi_B(\boldsymbol{J}')\, P(\boldsymbol{J}, \boldsymbol{J}'; \tau)\, \mathrm{d}\boldsymbol{J}'\mathrm{d}\boldsymbol{J}, \quad (1.4) \end{aligned}$$

and $K_J(0) = \langle \boldsymbol{J}^2\rangle = rd = rIkT$ according to (1.1).

The solution of Eq. (1.3) becomes possible only after its kernel $f(\boldsymbol{J}', \boldsymbol{J})$ has been specified on the basis of some physical model of collision. The latter are subdivided into adiabatic and non-adiabatic and described by alternative models: m- and J-diffusion [22]. Adiabatic (elastic) collisions do not change the magnitude of the angular momentum $J = \hbar j$, that is, j is conserved. Hence the only result of elastic collision is the deviation of the angular momentum vector $\boldsymbol{J}$ from its initial direction $\boldsymbol{J}'$ on angle α. In quantum theory this is performed by a linear transformation in the subspace of degenerate m-states (projections of j). That is why the model is known as m-diffusion. As far as non-adiabatic (inelastic) collisions are concerned, they induce transitions between both m- and j-states, changing the direction of $\boldsymbol{J}$, as well as its magnitude. Accordingly, it is

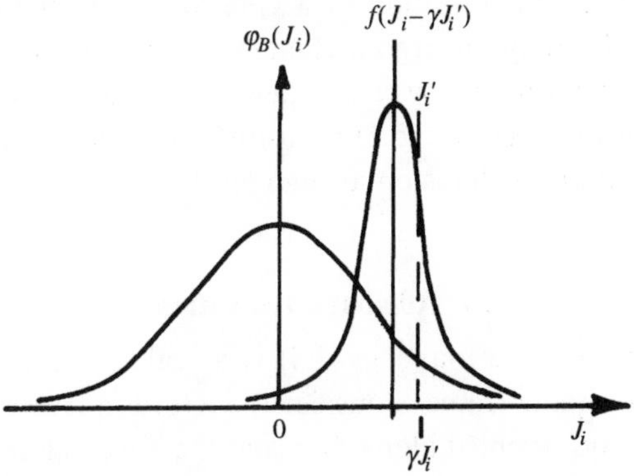

Fig. 1.2. One-dimensional Keilson–Storer kernel f in comparison with an equilibrium distribution φ_B.

assumed that

$$f(\boldsymbol{J}',\boldsymbol{J}) = \Phi(\alpha)\delta(J-J')/(J'J)^{\frac{1}{2}} \qquad \text{for m-diffusion } (J\tau_c \gg 1); \qquad (1.5)$$

$$f(\boldsymbol{J}',\boldsymbol{J}) = f\,(\boldsymbol{J}-\gamma\boldsymbol{J}') \qquad \text{for J-diffusion } (J\tau_c \ll 1). \qquad (1.6)$$

Here $\Phi(\alpha)$ is the density of distribution over α after collision and τ_c is the average collision time. The popular Keilson–Storer model, presented in Eq. (1.6) and Fig. 1.2, uses the single numerical parameter γ to characterize the strength of inelastic collisions. It will be discussed in Section 1.3.

The proposed specification of the kernel for m- and J-diffusion models is mathematically closed, physically clear and of quite general character. In particular, it takes into consideration that any collisions may be of arbitrary strength. The conventional m-diffusion model considers only strong collisions ($\Phi(\alpha) = 1/(2\pi)$), while J-diffusion considers either strong ($\gamma = 0$) or weak ($\gamma = 1$) collisions. Of course, the particular type of kernel used in (1.6) restricts the problem somewhat, but it does allow us to consider kernels with arbitrary $|\gamma| < 1$.

The validity conditions of the models given in (1.5) and (1.6) are also of great importance [4], though seldom taken into account. Transition between rotational states j and $j+1$ is induced only when the product of the transition frequency and the collision time τ_c is less than 1:

$$\omega_{j,j+1}\,\tau_c = \hbar j\tau_c/I = J\tau_c/I \ll 1. \qquad (1.7)$$

From a quasi-classical point of view, the Massey parameter (1.7) is a phase shift, which occurs during a collision and increases with rotation rate. Consequently, the duration of a collision being the same, rapidly rotating molecules only undergo a change in rotation plane orientation, while those rotating more slowly experience a change in rotation rate as well. This is shown in (1.5) and (1.6): J-diffusion occurs at low levels of rotational excitation, m-diffusion at high levels.

1.2 Adiabatic relaxation

Since the function to be found in (1.4) is a scalar, it can be calculated in a mobile coordinate frame. Consider the 'Gordon frame' (GF), where the z axis is always oriented along $\boldsymbol{J}(t)$, and the x axis along a molecule's axis. In the immobile frame the scalar product of Eq. (1.4) is a sum of $J_q(t)J_{-q}(0)$ over all projections ($q = 0, \pm 1$) whereas in GF it reduces to a single term with $q = 0$. In order to find $K_J(t)$ in the GF, it is sufficient to determine the average zth projection of the initial angular momentum $\langle J_0(t)\rangle = \overline{J_0(t)}$:

$$K_J(t) = \langle \sum_q (-1)^q J_q(t) J_{-q}(0)\rangle = \int J_0 \overline{J_0(t)} \varphi_B(J_0)\, \mathrm{d}J_0. \tag{1.8}$$

This projection is constant on a free path and changes only at collision moments (t_k) when rotation of the GF z axis takes place: $J_0(t_k + 0) = \sum_q T^1_{q0} J_q(t_k - 0)$. The collision operator $\hat{T}$ is expressed in terms of the operator $\hat{D}$, which rotates the coordinate frame [23]:

$$\hat{T} = \hat{D}(F, 0, 0,)\, \hat{D}\left(-\frac{\pi}{2},\ \alpha,\ \frac{\pi}{2}\right) \hat{D}\,(-F + \eta,\ 0, 0)\,. \tag{1.9}$$

It depends on both the angle α of the angular momentum vector rotation and other Euler angles F and η, which determine the molecule's axis shift. Besides, the angle F is also the azimuth of the change in angular momentum $\Delta\boldsymbol{J} = \boldsymbol{J}(t + 0) - \boldsymbol{J}(t - 0)$, which is the result of collision.

During adiabatic collision the molecule completes many rotations. Consequently $\Delta\boldsymbol{J}$ is oriented isotropically, and the operator (1.9) should be uniformly averaged over F:

$$\int_0^{2\pi} T^{\ell}_{q'q}(F, \alpha, \eta)\ \mathrm{d}F/2\pi = \delta_{q'q} D^{\ell}_{qq}\left(-\frac{\pi}{2}, \alpha, \eta + \frac{\pi}{2}\right).$$

This operation makes it diagonal. Therefore $\overline{J_0(t_k + 0)} = \overline{D^1_{00}\ (\alpha)\ J_0(t_k - 0)}$ is expressed through a $\hat{D}$ element, which is independent of η:

$$\overline{D^1_{00}\ (\alpha)} = \overline{\cos\alpha} = \lambda. \tag{1.10}$$

Here the averaging denoted by an overbar is performed with a two-dimensional probability density (1.5), i.e.

$$\lambda = \int \Phi(\alpha) \sin\alpha \cos\alpha \, \mathrm{d}\alpha.$$

The change of average z projection of the angular momentum per unit time is its change for a single collision $\Delta \boldsymbol{J} = (\overline{D^1_{00}} - 1)\overline{J_0(t)}$ multiplied by the collision frequency τ_0^{-1} :

$$\frac{\partial}{\partial t}\,\overline{J_0(t)} = -\left(1 - \overline{D^1_{00}}\right)\tau_0^{-1}\,\overline{J_0(t)} = -\frac{1-\lambda}{\tau_0}\,\overline{J_0(t)}.$$

After integrating this equation and substituting the result into (1.8), we find

$$K_J(t) = \int \varphi_B(J) J^2 \, \mathrm{d}J \, \mathrm{e}^{-t/\tau_J}, \tag{1.11}$$

where

$$1/\tau_J = (1-\lambda)/\tau_0. \tag{1.12}$$

1.3 The Keilson–Storer kernel

Because non-adiabatic collisions induce transitions between rotational levels, these levels do not participate in the relaxation process independently as in (1.11), but are correlated with each other. The degree of correlation is determined by the kernel of Eq. (1.3). A one-parameter model for such a kernel adopted in Eq. (1.6) meets the requirement formulated in (1.2). Mathematically it is suitable to solve integral equation (1.2) in a general way. The form of the kernel in Eq. (1.6) was first proposed by Keilson and Storer to describe the relaxation of the translational velocity [10]. Later it was employed in a number of other problems [24, 25], including the one under discussion [26, 27].

Owing to space isotropy of collision in the coordinate frame strongly coupled with a molecule's frame and hereafter referred to as the molecular system (MS), the kernel may be represented as

$$f\left(\boldsymbol{J} - \gamma \boldsymbol{J}'\right) = \prod_{i=1}^{r} f\left(J_i - \gamma J_i'\right). \tag{1.13}$$

Also, all factors should satisfy a stationarity condition

$$\int f\left(J_i - \gamma J_i'\right) \varphi_B\left(J_i'\right) \, \mathrm{d}J_i' = \varphi_B\left(J_i\right). \tag{1.14}$$

This is obtained from Eq. (1.2) after integrating both sides with respect to J. Having taken the Fourier transform (FT) of the equation, one can easily see that a particular form of Keilson–Storer kernel is defined by the corresponding equilibrium distribution [10, 24]:

$$f^F(z) = \varphi_B^F(z)/\varphi_B^F(\gamma z), \tag{1.15}$$

where $f^F = \text{FT}\, f$, and $\varphi_B^F = \text{FT}\, \varphi_B$. Since φ_B is the Gaussian distribution, f is also Gaussian, but with a lesser dispersion $(1-\gamma^2)\,d$:

$$f\left(J_i - \gamma J_i'\right) = \left[2\pi d\left(1-\gamma^2\right)\right]^{-1/2} \exp\left[-\left(J_i - \gamma J_i'\right)^2/2d\left(1-\gamma^2\right)\right]. \tag{1.16}$$

It is clear that the average value of the momentum after collision

$$\langle \boldsymbol{J} \rangle = \int \boldsymbol{J} f\left(\boldsymbol{J} - \gamma \boldsymbol{J}'\right)\, \mathrm{d}\boldsymbol{J} = \gamma \boldsymbol{J}' \tag{1.17}$$

is the centre of distribution which is Gaussian in both projections. The only parameter of the distribution γ determines both its width and the shift to the origin of coordinates (Fig. 1.2). Generally speaking $-1 \le \gamma < 1$. Three particular cases, which are the most important or often used, have to be considered separately.

(i) If collisions are weak, then γ is close to 1, while $f(\boldsymbol{J}-\gamma\boldsymbol{J}')$ is almost a δ-function. In this case $\boldsymbol{J}$ is slightly changed by collisions, and its subsequent values remain correlated for a long time (correlated process).

(ii) When $\gamma = 0$ an equilibrium distribution is established immediately after collision: $f(\boldsymbol{J}', \boldsymbol{J}) = \varphi_B(J)$. We consider these collisions to be strong. When angular momentum changes in this way the random process $\boldsymbol{J}(t)$ is non-correlated: each new value of $\boldsymbol{J}$ is independent of the preceding one.

(iii) At $\gamma = -1$ the process is anticorrelated since J changes only its sign but not its magnitude. This is the generalization of the well-known 'telegraph or dichotomous noise' [28] where only two values of random variable are available and they alternate with time. The anticorrelated process seems to be unrealistic in the gas phase but very useful for cage models of liquids described later.

From a physical point of view the change of the angular momentum per collision is

$$\Delta J = \int_{-\infty}^{\infty} M(t)\, \mathrm{d}t \approx M\tau_c \ge J, \tag{1.18}$$

where M is the torque 'switched' on and off by a perturbing particle.

The closer two particles pass, the greater is their interaction. Still, $\Delta J/J$ may turn out to be less than 1 even in the case of 'face to face' collision. In this limit collisions are weak, $\gamma \approx 1$ and the model of the correlated process fits the situation well. If close impacts produce a strong effect, then the influence of more distant paths is negligible, and the process approaches the non-correlated limit $\gamma \approx 0$.

From a mathematical perspective either of the two cases (correlated or non-correlated) considerably simplifies the situation [26]. Thus, it is not surprising that all non-adiabatic theories of rotational and orientational relaxation in gases are subdivided into two classes according to the type of collisions. Sack's 'model A' [26], referred to as 'Langevin model' in subsequent papers, falls into the first class (correlated or 'weak collisions' process) [29, 30, 12]. The second class includes Gordon's 'extended diffusion model' [8], [22] and Sack's 'model B' [26], later considered as a non-correlated or 'strong collision' process [29, 31, 32].

Though these are alternative models, they are both particular cases of the non-adiabatic impact theory of angular momentum relaxation in gases. Thus, we prefer to call them 'models of weak and strong collisions', as is usually done in analogous problems [13, 33].

1.4 Non-adiabatic relaxation

Using the Keilson–Storer kernel in Eq. (1.3) we obtain

$$\frac{\partial}{\partial \tau} P(\boldsymbol{J}, \boldsymbol{J}_0, \tau) = -\frac{1}{\tau_0}\left[P(\boldsymbol{J}, \boldsymbol{J}_0, \tau) - \int P(\boldsymbol{J}', \boldsymbol{J}_0, \tau) f(\boldsymbol{J} - \gamma \boldsymbol{J}')\, \mathrm{d}\boldsymbol{J}'\right]. \tag{1.19}$$

After multiplying the equation by $\boldsymbol{J}$ or by $\boldsymbol{J}\boldsymbol{J}_0\varphi_B$ and integrating, we obtain

$$\frac{\partial}{\partial \tau}\langle \boldsymbol{J}\rangle = -\frac{1}{\tau_J}\langle \boldsymbol{J}\rangle, \tag{1.20}$$

or

$$\frac{\partial}{\partial \tau} K_J(\tau) = -\frac{1}{\tau_J} K_J(\tau). \tag{1.21}$$

Here $\langle \boldsymbol{J}(\tau)\rangle = \int \boldsymbol{J} P(\boldsymbol{J}, \boldsymbol{J}_0, \tau)\, \mathrm{d}\boldsymbol{J}$ is the average value of the angular momentum vector equal to $\boldsymbol{J}_0$ for $t = 0$, $K_J(\tau)$ is the correlation function defined in (1.4) and

$$1/\tau_J = (1-\gamma)/\tau_0. \tag{1.22}$$

It is easily seen that impact non-adiabatic relaxation of angular momentum proceeds exponentially in the Keilson–Storer approximation. Let us dwell upon some particular cases.

In the case of weak collisions, the moment changes in small steps $\Delta J \approx (1-\gamma)J \ll J$, and the process is considered as diffusion in J-space. Formally, this means that the function $f(z)$ of width $[(1-\gamma^2)d]^{\frac{1}{2}}$ is narrow relative to $P(\boldsymbol{J}, \boldsymbol{J}', \tau)$. At $\tau \gg \tau_0$ the latter may be expanded at the point $\boldsymbol{J}$ up to terms of second-order with respect to $(\boldsymbol{J}' - \boldsymbol{J})$. Then at the limit $\gamma \to 1$, $\tau_0 \to 0$ with τ_J finite, the Feller equations turn into a Fokker–Planck equation

$$\dot{P} = r\frac{q_J}{I}\, P + \frac{q_J}{I}\,(\boldsymbol{J}\boldsymbol{\nabla}_J)\,P + D_J\; \boldsymbol{\nabla}_J^2 P. \tag{1.23}$$

Here

$$q_J = D_J/kT = I/\tau_J \tag{1.24}$$

is the 'mobility' in J-space, and D_J is the corresponding diffusion coefficient connected to the mobility by the Einstein relation. The solution of Eq. (1.23) has the following form:

$$\begin{aligned} P\left(\boldsymbol{J}, \boldsymbol{J}_0, \tau\right) \;&=\; \left\{2\pi I kT\left[1-\exp\left(-2\tau/\tau_J\right)\right]\right\}^{-r/2} \\ &\quad \exp\left(-\frac{[\boldsymbol{J}-\boldsymbol{J}_0\exp\left(-\tau/\tau_J\right)]^2}{2IkT[1-\exp\left(2\tau/\tau_J\right)]}\right). \end{aligned} \tag{1.25}$$

This shows the time evolution of the distribution over $\boldsymbol{J}$ when it was a δ-function at $t = 0$. The distribution spreads and shifts to $J = 0$, until it finally takes the equilibrium shape φ_B. Gradual transformation of the distribution is typical for correlated change in $\boldsymbol{J}(t)$ caused by weak collisions.

The above results were first obtained within the framework of Langevin phenomenology [30, 12]. At first sight, no assumptions concerning the strength of collisions have been made to obtain the original Langevin equation

$$\dot{\boldsymbol{J}} = -\zeta \boldsymbol{J}/I + \boldsymbol{m}(t). \tag{1.26}$$

Here $\boldsymbol{m}(t)$ is a random torque equal to zero on average while rotational friction is proportional to rotation rate J/I with a coefficient ζ. When averaged, Eq. (1.26) turns into Eq. (1.20). Hence

$$\zeta = I/\tau_J. \tag{1.27}$$

This relation is as general as the Langevin equation itself, i.e., it holds for collisions of any strength. When deriving Eq. (1.23) from Eq. (1.26), one has to assume in addition that, at short time intervals, the increment $\Delta \boldsymbol{J}$ is normally distributed and independent of its prior history [34, 35]. The density of the distribution is

$$\Phi(\Delta \boldsymbol{J}, \Delta t) = [4\pi k T \zeta \Delta t]^{-r/2} \exp\left[-(\Delta \boldsymbol{J})^2/4kT\zeta\Delta t\right]. \quad (1.28)$$

This assumption is easily verified, when one takes into consideration that

$$\Phi(\Delta \boldsymbol{J}, \Delta t) = P(\boldsymbol{J} + \Delta \boldsymbol{J}, \boldsymbol{J}, \Delta t), \quad \text{if} \quad \Delta t \ll \tau_J. \quad (1.29)$$

By substituting Eq. (1.25) into the above equation and taking into account Eq. (1.27) one can easily see that Eq. (1.29) holds for weak collisions.

The situation is different for strong collisions. For $\gamma = 0$, Eq. (1.19) is reduced to

$$\frac{\partial}{\partial \tau} P(\boldsymbol{J}, \boldsymbol{J}_0, \tau) = -\frac{1}{\tau_0} P(\boldsymbol{J}, \boldsymbol{J}_0, \tau) + \frac{1}{\tau_0} \varphi_B(\boldsymbol{J}). \quad (1.30)$$

Strictly speaking, the process of J-diffusion described by the above equation is not diffusion at all. The very first collision restores equilibrium in the whole J-space. In this sense, strong collisions represent the hopping mechanism of J-relaxation, which is the only alternative to the diffusion mechanism [20, 24, 25, 36]. Since the term 'J-diffusion' is so pervasive, we do not like to reject it. However, it should be understood in a wider sense and used to denote 'J-migration'. Then it remains valid for both weak and strong collisions. Still, it should be remembered that there is a considerable difference between these limits. For strong collision we obtain from Eq. (1.30)

$$P(\boldsymbol{J}, \boldsymbol{J}_0, \tau) = \delta(\boldsymbol{J} - \boldsymbol{J}_0)\, e^{-\tau/\tau_J} + \varphi_B(\boldsymbol{J}) \left[1 - e^{-\tau/\tau_J}\right], \quad (1.31)$$

where $\tau_J = \tau_0$. Equation (1.31) is qualitatively different from Eq. (1.25). As used in relation (1.29), it does not reproduce the Gaussian distribution (1.28). Hence, in the case of strong collisions $\Delta \boldsymbol{J}$ is not a Gaussian variable and the Markovian method used in [34] to derive (1.25) from (1.26) is inapplicable here.

In summary, the 'Langevin model' which addresses not Eq. (1.26) but rather solution (1.25) fits solely the limiting case of weak collisions. Only in this limit does rotational friction acquire the meaning of a 'mobility' in J-space, i.e.

$$\zeta = q_J \quad \text{at} \quad \gamma \to 1. \quad (1.32)$$

For strong collisions ($\gamma = 0$), ζ is still equal to I/τ_0, but q_J does not exist and the diffusion mechanism of rotational relaxation is replaced by a hopping mechanism.

The last particular case worthy of analysis is an anticorrelated process. The Keilson–Storer parameter γ, when negative, describes relaxation induced by collisions, which primarily change the direction of the angular momentum to its opposite. In other words, at $\gamma = -1$ the only result of a collision is that the direction of free rotation becomes reversed without changing the magnitude of the angular velocity. Substituting

$$f(\boldsymbol{J} - \gamma \boldsymbol{J}') = \delta(\boldsymbol{J} + \boldsymbol{J}') \tag{1.33}$$

into Eq. (1.19) we obtain

$$\frac{\partial}{\partial t} P(\boldsymbol{J}, \boldsymbol{J}_0, t) = -\frac{1}{\tau_0} P(\boldsymbol{J}, \boldsymbol{J}_0, t) + \frac{1}{\tau_0} P(\boldsymbol{J}_{\neg}, \boldsymbol{J}_0, t), \tag{1.34}$$

where $\boldsymbol{J}_{\neg} = -\boldsymbol{J}$. From the mathematical point of view, Eq. (1.34) is not closed, as it contains a 'new' conditional probability value, $P(\boldsymbol{J}_{\neg}, \boldsymbol{J}_0, t)$. However, we can write for the latter an additional equation following from Eq. (1.19) and Eq. (1.33), namely

$$\frac{\partial}{\partial t} P(\boldsymbol{J}_{\neg}, \boldsymbol{J}_0, t) = -\frac{1}{\tau_0} P(\boldsymbol{J}_{\neg}, \boldsymbol{J}_0, t) + \frac{1}{\tau_0} P(\boldsymbol{J}, \boldsymbol{J}_0, t). \tag{1.35}$$

The system of equations (1.34) and (1.35) is simple to solve by taking their Laplace transform with initial conditions

$$\begin{aligned} P(\boldsymbol{J}, \boldsymbol{J}_0, 0) &= \delta(\boldsymbol{J} - \boldsymbol{J}_0), \\ P(\boldsymbol{J}_{\neg}, \boldsymbol{J}_0, 0) &= \delta(\boldsymbol{J}_{\neg} - \boldsymbol{J}_0). \end{aligned} \tag{1.36}$$

Finally we have

$$\begin{aligned} P(\boldsymbol{J}, \boldsymbol{J}_0, t) &= \frac{1}{2}\delta(\boldsymbol{J}_{\neg} - \boldsymbol{J}_0)\left[1 - \exp\left(-\frac{2t}{\tau_0}\right)\right] \\ &+ \frac{1}{2}\delta(\boldsymbol{J} - \boldsymbol{J}_0)\left[1 + \exp\left(-\frac{2t}{\tau_0}\right)\right]. \end{aligned} \tag{1.37}$$

The physical properties of relaxation set by Eq. (1.37) are easy to discuss for its one-dimensional projection

$$\begin{aligned} P(J_x, J_{x0}, t) &= \frac{1}{2}\delta(J_x + J_{x0})\left[1 - \exp\left(-\frac{2t}{\tau_0}\right)\right] \\ &+ \frac{1}{2}\delta(J_x - J_{x0})\left[1 + \exp\left(-\frac{2t}{\tau_0}\right)\right], \end{aligned} \tag{1.38}$$

where $J_x = (\boldsymbol{J})_x = -(\boldsymbol{J}_{\neg})_x$. Eq. (1.38) describes exponential decay of the initial distribution centred at J_{x0}, and simultaneous growth of the number of particles having inverse x projection of angular momentum

$-J_{x0}$. At large times, both sub-ensembles tend to have an equal number of molecules

$$P(J_x, J_{x0}, \infty) \quad \longrightarrow \frac{1}{2}\delta(J_x + J_{x0}) + \frac{1}{2}\delta(J_x - J_{x0}). \tag{1.39}$$

Using solution (1.37) in definition (1.4), one has the angular momentum correlation function

$$K_J(t) = 2IkT \ \exp\left(-\frac{t}{\tau_J}\right) \tag{1.40}$$

with

$$\tau_J = \tau_0/2, \tag{1.41}$$

as follows from Eq. (1.22) at $\gamma = -1$. Comparison with Eq. (1.31) (when $\tau_J = \tau_0$) shows that reflection by collision accelerates relaxation.

1.5 General solution of the Feller equation

In order to determine the actual strength of collisions, it is desirable to have a general solution of the Feller equation that holds for any γ. Consider the problem of the eigenvalues of the integral operator $\hat{L}$, which appears in the Feller equation

$$\frac{\partial}{\partial\tau} \ P(\boldsymbol{J}, \boldsymbol{J}_0, \tau) = -\frac{1-\hat{L}}{\tau_0} \ P\left(\boldsymbol{J}', \boldsymbol{J}_0, \tau\right). \tag{1.42}$$

Determine eigenfunctions Ψ_λ of the operator $\hat{L}$ from the equation

$$\hat{L}\Psi_\lambda(\boldsymbol{J}') = \int f\left(\boldsymbol{J} - \gamma\boldsymbol{J}'\right) \Psi_\lambda\left(\boldsymbol{J}'\right) \ \mathrm{d}\boldsymbol{J}' = \lambda\Psi_\lambda(\boldsymbol{J}). \tag{1.43}$$

In view of multiplicity of Eq. (1.13) it is expedient to seek the solution in the form

$$\Psi_\lambda(\boldsymbol{J}) = \prod_{i=1}^{r} \chi_\lambda(J_i). \tag{1.44}$$

In this case Fourier transformation (1.43) gives for the case of two dimensions ($r = 2$)

$$f^F(u) f^F(v) \chi_\lambda^F(\gamma u) \chi_\lambda^F(\gamma v) = \lambda \chi_\lambda^F(u) \chi_\lambda^F(v), \tag{1.45}$$

where

$$\chi_\lambda^F(u) = \int \chi_\lambda(J_1) \ \mathrm{e}^{\mathrm{i}uJ_1} \ \mathrm{d}J_1 \quad \chi_\lambda^F(v) = \int \chi_\lambda(J_2) \ \mathrm{e}^{\mathrm{i}vJ_2} \ \mathrm{d}J_2.$$

In view of Eq. (1.15), we have

$$\chi_\lambda^F(\gamma u)\,\chi_\lambda^F(\gamma v) = \lambda(\gamma)\chi_\lambda^F(u)\chi_\lambda^F(v)\varphi_B^F(\gamma u)\varphi_B^F(\gamma v)/\varphi_B^F(u)\varphi_B^F(v). \tag{1.46}$$

After differentiating the expression with respect to γ and introducing new variables $U = \gamma u$ and $V = \gamma v$, we divide them

$$\frac{\partial \ln\left(\chi_\lambda^F/\varphi_B^F\right)}{\partial \ln U} = N_u, \frac{\partial \ln\left(\chi_\lambda^F/\varphi_B^F\right)}{\partial \ln V} = N_v, \tag{1.47}$$

where $N_u + N_v = \partial \ln \lambda / \partial \ln \gamma$. Integrating these equations, we may find the product of the solutions

$$\chi_\lambda^F(U)\,\chi_\lambda^F(V) = \text{const}\; U^{N_u} V^{N_v} \varphi_B^F(U)\varphi_B^F(V). \tag{1.48}$$

Substitution of this result into (1.46) gives

$$\lambda = \gamma^{N_u + N_v}. \tag{1.49}$$

It is seen from (1.42) and (1.43) that the quantity $(1-\lambda)/\tau_0$ has the meaning of a relaxation rate and is expected to be positive. Besides, if N_u and N_v are integers, then the original of χ^F is

$$\chi_{n_i}(J_i) \sim \frac{\mathrm{d}^{n_i}}{\mathrm{d}J_i^{n_i}}\,\varphi_B(J_i), \quad n_i = 0, 1, 2, \ldots. \tag{1.50}$$

Since $\varphi_B(J_i)$ is the Gaussian function (1.1),

$$\chi_{n_i}(J_i) \propto \varphi_B(J_i)\, H_{n_i}\left(\frac{J_i}{(2d)^{\frac{1}{2}}}\right), \qquad i = 1, \ldots, r;\; n_i = 0, 1, 2, \ldots \tag{1.51}$$

according to the definition of Hermite polynomials [37]. Since the latter constitute a complete set of functions, the general solution of the Feller equation may be expressed by

$$P(\boldsymbol{J}, \boldsymbol{J}_0, \tau) = \sum_{\{n_i\}} \mathrm{e}^{-\tau/\tau_0}\, \varphi_B(\boldsymbol{J}) \left(\prod_{i=1}^{r} a_{n_i} H_{n_i}\left[J_i/(2d)^{\frac{1}{2}}\right]\right) \exp\left(\frac{\tau}{\tau_0}\prod_{i=1}^{r}\gamma^{n_i}\right). \tag{1.52}$$

The expansion coefficients a_{n_i} are found from the initial condition and the orthogonality of H_{n_i} [37].

$$P(\boldsymbol{J}, \boldsymbol{J}_0, 0) = \delta(\boldsymbol{J} - \boldsymbol{J}_0).$$

Finally, the conditional probability takes the following form:

$$\begin{aligned} P(\boldsymbol{J}, \boldsymbol{J}_0, \tau) &= \exp\left(-\tau/\tau_0\right)\, \varphi_B(\boldsymbol{J}) \\ &\quad \times \sum_{\{n_i\}} \left(\frac{\prod_{i=1}^{r} H_{n_i}\left[J_{oi}/(2d)^{\frac{1}{2}}\right]\, H_{n_i}\left[J_i/(2d)^{\frac{1}{2}}\right]}{2^{n_i}(n_i)!}\right) \end{aligned}$$

$$\times \exp\left(\frac{\tau}{\tau_0}\prod_{i=1}^{r}\gamma^{n_i}\right). \tag{1.53}$$

Equation (1.25), which corresponds to weak collisions, Eq. (1.31) referring to strong ones and Eq. (1.37) for anticorrelated process are easily obtained from the above for $\gamma \to 1$, $\gamma = 0$ and $\gamma = -1$ respectively.

1.6 Angular momentum correlation functions

All correlation moments of $\boldsymbol{J}(t)$ for arbitrary γ may be calculated by employing formula (1.44). In particular, when used in (1.4), it yields (1.21). The fourth-order correlation function is

$$\langle J_{i_1}(t_1)J_{i_2}(t_2)J_{i_3}(t_3)J_{i_4}(t_4)\rangle = \int \mathbf{P}(\boldsymbol{J}_1, t_1, \ldots, \boldsymbol{J}_4, t_4) \prod_{m=1}^{4} \mathrm{d}\boldsymbol{J}_m \prod_{\alpha=1}^{r} (J_m)_\alpha^{\delta_{i_m\alpha}}, \tag{1.54}$$

$\mathbf{P}$ being multiplicative over $P(\boldsymbol{J}_k, \boldsymbol{J}_{k-1}, t)$. Higher order correlation functions $\langle \prod_{m=1}^{n} J_{i_m}(t_m)\rangle$ are found by analogy.

In practice, even the determination of a fourth-order correlation function requires a large amount of calculation. However, this procedure may be standardized by a graphical method, which performs the integration in Eq. (1.54) by using properties of Hermite polynomials [37]. Without going into details, we give the result

$$\begin{aligned}
&\langle J_{i_1}(t_1)J_{i_2}(t_2)J_{i_3}(t_3)J_{i_4}(t_4)\rangle \\
&= d^2\left[\left(\delta_{i_1i_3}\delta_{i_2i_4} + \delta_{i_1i_4}\delta_{i_2i_3}\right)\exp\left(-\frac{t_4 - \gamma(t_2 - t_3) - t_1}{\tau_J}\right)\right. \\
&\left.+\delta_{i_1i_2}\delta_{i_3i_4}\exp\left(-\frac{t_4 - t_3 + t_2 - t_1}{\tau_J}\right)\right].
\end{aligned} \tag{1.55}$$

Let $i_1 = i_2 = \alpha$, $i_3 = i_4 = \beta$, $t_1 = t_2 = 0$, $t_3 = t_4 = \tau$. Then, taking the summation over α and β, one can calculate the rotational energy correlation function:

$$\begin{aligned}
K_E(\tau) &= \langle J^2(\tau)J^2(0)\rangle/4I^2 \\
&= (r/4)\left[2(kT)^2\mathrm{e}^{-(1+\gamma)\tau/\tau_J} + r(kT)^2\right].
\end{aligned} \tag{1.56}$$

This result is analogous to that for the correlation function of kinetic energy, first derived in [10].

In order to find the correlation time $\tau_E = \tau_{J^2}$ of rotational energy, it is necessary to eliminate the constant component $K_E(\infty)$ from Eq. (1.56).

Introducing the corresponding cumulant [38]

$$\tilde{K}_E = \frac{\langle \delta J^2(t)\delta J^2(0)\rangle}{4I^2} = K_E(t) - K_E(\infty),$$

where $\delta J^2 = J^2 - \langle J^2\rangle$, we obtain by definition

$$\tau_E = \int_0^\infty \tilde{K}_E(\tau)\,\mathrm{d}\tau/\tilde{K}_E(0) = \tau_J/(1+\gamma). \tag{1.57}$$

In NMR theory the analogue of the relation (1.57) connects the times of longitudinal (T_1) and transverse (T_2) relaxation [39]. In the case of weak non-adiabatic interaction with a medium it turns out that $T_1 = T_2/2$. This also happens in a harmonic oscillator [40, 41] and in any two-level system. However, if the system is perturbed by strong collisions then $T_1 = T_2$ as for γ=0 [42]. Thus in non-adiabatic theory these times differ by not more than a factor 2 regardless of the type of system, or the type of perturbation, which may be either impact or a continuous process.

In the Keilson–Storer model of J-diffusion, non-adiabatic relaxation is assumed to extend to the whole energy spectrum of a rotator. Actually, for large J the relaxation becomes adiabatic. The considerable difference between the times appears in the adiabatic limit since $\tau_E = \infty$, while τ_J is defined by m-diffusion according to Eq. (1.12). As is seen from Eq. (1.5) and Eq. (1.6), both J- and m-diffusion are just approximations which hold for low- and high-excited rotational levels, respectively. In general $0 \le \tau_J/\tau_E \le 1+\gamma$.

It is clear that J-diffusion is a good approximation for rotational relaxation as a whole, if the centre of equilibrium distribution over J is within the limits of non-adiabatic theory. In the opposite case m-diffusion is preferable. Consequently, the J-diffusion model is applicable, if the following inequality holds:

$$\langle\omega\rangle\,\tau_c \propto \frac{\langle J\rangle\rho}{I\langle v\rangle} = \left(\frac{\pi^2\mu\rho^2}{16I}\right)^{\frac{1}{2}} < 1. \tag{1.58}$$

Here ρ is the radius of the effective cross-section, $\langle v\rangle$ is the average velocity of colliding particles, and μ is their reduced mass. When rotational relaxation of heavy molecules in a solution of light particles is considered, the above criterion is well satisfied. In the opposite case the situation is quite different. Even if the relaxation is induced by collisions of similar particles (as in a one-component system), the fraction of molecules which remain adiabatically isolated from the heat reservoir is fairly large. For such molecules energy relaxation is much slower than that of angular momentum, i.e. $\tau_E/\tau_J \gg 1$.

1.7 Impact processes with finite collision time

In impact theory the gas density is restricted by the criterion for the validity of the binary collision approximation. Roughly speaking, the time between collisions must be greater than their duration

$$\tau_0 \gg \tau_c. \tag{1.59}$$

However, only the left-hand side of the inequality has a clear, although qualitative, physical meaning. As far as collision time τ_c is concerned, its evaluation as $\rho/\langle v\rangle$ in Eq. (1.58) is rather arbitrary. Alternatively, it may be defined as the correlation time of the collisional processes which modulate the rotation. Using the mechanical equation of motion

$$\dot{\boldsymbol{J}} = \boldsymbol{M}, \tag{1.60}$$

rotational relaxation is considered as a molecule's response to perturbation by a random torque $\boldsymbol{M}(t)$. That is why the correlation time of $\boldsymbol{M}(t)$ seems to be the quantity to be found. However, this is not true. When averaged, Eq. (1.60) clearly shows that, at least in the impact approximation, decay of $\langle\boldsymbol{M}\rangle$ is not faster than that of $\langle\boldsymbol{J}\rangle$; indeed

$$\langle\boldsymbol{M}\rangle = \langle\dot{\boldsymbol{J}}\rangle = -\frac{1}{\tau_J}\langle\boldsymbol{J}\rangle = -\frac{\zeta}{I}\langle\boldsymbol{J}\rangle. \tag{1.61}$$

The above reasoning applies also to the correlation function of the torque. It follows from the stationarity of the random processes $\boldsymbol{J}(t)$ and $\boldsymbol{M}(t)$ that their correlation functions depend solely on difference of times arguments

$$K_M(t-t') = \langle\boldsymbol{M}(t)\boldsymbol{M}(t')\rangle = \langle\dot{\boldsymbol{J}}(t)\dot{\boldsymbol{J}}(t')\rangle = \frac{\mathrm{d}}{\mathrm{d}t}\frac{\mathrm{d}}{\mathrm{d}t'}K_J(t-t'). \tag{1.62}$$

Owing to this, $\mathrm{d}/\mathrm{d}t' = -\mathrm{d}/\mathrm{d}t$, and Eq. (1.62) turns into a simple and general relation which connects K_J to K_M:

$$\frac{\mathrm{d}^2}{\mathrm{d}t^2}K_J(t) = -K_M(t). \tag{1.63}$$

If $K_J(t)$ vanishes exponentially at times τ_J (Fig. 1.3), then the correlation function K_M behaves in a similar fashion, though opposite in sign (Fig. 1.4).

There are also other unusual peculiarities in K_M behaviour which follow from the definition of a stationary derivative $\dot{K}_J = \langle\boldsymbol{M}(t)\boldsymbol{J}(t')\rangle = -\langle\boldsymbol{J}(t)\boldsymbol{M}(t')\rangle$:

$$\dot{K}_J(\infty) = -\langle\boldsymbol{M}(0)\boldsymbol{J}(\infty)\rangle = -\langle\boldsymbol{M}(0)\rangle\cdot\langle\boldsymbol{J}(\infty)\rangle = 0, \tag{1.64}$$

$$\dot{K}_J(0) = -\langle\boldsymbol{M}(0)\boldsymbol{J}(0)\rangle = \langle\boldsymbol{M}(0)\boldsymbol{J}(0)\rangle = 0. \tag{1.65}$$

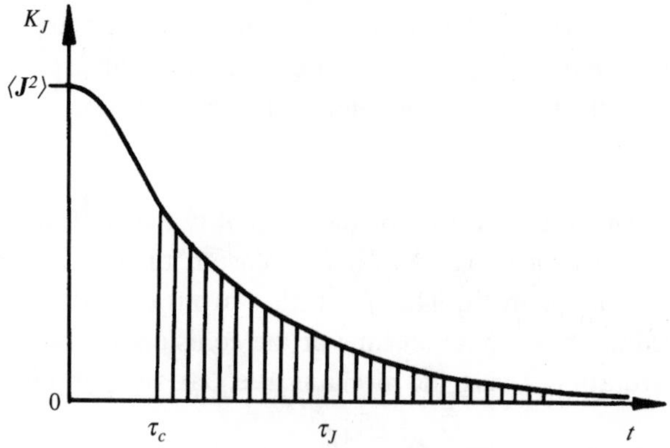

Fig. 1.3. The correlation functions of angular momentum. The long-time exponential asymptotics is shadowed.

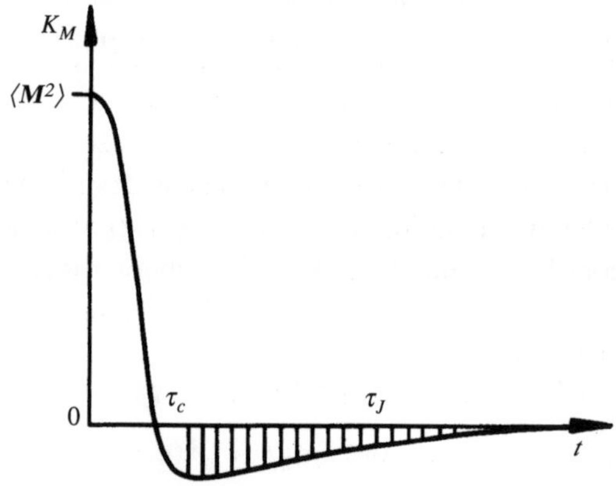

Fig. 1.4. The correlation functions of torque. The long-time exponential asymptotics is shadowed.

The first of the equations is a consequence of space isotropy ($\langle \boldsymbol{J} \rangle = 0$), while the second is a result of $\boldsymbol{J}(t)$ differentiability [43]. In the impact theory the latter equation does not hold since $\dot{K}_J(0)/K_J(0) = -1/\tau_J$. Therefore, if Eq. (1.65) is accepted, the impact approximation is inapplicable.

Consider the process proceeding on time scales comparable with τ_c, taking into account that $K_M(0) = \langle \boldsymbol{M}^2 \rangle > 0$ but at long times $K_M(t) < 0$.

Obviously the sign of $K_M(t)$ must reverse. Besides,

$$\int_0^\infty K_M(t)\,\mathrm{d}t = \int_0^\infty \langle \boldsymbol{M}(0)\dot{\boldsymbol{J}}(t)\rangle\,\mathrm{d}t = \langle \boldsymbol{M}(0)\boldsymbol{J}(\infty)\rangle - \langle \boldsymbol{M}(0)\boldsymbol{J}(0)\rangle = 0, \tag{1.66}$$

i.e. the areas under positive and negative branches of $K_M(t)$ are equal. Hence the corresponding correlation time τ_M is given by

$$\tau_M = \int_0^\infty K_M(t)\,\mathrm{d}t/K_M(0) = 0. \tag{1.67}$$

It is also important to note that

$$\int_0^\infty tK_M(t)\,\mathrm{d}t = -\int_0^\infty t\ddot{K}_J\,\mathrm{d}t = K_J|_0^\infty = -rd \tag{1.68a}$$

$$\int_0^\infty t^2K_M(t)\,\mathrm{d}t = -2rd\tau_J, \tag{1.68b}$$

where

$$\tau_J = \int_0^\infty K_J(t)\,\mathrm{d}t/rd. \tag{1.69}$$

Higher time moments of $K_M(t)$ are negative as well. This is physically accounted for by the 'anticorrelated' nature of successive collisions in a gas [44]. A collision 'from the front' is usually followed by a collision 'from the back' (Fig. 1.5). Owing to the opposite direction of collisions the product of the corresponding moments $\langle \boldsymbol{M}(0)\boldsymbol{M}(t)\rangle$ is negative, and it provides the main contribution to K_M for times of order τ_J. This is a manifestation of the correlated character of the interaction between a molecule and perturbers.

Taking the Fourier transform of Eq. (1.63) we get

$$g_M(\omega) = \omega^2 g(\omega), \tag{1.70}$$

where g_M and g are spectra of the corresponding processes. In the impact approximation, $g(\omega)$ takes a Lorentzian shape of width $1/\tau_J$. The finite duration of collisions produces no effect on the behaviour of rotational relaxation spectra near the maximum:

$$g(\omega) \approx \frac{r\,\mathrm{d}\tau_J}{\pi}\left[1-(\omega\tau_J)^2\right],\quad \omega\tau_J \ll 1.$$

On the contrary, according to Eq. (1.70) g_M becomes zero for $\omega = 0$, the width of a gap in the centre of the spectrum being about $1/\tau_J$. Note that the total width of a gas phase spectrum is much larger, namely $1/\tau_c$. This narrow gap in the centre of $g_M(\omega)$ points to the existence of intercollisional correlation. The same is valid for the spectrum of random

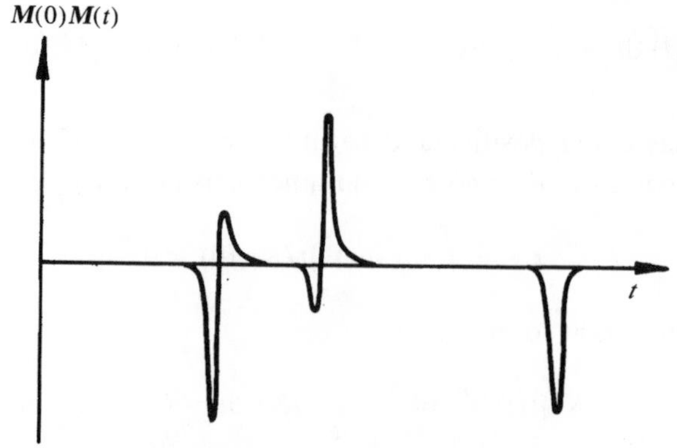

Fig. 1.5. Schematic presentation of the anticorrelation of subsequent collisions.

force modulating the translational velocity of the molecules. When theoretically calculated in the approximation of independent collisions, the spectrum contains no gap (Fig. 1.6).

For forbidden transitions in atoms and molecules this phenomenon may be experimentally observed in spectra induced by collisions. As is known, the selection rules on some transitions may be cancelled during collision. The perturbers are able to induce a dipole moment of transition having the opposite direction in successive collisions due to intercollisional correlation. Owing to this, the induced spectra do involve the gap (Fig. 1.7), the width of the latter being proportional to the gas density [46, 47]. Theorists consider intercollisional correlation to be responsible for the above phenomenon [48, 49, 50].

1.8 The memory function formalism

From the above discussion, it is apparent that the exponential asymptotic behaviour of $K_M(t)$ characterizes the correlation between collisions rather than collision itself. Hence the quantity τ_M defined in Eq. (1.67) cannot be considered as a collision time. To determine the true duration of collision let us transform Eq. (1.63) to the integral-differential equation as was done in [51]:

$$\dot{K}_J(t) = -\int_0^t R(t-t')K_J(t')\,\mathrm{d}t'. \tag{1.71}$$

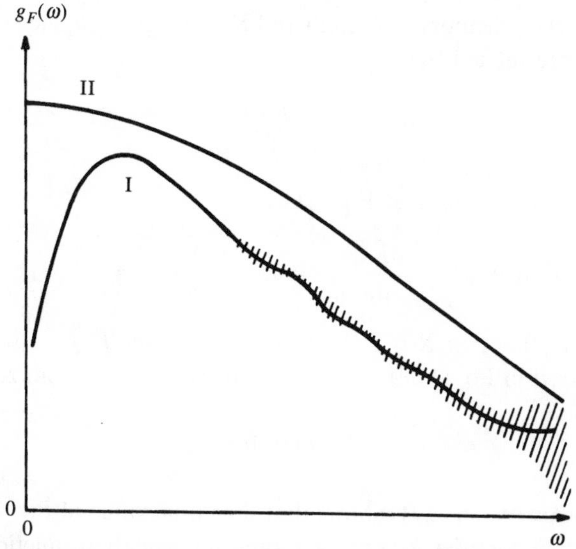

Fig. 1.6. The spectral manifestation of intercollision anticorrelation. Curve I is the spectral density of the force F acting in liquid [45] (the accuracy of MD calculation is shown), curve II the same but in the approximation of independent collisions.

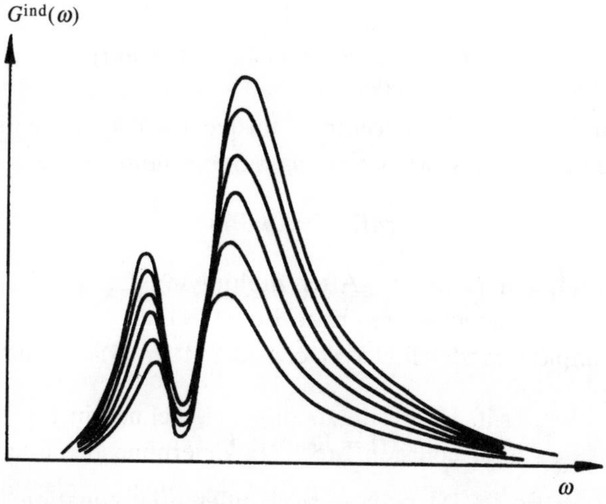

Fig. 1.7. Pressure-induced spectrum of H_2 in He [46] at room temperature. The curves correspond to different densities of He between 950 and 1465 amagat.

Thus, a single-valued connection is established between the kernel of this equation $R(t)$ (a 'memory function') and $K_M(t)$. Their Laplace transforms $\tilde{R}$ and $\tilde{K}_M$ are related by

$$\tilde{R}(p) = \frac{\tilde{K}_M(p)}{rd - \tilde{K}_M\,(p)/p}. \tag{1.72}$$

According to Eq. (1.66) and Eq. (1.68)

$$\tilde{K}_M(0) = 0, \quad \left.\frac{\mathrm{d}\tilde{K}_M}{\mathrm{d}p}\right|_0 = rd, \quad \left.\frac{\mathrm{d}^2\tilde{K}_M}{\mathrm{d}p^2}\right|_0 = -2rd\tau_J. \tag{1.73}$$

The memory function $R(t)$ has an advantage over $K_M(t)$, since, unlike what occurred in Eq. (1.66), the integral of $R(t)$ differs from zero:

$$\frac{1}{\tau_J} = \int_0^\infty R(t)\,\mathrm{d}t = R_0\tau_c. \tag{1.74}$$

Differentiation of Eq. (1.71) shows that the initial value of $R(t)$ is positive. Thus one may consider $R(t)$ to have typical correlation function properties: for $t = 0$ it is equal to the correspondingly determined dispersion R_0 and it decreases to zero in a time of the order τ_c.

This conclusion can be confirmed by an alternative derivation of Eq. (1.71). According to Mori [52], Eq. (1.71) may be obtained from a generalized Langevin equation:

$$\dot{\boldsymbol{J}} = -\int_0^t R(t-t')\boldsymbol{J}(t')\,\mathrm{d}t' + \boldsymbol{m}(t), \tag{1.75}$$

where $\boldsymbol{m}$ is the relevant 'centred' torque ($\langle \boldsymbol{m}\rangle = 0$) representing the sequence of collisions or continuous random noise, while

$$rdR(\tau) = \langle \boldsymbol{m}(0)\boldsymbol{m}(\tau)\rangle \tag{1.76}$$

is its correlation function. After multiplying Eq. (1.75) by $\boldsymbol{J}(0)$ and averaging the product we return to Eq. (1.71).

The simplest model for $R(\tau)$ considered in [51] is the exponential function

$$R(t) = R_0 \exp\left(-t/\tau_c\right). \tag{1.77}$$

In this case Eq. (1.71) reduces to a differential equation. Its rigorous solution with initial conditions $\dot{K}_J(0) = rd,\ K_J(0) = 0$ gives

$$K_J = rd\left(\frac{k_2}{k_2-k_1}\exp(-k_1 t) - \frac{k_1}{k_2-k_1}\exp(-k_2 t)\right), \tag{1.78}$$

where

$$k_1 = \left[1 - (1-4\kappa)^{\frac{1}{2}}\right]/2\tau_c,\ k_2 = \left[1 + (1-4\kappa)^{\frac{1}{2}}\right]/2\tau_c \tag{1.79}$$

and $\kappa = R_0\tau_c^2$. Using the above solution in Eq. (1.63) we have

$$K_M = \langle \boldsymbol{M}^2\rangle \left(\frac{k_2}{k_2-k_1}\ \exp\ (-k_2 t) - \frac{k_1}{k_2-k_1}\ \exp\ (-k_1 t)\right). \tag{1.80}$$

In analysing these equations, one should take into account that the dispersion of the torque $\langle \boldsymbol{M}^2\rangle$ is proportional to the fraction of time spent by a molecule in collision. We write that

$$\langle \boldsymbol{M}^2\rangle = \langle \boldsymbol{M}_c^2\rangle\ \frac{\tau_c}{\tau_0}, \tag{1.81}$$

where $\langle \boldsymbol{M}_c^2\rangle$ is the mean square torque during collision [20, 53]. Consequently

$$R_0 = \frac{\langle \boldsymbol{M}^2\rangle}{rd} = \frac{\langle \boldsymbol{M}_c^2\rangle\tau_c}{\langle \boldsymbol{J}^2\rangle\tau_0} = \frac{R_c\tau_c}{\tau_0} \tag{1.82}$$

and we note that $\kappa = R_c\tau_c^3/\tau_0$, as well as collisional frequency $1/\tau_0$, increases with the buffer gas density n.

When $\kappa < 1/4$ the roots (1.79) are real, and $K_J(t)$ decays monotonically with time. For low-density gas

$$\kappa = R_0\tau_c^2 \ll \frac{1}{4} \tag{1.83}$$

and the power expansion of the roots reduces Eq. (1.78) and Eq. (1.80) to the following

$$K_J \approx rd\ \left[(1+\kappa)\ \mathrm{e}^{-(t/\tau)} - \kappa\mathrm{e}^{-(t/\tau_c)(1-\kappa)}\right] \tag{1.84a}$$

$$K_M \approx \langle \boldsymbol{M}^2\rangle\ \left[\mathrm{e}^{-(t/\tau_c)(1-\kappa)}(1+\kappa) - \kappa\mathrm{e}^{-(t/\tau)}\right]. \tag{1.84b}$$

Here long-time asymptotic (collisional) rate

$$\frac{1}{\tau} = R_0\tau_c(1+\kappa) \tag{1.85}$$

and τ_c introduced in Eq. (1.77) does have the meaning of collision time. As in Fig. 1.4, $K_M(t)$ reverses its sign for $t \approx \tau_c\ \ln(1/\kappa)$ and slowly decays with the longer time τ.

Deviations from the impact (exponential) relaxation of angular momentum found in (1.84a) are rather insignificant. Taking account of finite collision time slows down the initial decay (at $t < \tau_c$) and accelerates the final stage (at $t > \tau_c$). The rate (1.85) is slightly greater than

$$\frac{1}{\tau_J} = R_0\,\tau_c = \frac{\langle(\boldsymbol{M}_c\tau_c)^2\rangle}{\langle \boldsymbol{J}^2\rangle\,\tau_0} = \frac{\langle(\Delta \boldsymbol{J})^2\rangle}{\langle \boldsymbol{J}^2\rangle}\,\frac{1}{\tau_0}. \tag{1.86}$$

In impact limit ($\kappa \to 0$) this difference disappears and $\tau = \tau_J$ must be the same as in Eq. (1.22). Therefore the relative mean-square change in the moment caused by a collision is

$$\frac{\langle(\Delta \boldsymbol{J})^2\rangle}{\langle \boldsymbol{J}^2\rangle} = 1 - \gamma \tag{1.87}$$

and the validity condition (1.83) for quasi-exponential relaxation may be expressed as

$$\kappa = \tau_c/\tau_J = (1-\gamma)\,\tau_c/\tau_0 \ll 1/4. \tag{1.88}$$

In the case of strong collisions, corresponding to $\gamma = 0$, inequality (1.88) is reduced to a conventional validity criterion of the binary theory (1.59). However, if collisions are weak ($\gamma \approx 1$), the actual criterion given in Eq. (1.88) is considerably weakened and Langevin phenomenology is valid at larger densities.

This conclusion is not unexpected. The molecule's response to weak collisions considered as a random process may be described by perturbation theory (with respect to interaction) if [53]

$$R_0\tau_c\tau_0 = \tau_0/\tau_J = 1 - \gamma \ll 1. \tag{1.89}$$

Markovian perturbation theory as well as impact theory describe solely the exponential asymptotic behaviour of rotational relaxation. However, it makes no difference to this theory whether the interaction with a medium is a sequence of pair collisions or a weak collective perturbation. Being binary, the impact theory holds when collisions are well separated ($\tau_c \ll \tau_0$) while the perturbation theory is broader. If it is valid, a new collision may start before the preceding one has been completed when $\tau_0 \ll \tau_c \ll \tau_J = \tau_0/(1-\gamma)$.

Inequality (1.88) defines the domain where rotational relaxation is quasi-exponential either due to the impact nature of the perturbation or because of its weakness. Beyond the limits of this domain, relaxation is quasi-periodic, and τ loses its meaning as the parameter for exponential asymptotic behaviour. The point is that, for $\kappa \gg 1/4$, Eq. (1.78) and Eq. (1.80) reduce to the following:

$$K_J(t) \approx rd\,\exp\left(-t/2\tau_c\right)\left[\cos\left(t\sqrt{R_0}\right) + \frac{\sin\left(t\sqrt{R_0}\right)}{2\tau_c\sqrt{R_0}}\right] \tag{1.90a}$$

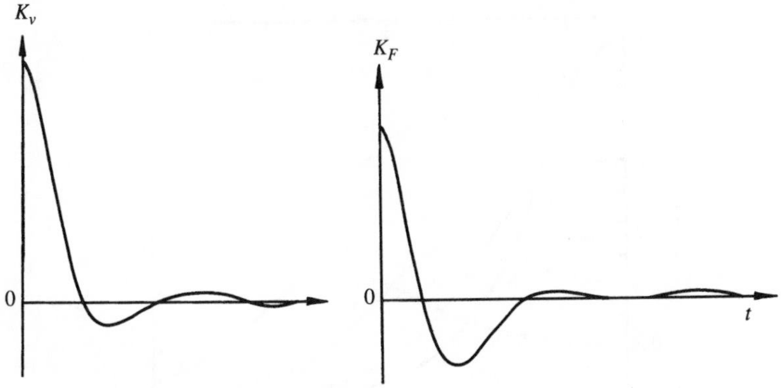

Fig. 1.8. MD calculation of autocorrelation functions of translational velocity (K_v) and the force (K_F) acting on a molecule in the liquid [45].

$$K_M(t) \approx \langle M^2 \rangle \exp\left(-t/2\tau_c\right) \left[\cos\left(t\sqrt{R_0}\right) - \frac{\sin(t\sqrt{R_0})}{2\tau_c\sqrt{R_0}}\right]. \quad (1.90b)$$

Judging by these results the angular momentum relaxation in a dense medium has the form of damped oscillations of frequency $\sqrt{R_0} = (R_c\tau_c/\tau_0)^{\frac{1}{2}}$ and decay decrement $1/(2\tau_c)$. This conclusion is quantitatively verified by computer experiments [45, 54, 55]. Most of them were concerned with calculations of the autocorrelation function of the translational velocity $\boldsymbol{v}(t)$. However the relation between $\boldsymbol{v}(t)$ and the force $\boldsymbol{F}(t)$ acting during collisions is the same as that between $\boldsymbol{\omega} = \boldsymbol{J}/I$ and $\boldsymbol{M}$. Therefore, the results are qualitatively similar. In Fig. 1.8 we show the correlation functions of the velocity and force for the liquid state density. Oscillations are clearly seen, which point to a regular character of collisions and non-Markovian nature of velocity changes.

The qualitative difference between low-density and high-density rotational relaxation is clearly reflected in the Fourier transform of the normalized angular momentum correlation function:

$$g(\omega) + \mathrm{i}f(\omega) = \frac{1}{\pi}\int_0^\infty \frac{K_J(t)}{K_J(0)} \exp(-\mathrm{i}\omega t)\, \mathrm{d}t. \quad (1.91)$$

Substituting Eq. (1.78) into Eq. (1.91), we find

$$g(\omega) = \frac{\tau_c\,\kappa}{\pi(\tilde{k}_1^2 + \tilde{\omega}^2)(\tilde{k}_2^2 + \tilde{\omega}^2)}, \quad f(\omega) = \frac{\tau_c\,\tilde{\omega}[(\kappa - 1) - \tilde{\omega}^2]}{\pi(\tilde{k}_1^2 + \tilde{\omega}^2)(\tilde{k}_2^2 + \tilde{\omega}^2)}, \quad (1.92)$$

with $\tilde{k}_1 = k_1\tau_c$, $\tilde{k}_2 = k_2\tau_c$, $\tilde{\omega} = \omega\tau_c$. At $\kappa \ll 1/4$ the spectra are bi-Lorentzian and one of the lines is much wider than the other ($\tilde{k}_1 \ll \tilde{k}_2$).

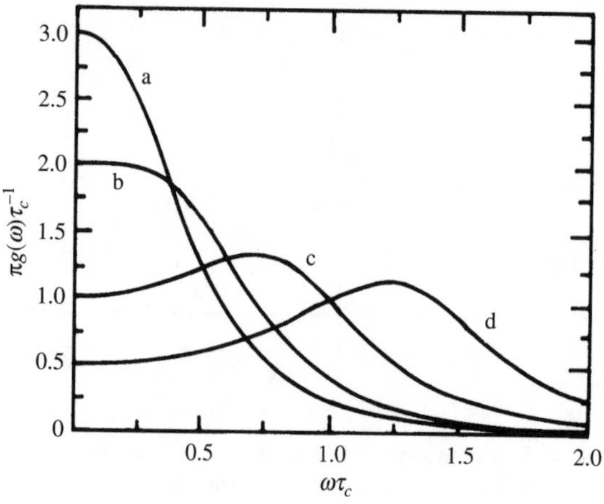

Fig. 1.9. Variation of the spectrum of rotational relaxation with $\tilde{\omega} = \omega\tau_c \geq 0$ and for values of $\kappa > 1/4$, viz. 1/3 (curve a), 1/2 (curve b), 1 (curve c), 2 (curve d).

The latter is negligible in the centre of the spectrum (at $\tilde{\omega} \ll 1$) which looks like a pure Lorentzian, as in the impact approximation with HWHH $1/\tau$. Only far wings are affected at finite collision time τ_c. When, however, $\kappa > 1/4$ ($k_2 = k_1^*$), the situation changes drastically. To describe it let us present the same formulae as

$$g(\omega) = \frac{\tau_c\,\kappa}{\pi[(\tilde{\omega}^2 - \kappa)^2 + \tilde{\omega}^2]}, \quad f(\omega) = \frac{\tau_c\,\tilde{\omega}[(\kappa - 1) - \tilde{\omega}^2]}{\pi[(\tilde{\omega}^2 - \kappa)^2 + \tilde{\omega}^2]}. \tag{1.93}$$

It is clear that the high-density spectra of rotational relaxation $g(\omega)$ have two maxima and these occur when

$$\tilde{\omega} = \pm\left(\kappa - \frac{1}{2}\right)^{1/2}. \tag{1.94}$$

Since the spectra are symmetrical, only their right-hand halves are shown in Fig. 1.9. The higher κ (gas density) is, the larger the splitting of the maxima and the deeper the well in the centre.

Similar results may be obtained not only for exponential $R(t)$ but for the Gaussian model of the kernel [56] and some others. It has recently been shown [57] that $K_J(t)$ changes sign at any κ when

$$R(t) = \frac{R_0}{(1 + t^2/\tau_c^2)^{3/2}} \tag{1.95}$$

is used as a kernel. The difference from the preceding results may be understood on the basis of a general relation following from Eq. (1.71)

$$g(\omega) = \frac{\tilde{R}'}{\pi[R'^2 + (\omega + \tilde{R}'')^2]}, \tag{1.96}$$

where $\tilde{R}' - i\tilde{R}'' = \int_0^\infty R(t) \exp(-i\omega t)\, dt$. According to this relation

$$\int_0^\infty K_J(t)t^2\, dt = -\lim_{\omega\to 0} \frac{\partial^2 g(\omega)}{\partial\omega^2} \propto -\kappa^2\overline{t^2}/\tau_c^2 + 2(1 - \kappa\bar{t}/\tau_c)^2, \tag{1.97}$$

where

$$\overline{t^2} = \int_0^\infty R(t)t^2\, dt/R_o\tau_c = -\lim_{\omega\to 0} \frac{\partial^2 \tilde{R}'(\omega)}{\partial\omega^2} \Big/ R_o\tau_c$$

$$\bar{t} = \int_0^\infty R(t)t\, dt/R_o\tau_c = \lim_{\omega\to 0} \frac{\partial \tilde{R}''(\omega)}{\partial\omega} \Big/ R_o\tau_c.$$

For exponential kernel (1.77) $\overline{t^2} = 2\bar{t}^2 = \tau_c^2$. Consequently $\partial^2 g(\omega)/\partial\omega^2|_0$ becomes positive only at $\kappa > 1/2$. For this condition, which is also seen from Eq. (1.94), $K_J(t)$ changes sign at sufficiently large t and the spectrum $g(\omega)$ is split into two lines with a minimum in between. In contrast, the second moment of the kernel (1.95) diverges and

$$\int_0^\infty K_J(t)\, t^2\, dt \propto -\int_0^\infty R(t)t^2\, dt/R_o\tau_c = -\infty \tag{1.98}$$

is negative for any κ. Thus even in the low-density limit $K_J(t)$ changes sign and approaches 0 from below. This is of course the peculiarity of slowly vanishing memory function (1.95) as well as the correlation function of the corresponding interparticle interaction (1.76). It is known that, for multipole–multiple interaction of rank k, the correlation function for binary collisions decays as t^{-k} [20]. When this interaction is responsible for line broadening in gases the latter is well described by binary (impact) theory only at $k > 3$ [58, 20]. At $k = 3$ the line shift diverges and at $k = 2$ the same happens with the width. This means that binary approach is impossible, however low the gas density. Analogous to this, rotational relaxation at $k = 3$ is not binary at any densities even if the memory function (1.95) is linear in n. Impact (Markovian) description of long-time rotational relaxation is possible only for $k > 3$ and small enough κ.

In dense fluids an additional complication appears, connected with the hydrodynamic nature of the slow collective motion that the molecule performs together with its neighbourhood. According to [59] it results in rather general asymptotic decay $K_J(t) \propto t^{-5/2}$ at times much larger than

10^{-12} s. According to Alder and Wainwright the velocity autocorrelation function decays even more slowly ($K_v(t) \propto t^{-3/2}$) for the same reason [60].

1.9 Non-Markovian differential theory

The theory of Section 1.8 is sometimes qualified as 'non-Markovian' since it accounts for non-exponential angular momentum relaxation, unlike impact theory which is 'Markovian' in this sense. However, it is not a unique non-Markovian generalization of impact theory. Not less known is a differential version of the theory

$$\dot{K}_J(t) = -\int_0^t R(t')\, dt'\; K_J(t). \tag{1.99}$$

It is often obtained from Eq. (1.71) when the kernel is assumed to relax much quicker than the solution to be found. Then it is nothing more than a low-density gas approximation to Eq. (1.71), valid when conditions (1.83) or (1.88) are met. For these conditions the differential theory is expected to be binary in collisions, and

$$\kappa = R_0\tau_c^2 = n\gamma\tau_c \tag{1.100}$$

is a linear function of the buffer gas density n as well as the long-time relaxation rate

$$1/\tau = \int_0^\infty R_0(t)\, dt = R_0\tau_c = n\gamma. \tag{1.101}$$

Although non-Markovian, the differential theory surely has Markovian asymptotics at sufficiently long times:

$$\dot{K}_J(t) = -\frac{1}{\tau} K_J(t) \text{ at } t \gg \tau_c. \tag{1.102}$$

In the impact approximation ($\tau_c = 0$) this equation is identical to Eq. (1.21), angular momentum relaxation is exponential at any times and $\tau = \tau_J$. In the non-Markovian approach there is always a difference between asymptotic decay time τ and angular momentum correlation time τ_J defined in Eq. (1.74). In integral (memory function) theory $R_0\tau_c$ is equal to $1/\tau_J$ whereas in differential theory it is $1/\tau$. We shall see that the difference between non-Markovian theories is not only in times but also in long-time relaxation kinetics, especially in dense media.

In the low-density limit Eq. (1.99) may be derived from the binary quantum theory of J-diffusion which is purely non-adiabatic ($\bar{\omega}\tau_c \ll 1$) and leads to the equation [57]

$$\dot{\mathbf{J}} = -n\hat{\mathbf{k}}(\tau)\mathbf{J} \tag{1.103}$$

with microscopically defined time-dependent 'rate constant'

$$\hat{\mathbf{k}}(\tau) = -\int_{-\infty}^{+\infty} \mathrm{d}t' \left\langle \frac{\partial}{\partial\tau} \exp\left(i \int_{t'-\tau}^{t'} \hat{\mathbf{L}}_i(t)\, \mathrm{d}t \right) \right\rangle, \tag{1.104}$$

where $\langle \ldots \rangle = \int_0^\infty 2\pi b\, \mathrm{d}b \int v f(v)\, \mathrm{d}v$ is an averaging over impact parameters of colliding particles and $\hat{\mathbf{L}}_i$ is a Liouvillian of the interparticle interaction. It is very likely that in the high-temperature (quasi-classical) limit $\mathbf{J}$ is an eigenvector of $\hat{\mathbf{k}}(\tau)$ as it is in the impact approximation [61] and hence

$$\hat{\mathbf{k}}(\mathbf{t})\mathbf{J} = k(t)\mathbf{J}.$$

Then the equation for $K_J = Sp[\rho\mathbf{J}(t)\mathbf{J}]$ simply follows from Eq. (1.103):

$$\dot{K}_J = -nk(t)K_J. \tag{1.105}$$

It is exactly the same as Eq. (1.99) with

$$R = n\dot{k}. \tag{1.106}$$

Using as an example $R(t)$ from Eq. (1.95), we can see from Eq. (1.106) and Eq. (1.100) that, in the low-density limit,

$$R(t) = \frac{n\gamma}{\tau_c(1 + t^2/\tau_c^2)^{3/2}}, \quad k(t) = \frac{\gamma\, t}{\tau_c(1 + t^2/\tau_c^2)^{1/2}}. \tag{1.107}$$

Although approximate, differential non-Markovian theory seems to be exact for the particular model of buffer gas atoms colliding with the molecule but not interacting among themselves. Such a model was first considered by Anderson and Talmen [63] in calculating shapes of atomic spectra in dense (or cooled) gases. However, it may be equally well implemented in estimating the multiparticle collision contribution to rotational relaxation [62, 57]. Of course this is a model of a rather artificial gas which cannot turn into a liquid or solid. Since the atoms move independently of each other, collisions are uncorrelated even if they occur simultaneously. If non-adiabatic interaction of a molecule with N atoms is additive ($\hat{\mathbf{L}} = \sum_i^N \hat{\mathbf{L}}_i$) and all terms commute at any time then averaging over the multiparticle distribution is partitioned into the product of one-particle distributions: $\langle \ldots \rangle_N = \prod_{i=1}^N \langle \ldots \rangle$. If so, an exact averaging of the N-particle evolution operator over all their trajectories becomes possible:

$$\mathbf{J}(t) = \left\langle \exp\left(i \int_0^t \hat{\mathbf{L}}(t')\, \mathrm{d}t' \right) \right\rangle_N \mathbf{J}(0) = \left[\left\langle \exp\left(i \int_0^t \hat{\mathbf{L}}_i(t')\, \mathrm{d}t' \right) \right\rangle \right]^N \mathbf{J}(0). \tag{1.108}$$

In the power expression obtained, a limiting transition is possible $N \to \infty$, $V \to \infty$ at $n = N/V = inv$ where V is the volume of a sample. A familiar result follows [63, 58, 20]:

$$\mathbf{J}(t) = \exp\left[-n \int_{-\infty}^{\infty} \left\langle \mathbf{I} - \exp\left(i \int_0^t \hat{\mathbf{L}}_i(\tau - t')\, \mathrm{d}\tau \right) \right\rangle \mathrm{d}t' \right] \mathbf{J}(0). \tag{1.109}$$

This is exactly an integral of Eq. (1.103) with $\hat{\mathbf{k}}(t)$ defined in Eq. (1.104). Hence the integral of differential equation (1.105)

$$K_J(t) = rd\ \exp\left(-n \int_0^t k(t')\, \mathrm{d}t' \right) \tag{1.110}$$

seems to be an exact result for an Anderson–Talmen model free of binary limitations (1.83) or (1.88). The authors of [62] believe that this conclusion is confirmed by their computer simulations of angular momentum relaxation in supra-critical nitrogen performed in a wide range of densities.

Using $k(t)$ from Eq. (1.107) as an example, we find from Eq. (1.110) that

$$K_J(t) = rd\ \exp\{-n\gamma\tau_c[(1 + t^2/\tau_c^2)^{\frac{1}{2}} - 1]\} \tag{1.111}$$

or

$$K_J(t) = rd \left[1 - \frac{\kappa}{2}\left(\frac{t}{\tau_c}\right)^2 + \frac{\kappa}{8}\left(\frac{t}{\tau_c}\right)^4 \left(1 + \frac{\kappa}{3}\right) \right] \quad \text{at } t \ll \tau_c \tag{1.112a}$$

$$K_J(t) = rd\ \exp\left[\kappa(1 - t/\tau_c)\right] = rd\,\mathrm{e}^{\tau_c/\tau}\mathrm{e}^{-t/\tau} \quad \text{at } t \gg \tau_c. \tag{1.112b}$$

Since relaxation is initially non-exponential, the true correlation time (1.69) does not coincide with $\tau = 1/n\gamma$, but is equal to

$$\tau_J = \tau\kappa\mathrm{e}^{\kappa}K_1(\kappa), \tag{1.113}$$

where $K_1(\kappa)$ is a modified Bessel function. As is seen from Fig. 1.10 and the expansions in $\kappa = \tau_c/\tau$

$$\tau_J/\tau = 1 + \kappa + \frac{1}{2}\kappa^2 \ln\kappa \quad \text{at } \kappa \ll 1, \tag{1.114a}$$

$$\tau_J/\tau = (\pi\kappa/2)^{\frac{1}{2}} \quad \text{at } \kappa \gg 1, \tag{1.114b}$$

the angular momentum relaxation increases more slowly with gas density than would be expected in binary impact theory. This means that collective interaction with atoms is less effective than the sum of independent collisions well separated in time (as at $\tau_c = 0$). As is seen from Eq. (1.112), $K_J(t)$ is positive at any t and exponentially approaches 0 at $t \to \infty$. The last stage is the longest and is dominant in the rarefied gas limit when the binary approximation is valid. In this approximation Eq. (1.114a) is reduced to linear relation $\tau_J = \tau(1+\kappa) = \tau + \tau_c$, which follows also from (1.85) and (1.86) and hence is the same for any non-Markovian theory of rotational relaxation. However, this conclusion follows from our choice of $k(t)$ and is not valid for other problems where it is different. For instance, quenching of an exciton by independently moving energy acceptors is also described rigorously by differential theory, but $k(t)$ decreases with time in contrast to Eq. (1.107). As a result, the integral life time is less than its asymptotic value and collective quenching at high concentrations is more effective than in the binary approximation [64].

Let us compare in detail the differential theory results with those obtained for rotational relaxation kinetics from the memory function formalism (integral theory). Using $R(t)$ from Eq. (1.107) as a kernel of Eq. (1.71) we can see that in the low-density limit

$$K_J(t) = rd\left[1 - \frac{\kappa}{2}\left(\frac{t}{\tau_c}\right)^2 + \frac{\kappa}{8}\left(\frac{t}{\tau_c}\right)^4(1+\kappa)\right] \quad \text{at } t \ll \tau_c \qquad (1.115)$$

differs from the short-time expansion (1.112a) only in second-order corrections with respect to small parameter κ. As is seen from Fig. 1.11, significant deviations appear at longer times since $K_J(t)$ in integral theory develops faster than in differential theory. The same was true for the model with exponential kernel: rate (1.85) is also greater than that found in Eq. (1.101). The principal difference between the two models appears at the very last stage of rotational relaxation. As was mentioned, in integral theory $K_J(t)$ does not approach 0 exponentially at any κ if the kernel (1.95) is used instead of (1.77). It inevitably follows from Eq. (1.98) that $K_J(t)$ has to be negative for large enough t. Although at low densities it changes sign, being very small (Fig. 1.12), this behaviour is qualitatively different to the exponential decay predicted by both differential and impact theories.

The difference is even more pronounced at $\kappa \gg 1$ when the functions

$$\tilde{R}' = \frac{x}{\tau}K_1(x), \quad \tilde{R}'' = \frac{x}{\tau}\left\{1 + \frac{\pi}{2}[L_1(x) - I_1(x)]\right\}$$

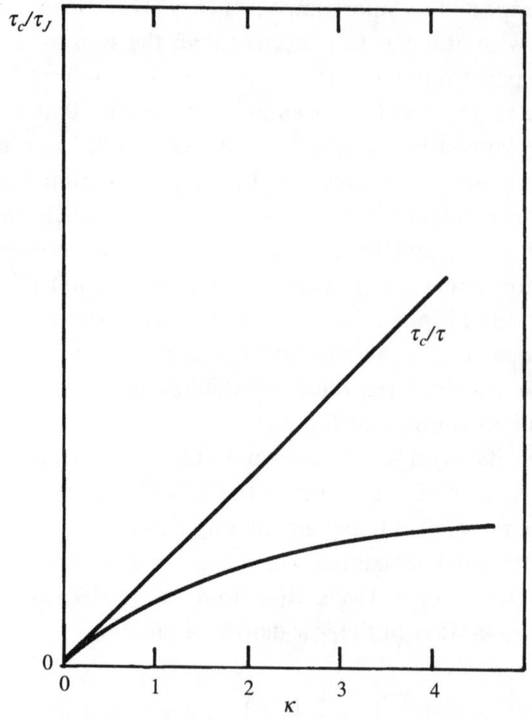

Fig. 1.10. The nonlinear dependence of inverse correlation time ($1/\tau_J$) on gas density n ($\kappa = \tau_c/\tau = n\gamma\tau_c$) in comparison with impact relaxation rate $1/\tau$ linear in n.

may be replaced by their high-frequency asymptotics at $x = \omega\tau_c \gg 1$ ($L_1(x)$ is a modified Struve function). Thus we get from Eq. (1.96)

$$g(\omega) = \frac{1}{\mathrm{i}(\omega - \omega_o^2/\omega) + (\pi\omega\tau_c/2\tau^2)^{\frac{1}{2}} \exp(-\omega\tau_c)}. \tag{1.116}$$

Restricting ourselves to frequencies near $\omega_o = 1/(\tau_c\tau)^{\frac{1}{2}}$, we approximate this expression by Lorentzian

$$g(\omega) = \frac{1}{\mathrm{i}(\omega - \omega_o) + \Gamma}$$

with

$$\Gamma = \frac{1}{\tau}\left(\pi\sqrt{\kappa}/2\right)^{1/2} \exp(-\sqrt{\kappa}). \tag{1.117}$$

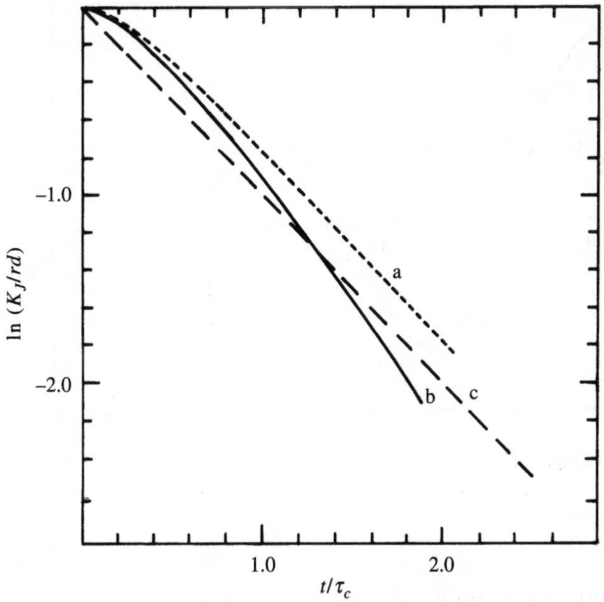

Fig. 1.11. The normalized angular momentum correlation function $K_J(t)/K_J(0)$ at $\kappa = 0.25$ in differential (curve a), integral (curve b) and impact (curve c) theories.

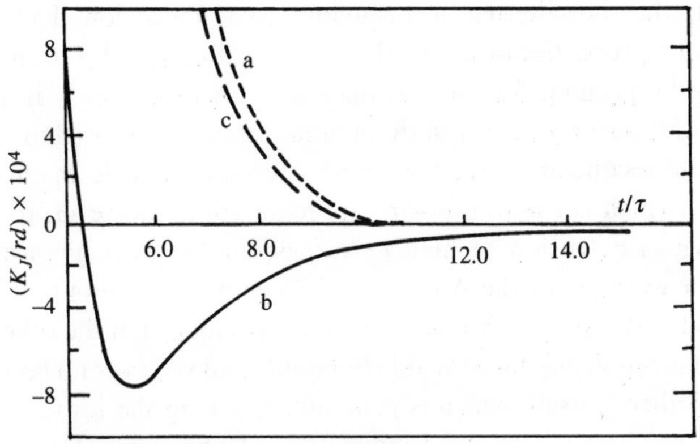

Fig. 1.12. Kinetics of rotational relaxation for the same parameters and designations as in Fig.1.11 but in the region of long times.

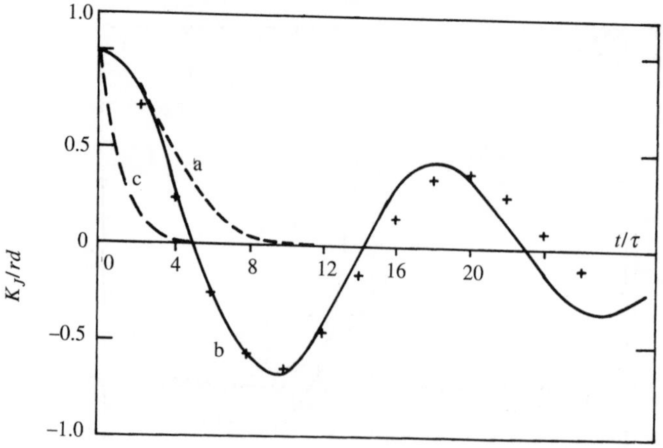

Fig. 1.13. Kinetics of rotational relaxation in differential (curve a), integral (curve b) and impact (curve c) theories at $\kappa = 10$ (+ + + + asymptotic result of Eq. (1.118)).

Hence we find by inverse Fourier transform of $g(\omega)$

$$K_J(t) = \exp(-\Gamma t) \cos(\omega_o t) \text{ at } \kappa \gg 1. \tag{1.118}$$

This result together with $K_J(t)$ calculated numerically for $\kappa = 10$ is given in Fig. 1.13.

Whereas the sign-alternating behaviour of $K_J(t)$ at low densities depends on the properties of a model kernel, oscillatory relaxation at high κ seems to be peculiar for integral (memory function) theory in general. In contrast, it never happens in differential theory. If we exclude kernels with infinite second moment then both theories coincide in the lowest order in κ, which is the true binary approximation. However, their phenomenological extension to high κ is questionable as it gives different results. The example of the Anderson–Talmen model shows that oscillations found with integral theory may be the artefacts. On the other hand, this model is physically inapplicable to liquids and perhaps no better than differential theory itself, which is generally speaking the low-density approximation. There is also the disadvantage of phenomenological choice of kernel $R(t)$ or the related time-dependent rate constant. Moreover, they are assumed to have the same shape at any densities. Even in cases when linear relation (1.100) is replaced by nonlinear dependence $\kappa(n)$ the shape remains unchanged, and this is a demerit which may be removed only by a consistent theory of collective motion.

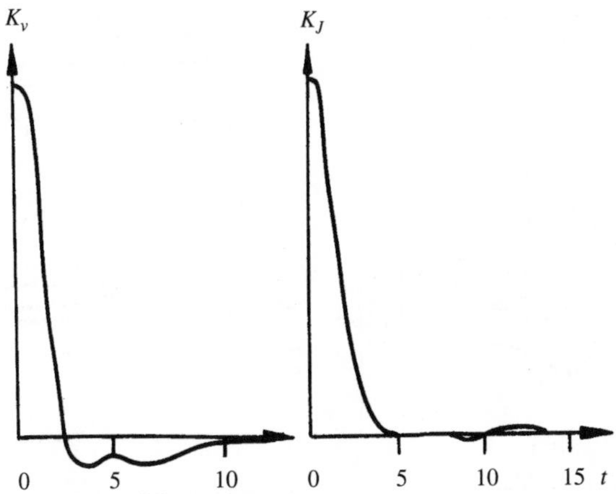

Fig. 1.14. Comparison of the MD calculations of the correlation functions of the translational velocity and angular momentum in liquid nitrogen [65]. The time is in units of 10^{-13} s.

1.10 Rotational kinetics and relaxation time

The effect produced by collisions on the translational velocity relaxation differs from that which they have on the rotational velocity. Rotational relaxation is induced by the anisotropic part of the interaction potential which decreases as the shape of a molecule becomes closer to a sphere. Thus rotational relaxation of spherical and diatomic molecules proceeds at a lower rate and remains exponential in a wider range of densities than does relaxation of translational velocity. In nitrogen, for instance, oscillations in rotational relaxation may be discerned solely near a triple point, and they are considerably weaker than those related to the correlation function of translational velocity (Fig. 1.14). The data presented in Fig. 1.14 are results of the first MD calculations of this type [65]. To date, these computer simulations remain among the best. They have been performed for 500 hard diatomics for nitrogen densities at the triple point and near the boiling point. The integral (memory function) theory was used with exponential and Gaussian kernels to fit to experimental data [65]. As R_0 was assumed to be equal to $\ddot{K}_J(0)/rd$ the optimal τ_c was found from Eq. (1.74). It turned out that near the triple point of nitrogen $\tau_c = 0.86 \times 10^{-13}$ s, while $\tau_J = 1.7 \times 10^{-13}$ s. Inequality (1.88) is obviously violated but still not inverted. However, in such liquids as F_2 and CCl_3F the value of τ_J measured by NMR relaxation is 0.95×10^{-13} s and 0.58×10^{-13} s, respectively [66, 67], which is less than τ_c. In a memory

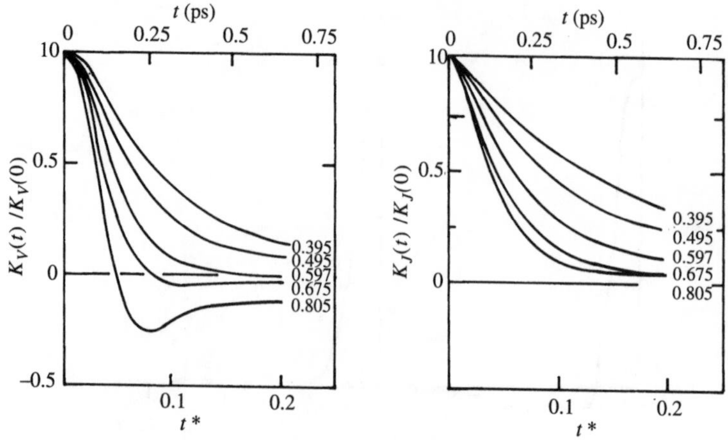

Fig. 1.15. Translational and angular velocity correlation functions for nitrogen. MD simulation data from [70], $T = 122$ K, densities are indicated in the figure. Reduced units: for time $t^* = (\epsilon/\sigma^2)^{\frac{1}{2}}$, for density $\rho^* = \rho\sigma^3$, m is the nitrogen mass, ϵ and σ are the parameters of the Lennard-Jones 12–6 site–site potential, $\epsilon/k = 36.4$ K, $\sigma = 3.32$ Å, top axis is time in picoseconds.

function formalism the above phenomenon is naturally accounted for by the fact that τ_J, being the integral characteristic of the oscillating relaxation, can be less than its decay time which is $2\tau_c$ according to Eq. (1.90).

The rotational correlation function $K_J(t)$ seems to be the most intriguing physical quantity for liquids composed of small molecules. Since it is unlikely to be measured experimentally we refer from time to time to the NMR measurements of τ_J, which is just the zeroth moment of K_J over time. Even these not very informative data are available for only a few cases. Moreover, being an integral, the magnitude of τ_J tells almost nothing about the shape or the negative tail of $K_J(t)$ associated with complicated collective phenomena of dynamical alignment. That is why MD simulation of $K_J(t)$ becomes of such great importance. Unfortunately, a very restricted number of investigations have been dedicated to this problem. For example, the authors of a detailed digest [68] restrict themselves to reviewing papers on translational correlation functions $K_v(t)$. Obviously, this may be due to the fact that MD simulations are very time-consuming and expensive and therefore rather rarely can be used to solve problems that do not have (or are at least not supposed to have) any practical significance. On the other hand, calculations of rotational correlation functions are connected with some principal problems

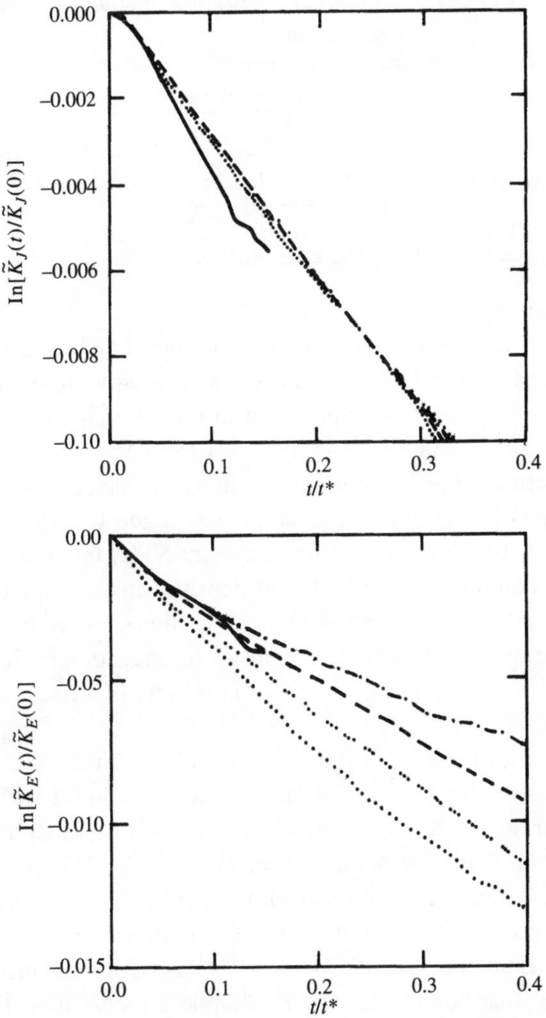

Fig. 1.16. Angular momentum and kinetic energy correlation functions for compressed nitrogen. MD simulation from [62]. $T = 300$ K. The lines are continuous (800 amagat), close dotted (600 amagat), sparse dotted (400 amagat), dashed (300 amagat) and pairwise dotted (200 amagat). Reduced time units are the same as in Fig. 1.15.

Table 1.1. *Cross-sections of rotational energy and angular momentum relaxation.*

	Density (Amagat)				
	200	300	400	600	800
σ_E (Å^2)	7.6	10.2	15.7	24.4	45.4
σ_J (Å^2)	10.1	13.5	20.8	32.3	60

of MD, e.g., the absence of isotropy due to boundary conditions [69].

However, interest in the MD calculations was greatly stimulated by study of orientational relaxation expounded in the next chapter. It is of particular importance for line shape analysis of FIR spectra and Rayleigh depolarized scattering [70]. The correlation functions calculated in [70] are shown in Fig. 1.15. In [62] an attempt was made to calculate K_J and the correlation function of rotational energy $K_E(t)$ for gaseous N_2 under isothermic compression up to liquid densities in the supercritical region (Fig. 1.16). Almost exponential relaxation allows one to find rates of angular momentum and rotational energy in the domain 200–800 amagat at $T = 300$ K. Cross-section representations of these data are listed in Table 1.1.

Experimental data on nitrogen obtained from spin–lattice relaxation time (T_1) in [71] also show that τ_J is monotonically reduced with condensation. Furthermore, when a gas turns into a liquid or when a liquid changes to the solid state, no breaks occur (Fig. 1.17). The change in density within the temperature interval under analysis is also shown in Fig. 1.17 for comparison. It cannot be ruled out that condensation of the medium results in increase in rotational relaxation rate primarily due to decrease in free volume. In the rigid sphere model used in [72] for nitrogen, this phenomenon is taken into account by introducing the factor $g(\eta)$ into the angular momentum relaxation rate

$$1/\tau_J = g(\eta) n \sigma_J \langle v \rangle, \tag{1.119}$$

where $\eta = \pi d_0^3 n/6$, and

$$g\,(\eta) = (2-\eta)/2(1-\eta)^3 \tag{1.120}$$

is taken from the conventional theory of simple liquids [73]. Nonlinear increase in rotational relaxation rate with density is provided by the finite

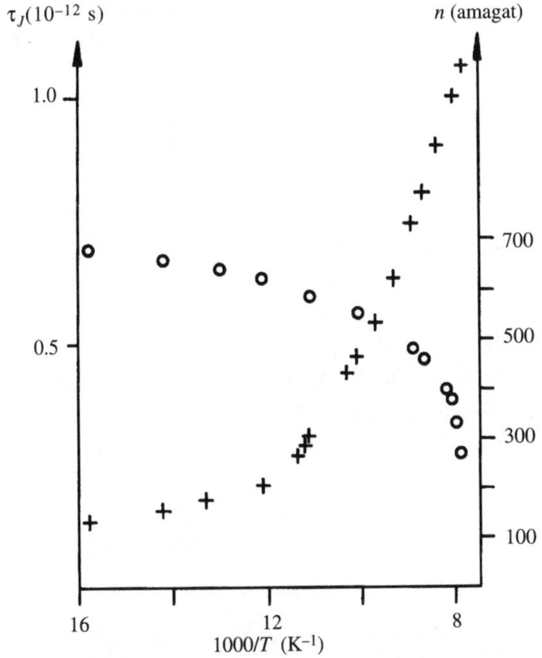

Fig. 1.17. The temperature-dependence of angular momentum relaxation time (+) in nitrogen [71] and accompanying density change due to cooling (0).

hard sphere diameter d_0 considered as a fitting parameter. This is in line with a collision rate of hard spheres that was proved to be exactly the product of the dilute-gas collision rate and the pair distribution function $g(d_0, n)$ of two hard spheres in contact [74]:

$$1/\tau_0 = g(d_0^3 n) n \pi d_0^2 \langle v \rangle.$$

This expression was shown to be valid for hard sphere systems of arbitrary density. Moreover, the free-path distribution scaled by $\lambda = \langle v \tau_0 \rangle$ is nearly density-independent and almost the same as in the limit of zero density [74].

Of course, the effect of excluded volume is opposite and greatly exceeds that shown in Fig. 1.10, which is produced by uncorrelated collective interaction. Unfortunately, neither of them results in sign-alternating behaviour of angular or translational momentum correlation functions. This does not have a simple explanation either in gas-like or solid-like models of liquids. As is clearly seen from MD calculations, even in

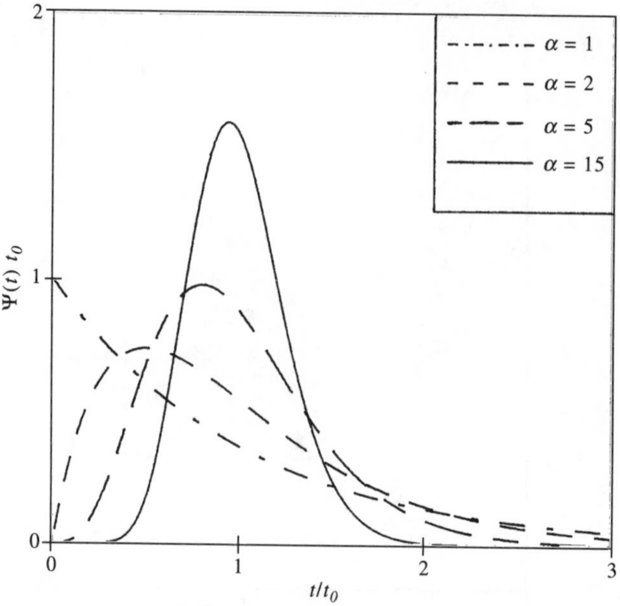

Fig. 1.18. Time distribution of free flights for $\alpha = 1, 2, 5, 15$ [77].

hard particle models, the correlation functions acquire negative values at long times [75]. In the latter case the effect must have an origin that is collective, but purely collisional.

The hypothesis proposed in [76] implies that collective effects change the collisional event statistics in a hard sphere liquid. The collisional distribution $\psi(t)$ dt is assumed to deviate from the famous Poissonian law $\psi(t) = \exp(-t/\tau_0)/\tau_0$ corresponding to the flow of events being uniform in time. The distribution

$$\psi_\alpha(t) = \frac{\alpha^\alpha}{t_o^\alpha \Gamma(\alpha)} t^{\alpha-1} \mathrm{e}^{-\alpha t/t_o} \tag{1.121}$$

is chosen as the simplest example in [77]. To obtain sign-alternating $K_J(t)$ or $K_v(t)$, an idealized inversion of corresponding momenta at any collision ('back reflection') must also be assumed [76, 75]. The negative loop in rotational or translational correlation functions appears if

(i) collisional distribution (1.121) exhibits an extremum at $t = t_e > 0$ and
(ii) Keilson–Storer parameter γ is negative and close to -1.

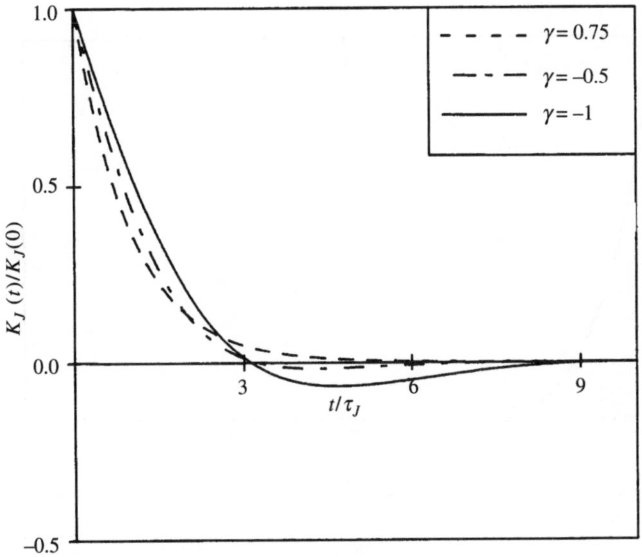

Fig. 1.19. Correlation function of angular momentum for $\alpha = 2$ and $\gamma = 0.75, -0.5, -1$ [77].

If the molecule moves without hindrance in a rigid-walled enclosure (the 'free enclosure'), as assumed in free volume theories, then 'rattling' back and forth is a free vibration, which could be considered as coherent in such a cell. The transfer time between opposite sides of the cell t_o is roughly the inverse frequency of the vibration. The maximum in the free-path distribution was found theoretically in many cells of different shape [74]. In model distribution (1.121) it appears at $\alpha \geq 2$ and shifts to t_o at $\alpha \to \infty$ (Fig. 1.18). At $\gamma \approx -1$ coherent vibration in a cell turns into translational velocity oscillation as well as a molecular libration (Fig. 1.19).

'Back reflection' of translational and rotational velocity is rather reasonable, but the extremum in the free-path time distribution was never found when collisional statistics were checked by computer simulation. Even in the hard-sphere solid the statistics only deviate slightly from Poissonian at the highest free-paths [74] in contrast to the prediction of free volume theories. The collisional statistics have recently been investigated by MD simulation of 108 hard spheres at reduced density $n/n_0 = 0.65$ (where n_0 is the density of closest packing) [75]. The obtained ratio $\overline{t^2}/\overline{t}^2 = 2.07$ was very close to 2, which is indirect evidence for uniform

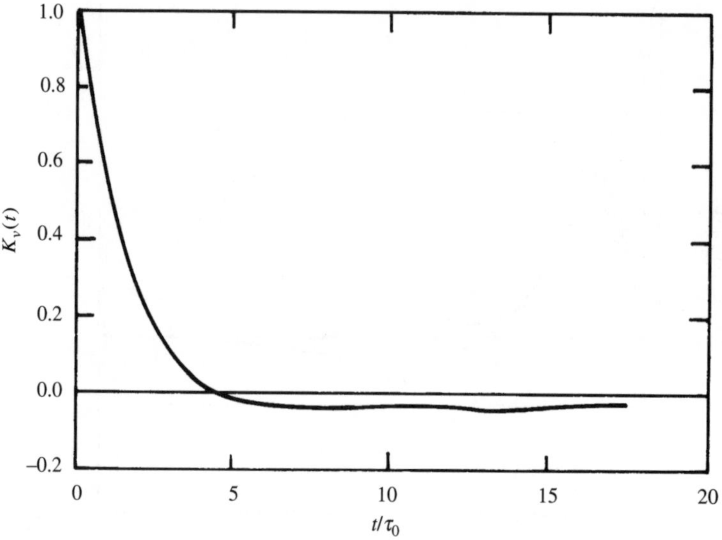

Fig. 1.20. Velocity autocorrelation function for the hard sphere fluid at a reduced density $\rho/\rho_0 = 0.65$ as a function of time measured in $\bar{t}$ units [75].

distribution in time of collisions ($\bar{t}$ is the averaged time interval between two successive collisions, $\overline{t^2}$ the averaged square of the same interval). The observed distribution was considered to be Poissonian with mean free path time $\tau_0 = \bar{t}$. At the same time the translational autocorrelation function has a negative loop, shown in Fig. 1.20, and crossover occurs at $t_e \approx 4.5\tau_0$.

This is an indication of the collective nature of the effect. Although collisions between hard spheres are instantaneous the model itself is not binary. Very careful analysis of the free-path distribution has been undertaken in an excellent old work [74]. It showed quite definite although small deviations from Poissonian statistics not only in solids, but also in a liquid hard-sphere system. The mean free-path λ is used as a scaling length to make a dimensionless free-path distribution, λp, as a function of a free-path length r/λ. In the zero-density limit this is an ideal exponential function $(\lambda p)_0$. In a one-dimensional system this is an exact result, i.e., $\lambda p/(\lambda p)_0 = 1$ at any density. In two dimensions the dense-fluid scaled free-path distributions agree quite well with each other, but not so well with the zero-density scaled distribution, which is represented by a horizontal line (Fig. 1.21(a)). The maximum deviation is about

(a)

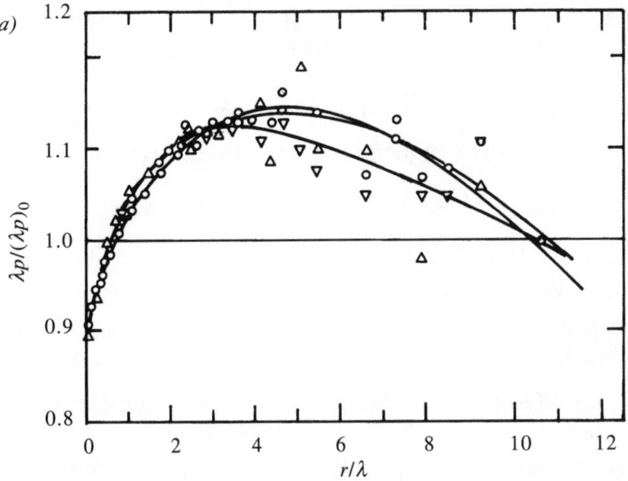

(b)

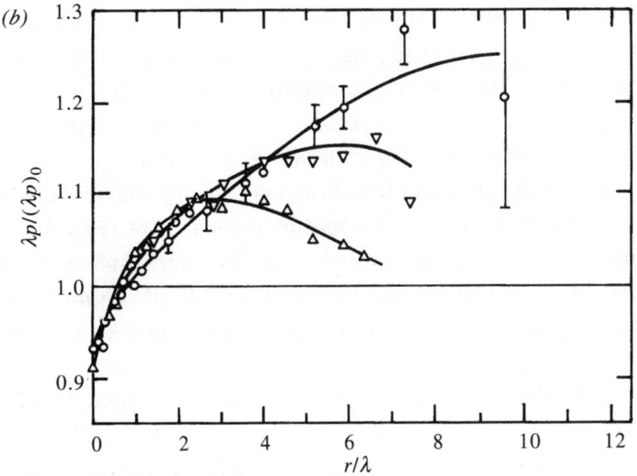

Fig. 1.21. Ratio of free-path distribution λp to the scaled zero-density free-path distribution plotted as a function of reduced free-path length r/λ for two-dimensional (a) and three-dimensional (b) liquids. Circles, inverted triangles and upright triangles refer to reduced volumes V/V_0 of 1.6, 2, and 3, respectively (V_0 is the volume of the system at close packing) [74].

12% and occurs at 3–5 mean free paths. In three dimensions, free-path distributions do not agree even with each other (Fig. 1.21(*b*)). Moreover, deviations from the zero-density scaled distribution are larger, as much as 22%, and occur at free-path lengths which depend much more on density. Semiquantitatively these deviations may be taken in account by a small correction to the exponential distribution of free paths or times:

$$\psi(t) = (1 - \aleph)\exp(-t/\tau_0)/\tau_0 + \aleph\psi_\alpha(t). \tag{1.122}$$

According to Table II in [74] the fitting parameters $\aleph \approx 0.08$ and $t_0 \approx 4\tau_0$ are weakly dependent on density in the interval $1.5 < V/V_0 < 3.0$. At $\aleph = 0$ the distribution (1.122) coincides with the Poissonian law whereas at $\aleph = 1$ it reduces to one of the family (1.121). At $\alpha \geq 2$ it compels $K_J(t)$ to change sign at $t \approx t_0$. In reality $\aleph$ is at the most 0.1, which is hopefully enough to explain the small negative loop in the tail of $K_J(t)$. The more pronounced this loop is, the smaller τ_J is in comparison with τ_0.

In a solid medium there is an additional reason for τ_J being anomalously short. Interaction with the neighbourhood, which is different from zero in a crystal lattice, should be taken into consideration in (1.60). This is achieved by introduction of a regular angle-dependent potential in which a molecule must rotate. If potential barriers cannot be overcome, then rotation is transformed into libration, and the motion becomes finite in the angular space. Consider a one-dimensional case (Fig. 1.22). In this case rotation angle θ is connected to angular momentum J by the relation $\dot{\theta} = J/I$, which is analogous to (1.60). If librations occur in the infinitely deep well, $\theta(t)$ is a stationary process, and $\tau_J = 0$ for the same reason that makes τ_M become 0 in Eq. (1.67). However, it is not so because the barriers are of finite height U_0 and may be overcome. In this case [78]

$$\tau_J = \tau_J^0 \, e^{-U_0/kT}, \tag{1.123}$$

where τ_J^0 is the angular momentum relaxation time during motion over the barrier. The exponential factor in (1.123) is an additional independent reason for reduction of τ_J in glasses and to some extent in liquids.

The activated character of the dependence $\tau_J(T)$ shown in (1.123) is often considered as a feature suggesting a quasicrystal model of the liquid. Data taken from liquid–vapour co-existence curves are frequently analysed in coordinates $\ln \tau_J$ from $1/T$ in order to determine U_0. The point that $\tau_J(n, T)$ is a function not only of the temperature T, but also of the density n is ignored. The density along the co-existence curve is

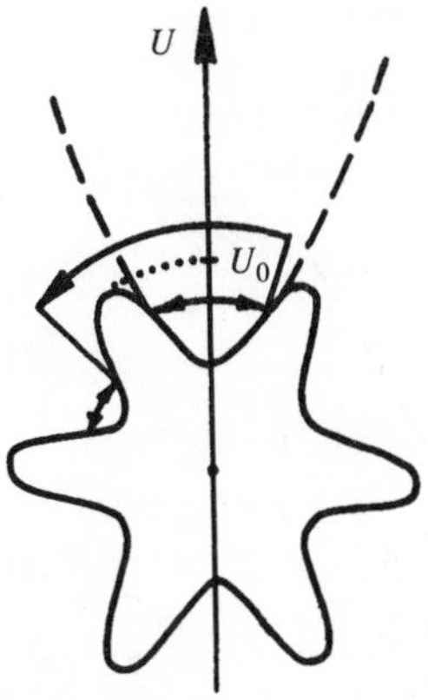

Fig. 1.22. Angular-dependent potential U for one-dimensional libration over barriers of height U_0. The arrows show the way of libration below barriers and random translations from one well to another due to high-energy fluctuations. The broken line presents the approximation of the parabolic well valid at the bottom.

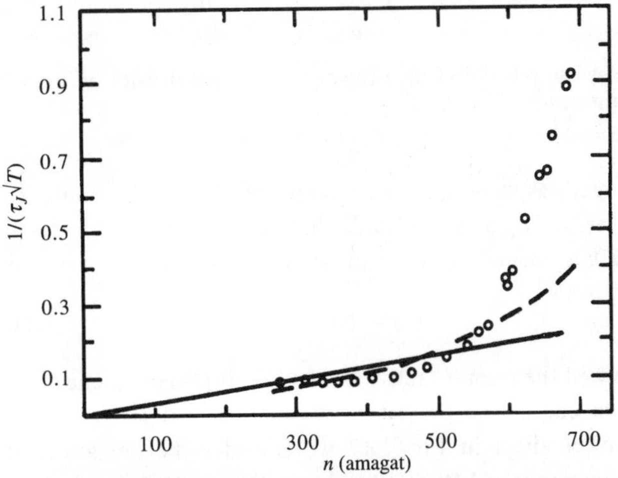

Fig. 1.23. Density-dependence of angular momentum relaxation rate. Points correspond to experimental data presented in Fig. 1.17. The straight solid line is a binary estimation of this rate with the cross-section $\sigma_J = 3 \times 10^{-15}$ cm^2 and the broken curve presents the result obtained in the rough-sphere approximation used in [72, 80].

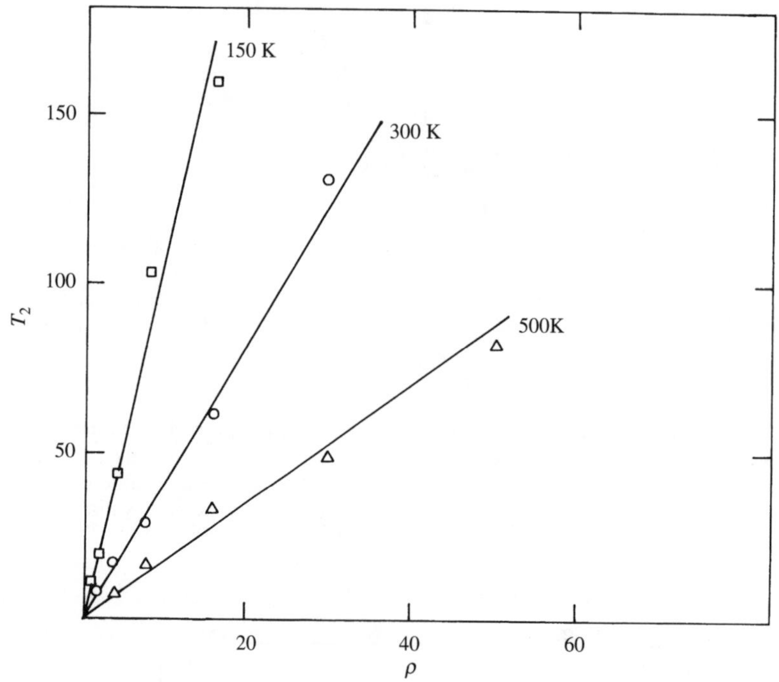

Fig. 1.24. Transverse relaxation time (in ms) versus gas density (in amagat) at various temperatures for $^{15}N_2$ [81].

abruptly changed, especially in the vicinity of the critical point. It was shown in [79] with SF_6 as an example that, with allowance for the actual change in $n(T)$, a normal binary estimate of the frequency of collisions

$$1/\tau_J = n\sigma_J \langle v \rangle \tag{1.124}$$

imitates quite well the pseudo-activated temperature-dependence τ_J along the co-existence curve.

Using the data given in Fig. 1.17 we consider the deviation of the isothermic dependence of $1/\tau_J(\eta)$ from linear (binary) relationship (1.124). The dependence of $1/(\tau_J\langle v \rangle)$ on η in a liquid is presented in Fig. 1.23. The experimental results practically coincide with a straight line corresponding to a binary approximation up to a critical point. Hence the impact approximation is not too bad even for moderately condensed gases. However, the abrupt increase in $1/\tau_J$ observed in the cryogen liquid is too sharp to be described even with the hard-sphere correction

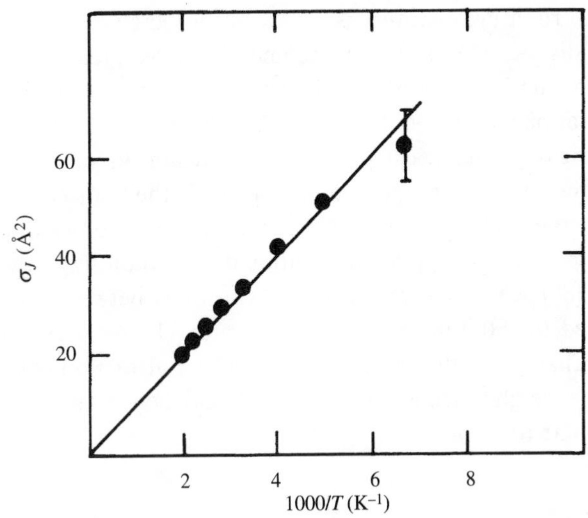

Fig. 1.25. Temperature-dependence of rotational relaxation cross-section from [81]. For the lowest temperature point the experimental uncertainty is indicated, the latter being the biggest one over the whole set of measurements.

factor (1.120). The latter serves well for high-temperature compressed gas, but for real liquids it is better to use actual experimental data for τ_J when they are available.

To extract information about τ_J from NMR data, the transverse relaxation time T_2 may be used as well as the longitudinal time T_1. For gaseous nitrogen it was done first with T_2 in [81] and confirmed later [82] when T_1 was measured and used for the same goal. The NMR linewidth of $^{15}N_2$ is the inverse of T_2, and the theory, relating to T_2 to τ_J, is well known [39, 83]. For the case of diatomic and linear molecules the formula is

$$T_2^{-1} = \frac{4}{3}(IkT/\hbar^2)(2\pi C)^2\tau_J.$$

The value of the magnetic hyperfine interaction constant $C = 22.00$ kHz is supposed to be reliably measured in the molecular beam method [71]. Experimental data for $^{15}N_2$ are shown in Fig. 1.24, which depicts the density-dependence of $T_2 = (2\pi\Delta\nu_{1/2})^{-1}$ at several temperatures. The fact that the dependences $T_2(\rho)$ are linear until 200 amagat proves that binary estimation of the rotational relaxation rate is valid within these limits and that Eq. (1.124) may be used to estimate cross-section σ_J from

τ_J values found from NMR data. The results are shown in Fig. 1.25. The cross-section at room temperature is 34 Å^2, but it increases as $1/T$ when temperature falls. At $T = 150$ K it becomes 1.5 times greater than that calculated from the Van der Waals diameter $d_0 = 3.7$ Å. This fact casts doubt on the applicability of the hard-sphere model to τ_J calculations at least in the gas phase. Most probably, J-diffusion in gases is chiefly induced by attractive long-range interactions while the 'molecular size' is determined by the repulsive potential branch. The difficulties arising in the hard-sphere model as applied to rotational relaxation have also been mentioned in [84] when the transition cross-sections between rotational levels, measured for HCl in argon, amount up to 115 Å^2. However, the situation may change for the better in liquids where attraction with many particles in the neighbourhood is smoothed and hence less effectively hinders molecular rotation.

2

Orientational relaxation in dense media

Information on orientational relaxation may be obtained by a wide range of techniques. Dielectric relaxation and magnetic resonance, neutron and light scattering, infrared spectroscopy and fluorescence depolarization are widely used [8, 85]. These different experimental probes of the phenomenon characterize it in different ways. The advantage of spectroscopic investigations is that they give information on relaxation times as well as on the corresponding correlation functions and their spectra. In particular, by combining the information in an absorption spectrum with that obtained from Raman scattering, one can determine the two lowest correlation functions of a molecule's axis position. A complete description of orientational relaxation is given by the infinite set of these functions.

The orientation of linear rotators in space is defined by a single vector directed along a molecular axis. The orientation of this vector and the angular momentum may be specified within the limits set by the uncertainty relation. In a rarefied gas angular momentum is well conserved at least during the free path. In a dense liquid it is a molecule's orientation that is kept fixed to a first approximation. Since collisions in dense gas and liquid change the direction and rate of rotation too often, the rotation turns into a process of small random walks of the molecular axis. Consequently, reorientation of molecules in a liquid may be considered as diffusion of the symmetry axis in angular space, as was first done by Debye [1].

The Debye phenomenology is consistent with both gas-like and solid-like model representations of the reorientation mechanism. Reorientation may result either from free rotation paths or from jumps over libration barriers [86]. Primary importance is attached to the resulting angle of reorientation, which should be small in an elementary step. If it is small enough, perturbation theory with respect to this parameter may be used to describe orientational relaxation ('rotational diffusion') in dense media. Perturbation calculation of correlation functions and their spectra for linear and spherical rotators are discussed here before a traditional description of rotational diffusion in terms of distribution functions in the angular space.

The results of the Debye theory reproduced in the lowest order of perturbation theory are universal. Only higher order corrections are peculiar to the specific models of molecular motion. We have shown in conclusion how to discriminate the models by comparing deviations from Debye theory with available experimental data.

2.1 Correlation characteristics of rotational relaxation

If the resolving capacity of the instruments is ideal then vibrational–rotational absorption and Raman spectra make it possible in principle to divide and study separately vibrational and orientational relaxation of molecules in gases and liquids. First one transforms the observed spectrum of infrared absorption F^{IR} and that of Raman scattering F^{R} into spectral functions

$$I^{\mathrm{IR}}(\omega) \propto F^{\mathrm{IR}}(\omega) \exp\left(-\hbar\omega/2kT\right), \tag{2.1}$$

$$I^{\mathrm{R}}(\omega) \propto F^{\mathrm{R}}(\omega) \exp\left(-\hbar\omega/2kT\right), \tag{2.2}$$

which are symmetrical in frequency and normalized to 1. By performing Raman scattering experiments with polarized light one may determine the spectral function of scattered light of the same polarization as that of the incident light $I_{\parallel}^{\mathrm{R}}$ and that having a polarization perpendicular to the exciting light $I_{\perp}^{\mathrm{R}}$. Using this information, the contributions of isotropic and anisotropic scattering may be separated [2, 87]:

$$I_{\mathrm{is}}^{\mathrm{R}} = I_{\parallel}^{\mathrm{R}} - \frac{4}{3} I_{\perp}^{\mathrm{R}}, \tag{2.3}$$

$$I_{\mathrm{an}}^{\mathrm{R}} = I_{\perp}^{\mathrm{R}}. \tag{2.4}$$

The following correlation functions are found by the Fourier transformation of spectral functions (2.1), (2.3) and (2.4):

$$K_0 = \mathrm{FT}\, I_{\mathrm{is}}^{\mathrm{R}}, \tag{2.5}$$

$$K_1 K_0 = \mathrm{FT}\, I^{\mathrm{IR}}, \tag{2.6}$$

$$K_2 K_0 = \mathrm{FT}\, I_{\mathrm{an}}^{\mathrm{R}}, \tag{2.7}$$

where $K_0(t)$ characterizes solely vibrational relaxation and $K_1(t)$ and $K_2(t)$ characterize orientational relaxation. The latter may be calculated by simple division of (2.6) and (2.7) by (2.5). The above procedure is the only correct method for experimental determination of true correlation functions from infrared and Raman spectra. Each of the autocorrelation functions considered below is an equilibrium average $\langle \ldots \rangle$ of a binary product of the corresponding molecular characteristic [8]:

$$K_0 = \frac{\langle \bar{\alpha}(0)\bar{\alpha}(t)\rangle}{\langle \bar{\alpha}^2\rangle}, \quad K_1 = \frac{\langle \boldsymbol{\mu}(t)\cdot\boldsymbol{\mu}(0)\rangle}{\langle \boldsymbol{\mu}^2\rangle}, \quad K_2 = \frac{\langle S_P\hat{\beta}(t)\hat{\beta}(0)\rangle}{\langle S_P\hat{\beta}^2\rangle}. \tag{2.8}$$

Here $\bar{\alpha} = S_P\hat{\alpha}/3$ is the average value of the polarization tensor of the molecule, $\hat{\beta} = \hat{\alpha} - \hat{I}\bar{\alpha}$ being its anisotropy, and $\boldsymbol{\mu}$ the dipole moment of the molecule. We assume that the concentration of active molecules in the gas mixture or liquid solution is so small that intermolecular coupling may be neglected.

Although K_1 and K_2 are defined by physical quantities of different nature, their time evolution is universally determined by orientational relaxation. This discussion is restricted to linear molecules and vibrations of spherical rotators for which $\hat{\beta}$ is a symmetric tensor: $\beta_{ik} = \beta_{ki}$. In this case the following relation holds

$$\langle S_P\,\hat{\beta}(0)\,\hat{\beta}(t)\rangle \propto \left\langle \frac{3}{2}\ [\boldsymbol{u}(0)\cdot\boldsymbol{u}(t)]^2 - \frac{1}{2}\right\rangle . \tag{2.9}$$

Here $\boldsymbol{u}$ is a unit vector oriented along the rotational symmetry axis, while in a spherical molecule it is an arbitrary vector rigidly connected to the molecular frame. The scalar product $\boldsymbol{u}(t)\cdot\boldsymbol{u}(0)$ is $\cos\theta(t)$ in classical theory, where $\theta(t)$ is the angle of $\boldsymbol{u}$ reorientation with respect to its initial position. It can be easily seen that both orientational correlation functions are the average values of the corresponding Legendre polynomials:

$$K_\ell = \langle P_\ell\ [\boldsymbol{u}(t)\cdot\boldsymbol{u}(0)]\rangle, \quad \ell = 1, 2. \tag{2.10}$$

Using the summation theorem for spherical harmonics, these correlation functions may be represented as scalar products

$$K_\ell = \langle [d^+d(t)]\rangle = \left\langle \sum_{q=-\ell}^{\ell} (-1)^q\, d_q^\ell\, d_{-q}^\ell\,(t)\right\rangle , \tag{2.11}$$

where

$$d_q^\ell(t) = \left(\frac{4\pi}{2\ell+1}\right)^{\frac{1}{2}}\ Y_q^\ell\,(\boldsymbol{u}(t)) \tag{2.12}$$

and d^+ is the Hermitian conjugate of d.

Correlation functions are sometimes believed to be more reliable than their spectra

$$G_\ell(\omega) = \frac{\mathrm{Re}}{\pi}\left[\int_0^\infty K_\ell(t)\,\mathrm{e}^{-i\omega t}\,\mathrm{d}t\right] = \frac{1}{2\pi}\int_{-\infty}^{+\infty} K_\ell(t)\,\mathrm{e}^{-i\omega t}\,\mathrm{d}t, \qquad (2.13)$$

where $K(t) = K^*(-t)$. From the experimental point of view this assumption is justified to some extent [88] though changes observed in spectra under condensation are more pronounced and easier to interpret. Besides, double Fourier transformation which relates $F(\omega)$ to $G(\omega)$ is sometimes unnecessary. In particular, $G_0(\omega)$ simply coincides with $I_{\mathrm{is}}^{\mathrm{R}}(\omega)$. Moreover, the width of the isotropic scattering spectrum usually decreases significantly with increase in density [89] and in many liquids one may take K_0 equal to 1 in Eq. (2.6) and Eq. (2.7) without significant error. In this approximation the observed spectral functions I^{IR} and $I_{\mathrm{an}}^{\mathrm{R}}$ are the desired G_1 and G_2.

The same information may be obtained from purely rotational far infrared spectroscopy (FIR) and depolarized Rayleigh spectra. Dielectric relaxation measurements are also used for the same goal, most successfully in combination with far-infrared data. The absorption coefficient of a periodic electric field

$$a(\omega) = \frac{\omega\epsilon''(\omega)}{cn(\omega)} \qquad (2.14)$$

is expressed via its refractive index $n(\omega) = \mathrm{Re}[\epsilon^{1/2}(\omega)]$ and the imaginary part of the complex permittivity $\epsilon(\omega) = \epsilon'(\omega) - \mathrm{i}\epsilon''(\omega)$ [90]. The latter is related to the lowest orientational correlation function:

$$\frac{\epsilon(\omega) - \epsilon_\infty}{\epsilon_s - \epsilon_\infty} = 1 - \mathrm{i}\omega\int_0^\infty K_1(t)\mathrm{e}^{-\mathrm{i}\omega t}\,\mathrm{d}t, \qquad (2.15)$$

where $\epsilon_s = \epsilon(0)$ is static and ϵ_∞ is high-frequency (optical) permittivity. This relation combined with Eq. (2.14) yields

$$a(\omega) = \pi A(\omega)\omega^2 G_1(\omega), \qquad (2.16)$$

where $A(\omega) = (\epsilon_s - \epsilon_\infty)/cn(\omega)$ is a slow function of ω, which tends to constant value

$$A_\infty = (\epsilon_s - \epsilon_\infty)/c\epsilon_\infty^{1/2} \quad \text{at} \quad \omega \to \infty.$$

Since the absorption coefficient $a(\omega)$ is roughly speaking $G_1(\omega)$ multiplied by ω^2, the centrum of the spectrum is suppressed to 0 whereas the shape of the spectral wings is greatly pronounced. The same happens with the Stokes wing ($\omega > 0$) of the depolarized Rayleigh spectrum

$$R(\omega) = \hbar\omega(1 - \mathrm{e}^{-\hbar\omega/kT})G_2(\omega).$$

In the high-temperature quasi-classical limit ($\hbar\omega \ll 1$) it relates to $G_2(\omega)$ the same way as $a(\omega)$ does to $G_1(\omega)$:

$$R(\omega) = \frac{(\hbar\omega)^2}{kT} G_2(\omega). \tag{2.17}$$

Hence Rayleigh scattering gives the same information about non-polar molecules as dielectric measurements do on polar molecules.

As far as indirect methods are concerned (for instance, that of magnetic resonance), they measure solely the correlation times of orientational relaxation, which are integral characteristics of the process:

$$\tau_{\theta}, \ell = \int_0^{\infty} K_\ell(t)\, \mathrm{d}t = \pi G_\ell(0). \tag{2.18}$$

In this case one determines the spectral intensity solely in the centre, not over the whole frequency range. Therefore the analysis often refers not to the spectrum as a whole, but to relaxation times $\tau_{\theta,1}$ or $\tau_{\theta,2}$ and their dependence on rotational relaxation time τ_J [85]. This dependence contains much information and can be easier to interpret. It enables one to determine when free rotation turns into rotational diffusion.

Of course, knowledge of the entire spectrum does provide more information. If the shape of the wings of $G_\ell(\omega)$ is established correctly, then not only the value of τ_J but also angular momentum correlation function $K_J(t)$ may be determined. Thus, in order to obtain full information from the optical spectra of liquids, it is necessary to use their periphery as well as the central Lorentzian part of the spectrum. In terms of correlation functions this means that the initial non-exponential relaxation, which characterizes the system's behaviour during free rotation, is of no less importance than its long-time exponential behaviour. Therefore, we pay special attention to how dynamic effects may be taken into account in the theory of orientational relaxation.

2.2 Stochastic perturbation theory

The change in a molecule's orientation in space as a result of rotation is described by the dynamic equation of motion

$$\dot{d}_q^\ell = -i \sum_{q'=-\ell}^{\ell} \frac{\left(\boldsymbol{J} \cdot \hat{\boldsymbol{L}}\right)_{q'q}}{I} d_{q'}^\ell, \tag{2.19}$$

where $\hat{\boldsymbol{L}}$ is the rotation operator [23]. If $\boldsymbol{J}(t)$ is considered as a random perturbation of a mechanical system, then $d_q^\ell(t)$ is the system's response. The problem is that of how to find the correlation characteristics of the response, when the stochastic properties of the perturbation are known. Only by perturbation theory may the above problem be approximately solved without going into detailed description of random perturbation affecting the rotator. It makes no difference, for instance, whether perturbation is by impact or not. The only requirement is that it must be weak enough, i.e., it should induce slow orientational relaxation. In this case the equation of motion becomes

$$\dot{d}_q^\ell = \sum_{\mu q'} \frac{(-1)^\mu}{I} J_{-\mu}(t) \, \left[\ell/(\ell+1)\right]^{\frac{1}{2}} C_{\ell q' 1 \mu}^{\ell q} d_{q'}^\ell = \sum_{q'} H_{q'q}(t)\, d_{q'}^\ell(t), \tag{2.20}$$

where $J_\mu(t)$ may be considered as a fluctuating random perturbation ($C_{\ell q' 1\mu}^{\ell q}$ are Clebsch–Gordan coefficients [23]).

When perturbation is fast enough in comparison with the molecular response to it, the averaging procedure proposed in [91] is justified. After substitution of a formal solution of Eq. (2.20)

$$d_q^\ell(t) = \sum_{q'} \int_0^t H_{q'q}(t') d_{q'}^\ell(t')\, \mathrm{d}t'$$

into the right-hand side of that very equation, it may be averaged over all realizations of $J_\mu(t)$ using decoupling of the type

$$\langle J_\mu(t)\, J_\nu(t')\, d_q^\ell(t')\rangle = \langle J_\mu(t)\, J_\nu(t')\rangle\, \langle d_q^\ell(t')\rangle. \tag{2.21}$$

Note that

$$\langle J_\mu(t)\, J_\nu(t')\rangle = K_J\,(t-t')\,\delta_{\mu\nu}/r$$

and $\langle J_\mu\rangle = 0$. This procedure results in the integro-differential equation

$$\langle \dot{d}_q^\ell(t)\rangle = -\frac{\ell(\ell+1)}{r I^2} \int_0^t K_J\,(t-t')\, \langle d_q^\ell(t')\rangle\, \mathrm{d}t'. \tag{2.22}$$

The price of the decoupling approximation is unknown *a priori*, but a number of examples considered in [20] showed that it usually coincides with a generally accepted application criterion for standard perturbation theory [39]

$$\tau_{\theta,\ell} \gg \tau_J . \tag{2.23}$$

This means that the theory may be applied only to dense media where rotational relaxation proceeds at a higher rate than does orientational relaxation.

Multiplying (2.22) by $(-1)^q\, d^{\ell}_{-q}(0)$ and summing over q according to (2.11), we obtain an equation that connects the correlation functions of the perturbation to those of the response:

$$\dot{K}_\ell(t) = -\frac{\ell(\ell+1)}{r I^2} \int_0^t K_J(t-t')K_\ell(t')\,\mathrm{d}t'. \tag{2.24}$$

It is commonly believed that $K_\ell(t')$ may be carried outside the integral without lack of accuracy if inequality (2.23) is satisfied. This is the same way that was used in Chapter 1 to obtain the non-Markovian differential equation

$$\dot{K}_\ell(t) = -\frac{\ell(\ell+1)}{r I^2} \int_0^t K_J(t')\,\mathrm{d}t'\; K_\ell(t). \tag{2.25}$$

However, this equation still differs from a basic kinetic equation of the standard (Markovian) perturbation theory [39].

The Markovian theory is obtained when the integration over time in Eq. (2.25) is extended to infinity:

$$\dot{K}_\ell = -\Gamma_\ell K_\ell. \tag{2.26}$$

The relaxation rate

$$\Gamma_\ell = \frac{\ell(\ell+1)}{r I^2} \int_0^\infty K_J(t')\,\mathrm{d}t' = \ell(\ell+1)kT\tau_J/I \tag{2.27}$$

is actually the width of the corresponding spectrum $G_\ell(\omega)$, which is Lorentzian in this limit. Since $1/\Gamma_\ell = \pi G_\ell(0) = \tau_{\theta,\ell}$, Eq. (2.27) may be presented as the famous Hubbard relation [30, 12]

$$\tau_{\theta,\ell}\, \tau_J = I/\ell(\ell+1)kT. \tag{2.28}$$

This connection between the correlation time of perturbation and that of response is a very general result independent of a model of molecular motion. It is valid not only when a molecule is perturbed by a sequence of 'instantaneous' collisions (as in a gas), but also when it is subjected to perturbations that are continuous in time (caused by the nearest neighbourhood in liquids or solids). By the way, it follows from Eq. (2.28) that

$$\frac{\tau_{\theta,1}}{\tau_{\theta,2}} = 3. \tag{2.29}$$

This ratio of orientational relaxation times is sometimes used to identify the situation corresponding to perturbation theory [85].

According to Eq. (2.23) and Eq. (2.28), perturbation theory holds if

$$\mathscr{H}_\ell = \frac{\ell(\ell+1)kT}{I}\,\tau_J^2 = \frac{\ell(\ell+1)}{r}\bar{\omega}^2\,\tau_J^2 \ll 1, \tag{2.30}$$

where $\bar{\omega}$ is the root mean square frequency of a molecule's rotation

$$\bar{\omega} = (\langle\omega^2\rangle)^{\frac{1}{2}} = (rkT/I)^{\frac{1}{2}}. \tag{2.31}$$

It is easy to ascertain the physical meaning of Eq. (2.30). It consists of the requirement that the angular displacement of a molecule's axes during free rotation should be small. In other words, in the framework of perturbation theory a molecule fails to complete a full rotation cycle during the time of a free path. Reorientation is accomplished by a sequence of small turns (much less than 2π), their frequency and direction being changed after each collision. This is Brownian motion in angular space. At time intervals larger than τ_J such motion may be described in terms of Debye rotational diffusion [1] that we will discuss later.

The Hubbard relation, as well as Eq. (2.27), is a particular case of a more general result of perturbation theory, namely

$$\Gamma = M_2\tau_\omega, \tag{2.32}$$

where $M_2 = \langle\Delta\omega^2\rangle$ is a second moment of the random frequency shift $\Delta\omega(t)$, and τ_ω is the correlation time of this process. This result is known as the famous 'motional narrowing' effect [9, 39] as the spectral width Γ decreases when frequency modulation becomes faster. If only the spectrum has a finite second moment it always transforms to a narrow Lorentzian contour in the perturbation theory limit ($\mathscr{H} \ll 1$) [92]. In particular it happens with rotational bands of numerous vibrational modes of spherical molecules in cryogenic solutions [93]. When second moments M_2 of these branches were calculated, they were found to differ even in order of magnitude as a result of Coriolis interaction. However, in liquids all of them turn into Lorentzian lines of width Γ, which must be linear in M_2 according to Eq. (2.32). This linearity was confirmed experimentally in [93] and used to deduce $\tau_\omega \approx \tau_J$ from the slope of the straight line in Fig. 2.1. The values of τ_J as well as $\tau_J^* = \tau_J(kT/I)^{\frac{1}{2}}$ were

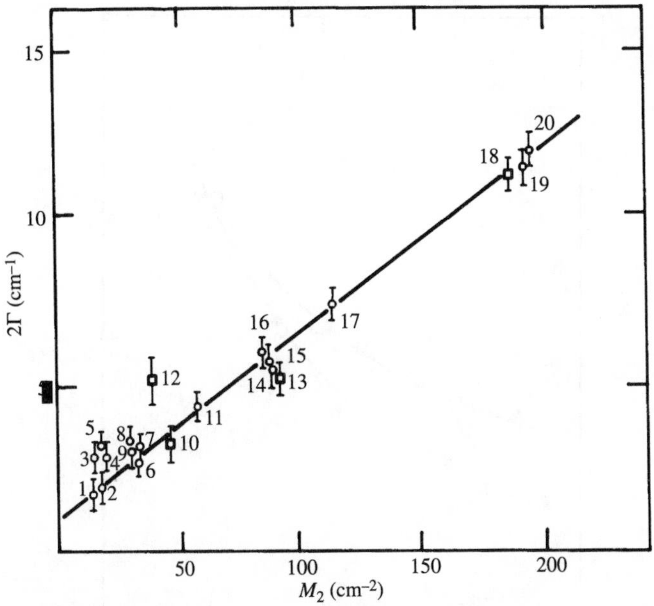

Fig. 2.1. The dependence of the experimental full width 2Γ on the second moment M_2 of perpendicular (○) and parallel (□) bands in the spectrum of CF_3Cl in liquid argon at 90 K. The numbers 1–20 have the following significance: (1) ν_4; (2)$\nu_1+\nu_4$; (3)$2\nu_1+\nu_4$; (4)$\nu_2+\nu_6$; (5)$3\nu_4$; (6)$\nu_3+\nu_6$; (7)$\nu_4+\nu_5$; (8)$2\nu_3+\nu_6$; (9)$\nu_2+\nu_6$; (10)$\nu_5+\nu_6$; (11)$\nu_2+2\nu_6$; (12)$\nu_4+\nu_6$; (13)$\nu_4+\nu_5$; (14)ν_5; (15)$\nu_2+\nu_5$; (16)$2\nu_3+\nu_5$; (17)$\nu_4+\nu_6$; (18)ν_4; (19)$2\nu_4$; (20)$\nu_1+2\nu_4$.

found for many solute molecules in their respective solvents as explained in Table 2.1. There one may see that the values of τ_J from optical experiments are in agreement with values found by nuclear magnetic relaxation experiments [94, 95]. Furthermore we see from Fig. 2.2 that the temperature-dependence of τ_J found by the above two methods is the same for the two molecules CF_4 and SiF_4 over a wide temperature range. It is similar to that shown in Fig. 1.25 for nitrogen: rotational relaxation rate and cross-section decrease with temperature.

It appears from Table 2.1 and Fig. 2.2 that, at least in the liquids studied in [93],

$$\mathscr{H}_1 = 2(\tau_J^*)^2 \approx 0.03 \ll 1, \tag{2.33}$$

i.e., the condition (2.30) for the applicability of perturbation theory is well fulfilled. Hence not only the Hubbard relation, which is a consequence of the simplest Markovian perturbation theory, but also the non-Markovian

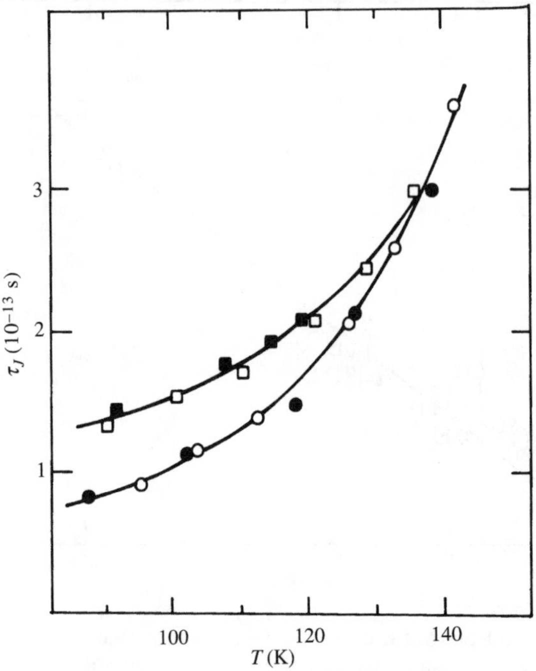

Fig. 2.2. The temperature-dependence of τ_J in SiF_4 obtained in [93] by two different methods, viz. IR spectroscopy (•) and NMR (○). Squares are the same for CF_4.

versions of this theory formulated in Eq. (2.24) and Eq. (2.25) are justified. They constitute a reliable base for formal description of J-diffusion in dense gases and liquids.

It should be noted in conclusion that, by virtue of the impact condition ($\tau_J \gg \tau_c$) and adiabaticity of the m-diffusion model ($\bar{\omega}\tau_c \gg 1$), the latter relates to the limit alternative to the perturbation theory:

$$\bar{\omega}\,\tau_J \gg \bar{\omega}\tau_c \gg 1. \tag{2.34}$$

Consequently, the m-diffusion model does not extend to the domain where the Hubbard relation holds. Therefore, the J-diffusion model is the only realistic description of rotational diffusion within the framework of impact theory.

Table 2.1. *Angular momentum correlation times τ_J of various molecules in liquefied gases [93].*

Molecule	Solvent	Temperature K	τ_J (10^{-12} s)	τ_J^*	τ_J by other methods (10^{-12} s)
CF_4	N_2	77	0.14	0.12	0.13 (liquid 90 K)
CF_4	Ar	90	0.14	0.13	0.14 (NMR 90 K)
SiF_4	Ar	90	0.09	0.07	0.09 (NMR 90 K)
SF_6	O_2	90	0.21	0.13	0.22 (solid 90 K)
SF_6	Ar	90	0.19	0.12	
NF_3		90	0.16	0.15	0.11 (liquid 78 K)
CHF_3	O_2	90	0.13	0.12	
CDF_3	O_2	90	0.13	0.12	
CF_3Cl	Ar	90	0.16	0.15	
CF_3I	Ar	90	0.13	0.12	

2.3 Rotational diffusion

Correlation functions (2.11) may be considered as ℓ-order moments of the two-dimensional probability $\mathbf{P}$:

$$K_\ell(t) = \frac{4\pi}{2\ell+1} \int \mathbf{P}(\boldsymbol{u}, \boldsymbol{u}_0, t) \sum_{q=-\ell}^{\ell} (-1)^q \times Y_q^\ell(\boldsymbol{u})\, Y_{-q}^\ell(\boldsymbol{u}_0)\, \mathrm{d}\boldsymbol{u}\, \mathrm{d}\boldsymbol{u}_0 . \quad (2.35)$$

$\mathbf{P}$ defines the orientation of the axis of the molecule ($\boldsymbol{u}$) at time t, provided that $\boldsymbol{u} = \boldsymbol{u}_0$ at $t = 0$. The opposite is also true: if all $K_\ell(t)$ are known, then $\mathbf{P}(\boldsymbol{u}, \boldsymbol{u}_0, t)$ is unambiguously defined by the series

$$\mathbf{P}(\boldsymbol{u}, \boldsymbol{u}_0, t) = \frac{1}{4\pi} \sum_{\ell q} K_\ell(t)\, Y_q^\ell(\boldsymbol{u})\, Y_{-q}^\ell(\boldsymbol{u}_0)\, (-1)^q . \quad (2.36)$$

It is clear from the preceding section that, until inequality (2.30) holds, the moments satisfy Eq. (2.24). Using this fact in Eq. (2.36) and the eigenvalue equation

$$\ell(\ell+1) Y^\ell(\boldsymbol{u}) = \boldsymbol{L}^2\, Y^\ell(\boldsymbol{u}) = \Delta\, Y^\ell(\boldsymbol{u}),$$

we get

$$\frac{\partial \mathbf{P}}{\partial t} = \frac{1}{r I^2} \int_0^t K_J(t-t')\, \Delta \mathbf{P}(\boldsymbol{u}, \boldsymbol{u}_0, t')\, \mathrm{d}t' . \quad (2.37)$$

It should be noted, however, that $\mathscr{H}_\ell$ increases quadratically with ℓ. Starting with $\ell = L$ such that $\Gamma_L \tau_J = 1$, inequality (2.30) is reversed, and Eq. (2.24) is no longer valid. Consequently, in series (2.36) only the first L terms, which decay at rates $\Gamma_\ell \ll 1/\tau_J$, are correctly estimated by perturbation theory. Eq. (2.37) holds solely for times

$$t \gg \tau_J , \tag{2.38}$$

when the contribution of K_ℓ with $\ell > L$ is neglected. However, in this time range there is no necessity to keep the probability $\mathbf{P}$ inside the integral on the right-hand side of Eq. (2.37). In this case the equation may be rewritten as

$$\frac{\partial \mathbf{P}}{\partial t} = \frac{1}{rI^2} \int_0^t K_J(t')\, dt' \, \Delta \mathbf{P}, \tag{2.39}$$

which is the equation used by Steele [96]. The difference between Eq. (2.37) and Eq. (2.39) is similar to that between Eq. (2.24) and Eq. (2.25).

Eq. (2.39) may be considered as the generalized diffusion equation in angular space with a diffusion coefficient that varies in time as

$$D(t) = \frac{1}{rI^2} \int_0^t K_J(t')\, dt' . \tag{2.40}$$

Though the applicability of Eq. (2.39) is restricted, it has certain advantages over the conventional equation of orientational diffusion proposed by Debye [1]

$$\frac{\partial \mathbf{P}}{\partial t} - D_\theta \, \Delta \mathbf{P} = 0 , \tag{2.41}$$

where

$$D_\theta = \frac{D(\infty)}{rI^2} = \frac{\langle \omega^2 \rangle}{r} \tau_J . \tag{2.42}$$

In particular the use of solution of Eq. (2.37) or Eq. (2.39) allows one to calculate the moments in (2.35) preserving fast vanishing components with $\ell < L$. Consequently, the time-dependent diffusion coefficient $D(t)$ makes it possible to recover information about the periphery of the spectrum $G_\ell(\omega)$.

This information is lost in the Debye theory. If one uses the solution of Eq. (2.41) in Eq. (2.35), then any $K_\ell = \exp(-t/\tau_{\theta,\ell})$ decays exponentially with a rate

$$1/\tau_{\theta,\ell} = \ell\,(\ell + 1) D_\theta . \tag{2.43}$$

This is nothing other than a long-time asymptotics of orientational relaxation valid at $t \gg \tau_J$. Sometimes deviations from Debye's relaxation were also found at $t \gg \tau_{\theta,\ell}$. This is an effect of long-range Coulomb and dipolar forces on the collective tumbling of rigid linear dipoles [97]. In dense solutions it slows down the final decay of the correlation function.

The microscopic meaning of D_θ is not specified in Eq. (2.43), just as in any phenomenological theory. However, the relationship between the Debye description of orientational relaxation and Langevin description of the angular momentum relaxation may be revealed. It has been ascertained that the parameters D_θ and ζ of both theories are related. This becomes evident, if we consider how an external orientating field affects the rotational motion of a molecular dipole. When the field is switched on the torque affecting a molecule is on the average different from zero: $\langle \boldsymbol{m} \rangle = \boldsymbol{m}_0$. According to Eq. (1.26) the stationary drift in the angular space of the linear molecule axis (dipole) has a rate proportional to $\boldsymbol{m}_0$:

$$\langle \omega \rangle = \langle \boldsymbol{J} \rangle / I = \zeta^{-1} \boldsymbol{m}_0 = q_\theta \, \boldsymbol{m}_0 \,, \tag{2.44}$$

where q_θ is the mobility in angular space. As is any mobility, it is connected to the corresponding diffusion coefficient by the Einstein relation

$$q_\theta \;=\; D_\theta / kT . \tag{2.45}$$

It follows from Eq. (2.44) and Eq. (2.32) that the mobilities in J- and θ-spaces (when both of them have any meaning) are inversely proportional to each other:

$$q_\theta = 1/\zeta = 1/q_J \,. \tag{2.46}$$

Since q_θ is expressed via D_θ in Eq. (2.45) and ζ is expressed via τ_J according to Eq. (1.27), Eq. (2.46) reduces to

$$\tau_J / D_\theta \;=\; I / kT \,. \tag{2.47}$$

In view of (2.43) this result is identical to the Hubbard relation (2.28).

Relations (2.46) and (2.47) are equivalent formulations of the fact that, in a dense medium, increase in frequency of collisions retards molecular reorientation. As this fact was established by Hubbard within Langevin phenomenology [30] it is compatible with any sort of molecule–neighbourhood interaction (binary or collective) that results in diffusion of angular momentum. In the gas phase it is related to weak collisions only. On the other hand, the perturbation theory derivation of the Hubbard relation shows that it is valid for dense media but only for collisions of arbitrary strength. Hence the Hubbard relation has a more general and universal character than that originally accredited to it.

2.4 Reorientation in the impact approximation

Markovian theory of orientational relaxation implies that it is exponential from the very beginning but actually Eq. (2.26) holds for $t \gg \tau_J$ only. If any non-Markovian equations, either (2.24) or (2.25), are used instead, then the exponential asymptotic behaviour is preceded by a short dynamic stage which accounts for the inertial effects (at $t \leq \tau_J$) and collisions (at $t \leq \tau_c$).

Let us start from the simplest model, which is impact relaxation of angular momentum. According to Eq. (1.21) it proceeds exponentially with relaxation time (1.22):

$$K_J = rIkT e^{-t/\tau_J}. \tag{2.48}$$

With this kernel the exact solution of Eq. (2.25) is similar to that of Eq. (1.71) with a kernel Eq. (1.77):

$$K_\ell(t) = e^{-t/2\tau_J}\left(\cosh\left[t(1-4\mathscr{K}_\ell)^{\frac{1}{2}}/2\tau_J\right] + \frac{\sinh\left[t(1-4\mathscr{K}_\ell)^{\frac{1}{2}}/2\tau_J\right]}{(1-4\mathscr{K}_\ell)^{\frac{1}{2}}}\right). \tag{2.49}$$

The only distinction between Eq. (1.78) and Eq. (2.49) is that $K_\ell(0) = 1$, τ_c is replaced by τ_J, and $\mathscr{K}_\ell$ is substituted for κ. We may also solve Eq. (2.25) by substitution of Eq. (2.48). In this case we find

$$K_\ell(t) = \exp\left[-\mathscr{K}_\ell\left(e^{-t/\tau_J} + t/\tau_J - 1\right)\right]. \tag{2.50}$$

For one-dimensional rotation ($r = 1$), orientational correlation functions were rigorously calculated in the impact theory for both strong and weak collisions [98, 99]. It turns out in the case of weak collisions that the exact solution, which holds for any $\mathscr{K}_\ell$, happens to coincide with what is obtained in Eq. (2.50). Consequently, the accuracy of the perturbation theory is characterized by the difference between Eq. (2.49) and Eq. (2.50), at least in this particular case. The degree of agreement between approximate and exact solutions is readily determined by representing them as a time expansion

$$K_\ell(t) = 1 - \tilde{I}_2(\mathscr{K}_\ell)\frac{t^2}{2!\tau_J^2} + \tilde{I}_3(\mathscr{K}_\ell)\frac{t^3}{3!\tau_J^3} + \tilde{I}_4(\mathscr{K}_\ell)\frac{t^4}{4!\tau_J^4} + \cdots, \tag{2.51}$$

where the coefficients are polynomials of $\mathscr{K}_\ell$. The latter are the same for both solutions solely in the lowest order of $\mathscr{K}_\ell$ when

$$\tilde{I}_2 = \tilde{I}_3 = -\tilde{I}_4 = \mathscr{K}_\ell\,. \tag{2.52}$$

By keeping higher powers of $\mathscr{K}_\ell$ in these polynomials, one exceeds the accuracy of perturbation theory. In other words, there is no use nor necessity in seeking non-Markovian solutions of higher order in $\mathscr{H}_\ell$. With the accuracy assured in perturbation theory, both non-Markovian results of Eq. (2.49) and Eq. (2.50) reduce to

$$K_\ell = \mathrm{e}^{-\Gamma_\ell t} - \mathscr{H}_\ell \mathrm{e}^{-t/\tau_J}, \qquad \mathscr{H}_\ell \ll 1. \tag{2.53}$$

The first component in expression (2.53) corresponds to the long-time behaviour of $K_\ell(t)$ described by Markovian perturbation theory, while the second term introduces a correction for times less than τ_J. Within this time interval (before the first collision occurs) the system should display the dynamic properties of free rotation ('inertial effects').

According to the uncertainty principle the non-exponential short-time behaviour of K_ℓ determines the deviation of the high-frequency spectral wings from Lorentzian shape. The actual spectrum obtained by substitution of Eq. (2.53) into Eq. (2.13) is bi-Lorentzian:

$$G_\ell(\omega) = \frac{\Gamma_\ell}{\pi(\omega^2 + \Gamma_\ell^2)(1 + \omega^2\tau_J^2)}. \tag{2.54}$$

At $\ell = 1$ it coincides with the well-known Rocard formula for the spectrum of dipole relaxation [90]:

$$G_1(\omega) = \frac{\tau_D}{\pi(1 + \omega^2\tau_D^2)(1 + \omega^2\tau_J^2)}. \tag{2.55}$$

It is valid for spherical, linear and symmetric rotators and is applicable in both IR and FIR regions. In the latter case we find from Eq. (2.16) and Eq. (2.55) that

$$\frac{a(\omega)}{A(\omega)} = \frac{\omega^2\tau_D}{\pi(1 + \omega^2\tau_D^2)(1 + \omega^2\tau_J^2)}. \tag{2.56}$$

In Markovian approximation ($\tau_J = 0$) this quantity approaches the famous Debye plateau shown in Fig. 2.3 whereas non-Markovian absorption coefficient (2.56) tends to 0 when $\omega \to 0$ as it is in reality. This is an advantage of the Rocard formula that eliminates the discrepancy between theory and experiment by taking into account inertial effects. As is seen from Eq. (2.56) and the Hubbard relation (2.28)

$$a(\omega) = \frac{2kT}{I\tau_J}\frac{A_\infty}{\omega^2} \quad \text{at} \quad \omega^2\tau_D^2 \gg \omega^2\tau_J^2 \gg 1. \tag{2.57}$$

This explains the 'return to transparency' in the high-frequency region. However, the position and height of the absorption maximum (see Fig. 2.3) remain incomprehensible in the impact approximation. The

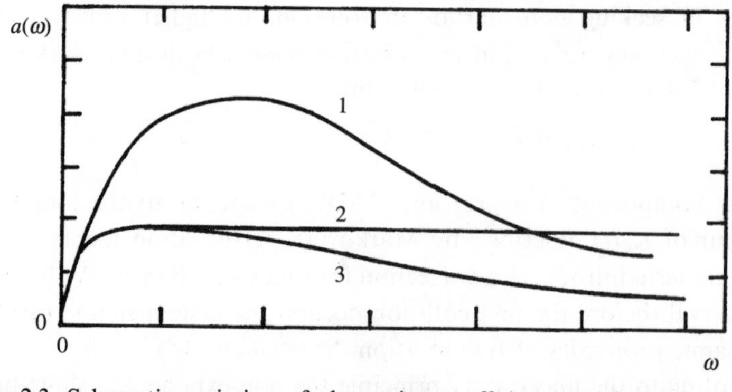

Fig. 2.3. Schematic variation of absorption coefficient as a function of the frequency of absorbed radiation: experimental curve (1) and theoretical curves in Debye approximation (2) and impact approximation (3) (Rocard formula).

super-absorption around the maximum discovered by Poley [100] was a subject of numerous experimental and theoretical works but has not been properly understood until recently. For a reasonable explanation of Poley absorption one must search outside the impact approximation, as will be done in the next section.

The central Lorentzian part of the IR spectrum (2.55) has the same shape as in the classical Debye theory and may be of various origins. The impact mechanism of reorientation can be confirmed judging by the shape of the wings only. The inertial effects show themselves in the asymptotic relation

$$\omega^4 G_1(\omega) = \frac{2kT}{\pi I \tau_J} . \tag{2.58}$$

Its experimental confirmation provides information about the free rotation time τ_J. However, this is very difficult to do in the Debye case. From one side the density must be high enough to reach the perturbation theory (rotational diffusion) region where $\mathscr{K}_1 \ll 1$. From the other side it must not be too high to preserve the impact description of rotational relaxation which is valid at $\kappa \ll 1$. The two conditions are mutually contradictory. The validity condition of perturbation theory

$$\mathscr{K}_1 = \frac{2}{r}\, \bar{\omega}^2 \tau_J^2 = \frac{2\, \bar{\omega}^2 \tau_c^2}{r\, \kappa^2} \ll 1 \tag{2.59}$$

is hardly compatible with inequality (1.83), which guarantees quasi-exponential (impact) rotational relaxation. For a typical collision time $\tau_c = 10^{-13}$ s and rotational frequency $\bar{\omega} = 10^{12}$ s^{-1} perturbation theory starts to be valid right after the impact approximation becomes unfounded. Only heavy molecules ($\bar{\omega} < 10^{11}$ s^{-1}) may reach the Debye limit in a relatively rarefied gas where the impact approximation remains valid.

The numerous attempts to extend the integral theory to rarefied gas (impact) region must not be passed over in silence. Eq. (2.24) may be considered as a particular case of the first equation in a Zwanzig–Mori chain [101, 52]

$$\dot{K}_\ell(t) = -\int_0^t F(t-t')K_\ell(t)\,\mathrm{d}t' . \tag{2.60}$$

In this very case $F(t) = [\ell(\ell+1)/(rI^2)]\; K_J(t)$ because the condition $\mathscr{H}_\ell \ll 1$ is met. However, the equation may serve even at $\mathscr{H}_\ell > 1$ if its kernel $F(t)$ is changed to a properly defined one. Even in the extreme case of free rotation ($\mathscr{H}_\ell = \infty$), the appropriate kernel $F_0(t)$ as was found which obeys Eq. (2.60) containing a given K_ℓ. To obtain it alternatively, all the rest of the Mori chain must be summed up. None of these ways are feasible as soon as collisions are taken into account. Some phenomenological approaches were proposed instead. Taking $F_0(t)$ as known, one can use as a kernel $F(t) = F_0(t)f(t)$ where $f(t)$ is taken exponential (at $\kappa \ll 1$), Gaussian (at $\kappa \gg 1$) [102] or equal to $K_J(t)$ [103]. Exponential $f(t)$ is unable to reproduce $K_2(t)$ as was found experimentally from light scattering in N_2 at room temperatures and $n < 135$ amagat ($\mathscr{H}_2 > 2$) [102]. Better results were achieved at $T = 150$ K, $n = 585$ amagat for $F(t) = F_0(t)K_J(t)$, and $K_J(t)$ found from Eq. (1.71) with Gaussian kernel [103]. However, this kernel does not reduce to $K_J(t)$ at $\mathscr{H}_\ell \ll 1$. Therefore, it may not be considered even as interpolation between the free rotation limit and perturbation theory. As no regular ways to find a true $F(t)$ for large $\mathscr{H}_\ell$ are known, one can gain no advantages of the Mori approach in comparison with perturbation theory.

While the Rocard formula as well as the perturbation theory is inapplicable to rarefied gases ($\mathscr{H}_\ell \gg 1$), its high-frequency asymptotics (2.58) may be a subject of experimental study since it is of more general origin. The shape of the wings is determined by the short-time expansion (2.51), which is valid at any $\mathscr{H}_\ell$. Since in the impact approximation all derivatives of K_ℓ higher than second have a break at $t = 0$, the definition of the expansion coefficients implies they are taken from the right:

$$\tilde{I}_{2n+1} = \tau_J^{2n+1}\,\mathrm{Im}\left((-\mathrm{i})^{2n+1} \left.\frac{\mathrm{d}^{2n+1}K_\ell}{\mathrm{d}t^{2n+1}}\right|_{0+0} \right),$$

$$\tilde{I}_{2n} = \tau_J^{2n}\,\mathrm{Re}\left((-\mathrm{i})^{2n} \left.\frac{\mathrm{d}^{2n}K_\ell}{\mathrm{d}t^{2n}}\right|_{0+0} \right). \tag{2.61}$$

Using the impact approximation presented in Chapter 6, they may easily be found for any rotational band even if rotational–vibrational interaction is nonlinear in J. In 1954 P. W. Anderson proved as a theorem [104] that expansion of the spectral wings in inverse powers of frequency is controlled by successive odd derivatives of the correlation function at the origin. In impact approximation the lowest non-zero derivative of this type is the third and therefore asymptotics $G_\ell(\omega)$ is described by the power expansion [20]

$$G_\ell(\omega) = \frac{1}{\pi}\left[\frac{\tilde{I}_3}{\omega^4\tau_J^3} + \frac{\tilde{I}_5}{\omega^6\tau_J^5} + \cdots\right] \tag{2.62}$$

for $\omega \gg 1/\tau_J$. Eq. (2.58) is a particular case of this general formula. The latter may be equally well used to investigate the wing beyond the edge of IR spectra peculiar to some vibrational modes. Such an experiment was performed in [105] at such low pressures of buffer gases that the rotational structure of the CO_2 vibrational spectrum was well resolved. Of course, perturbation theory is not valid here ($\mathscr{H}_1 \gg 1$) but the impact approximation works well instead. This is enough to obtain experimental confirmation of Eq. (2.58), shown in Fig. 2.4. It provides the means of measuring $1/\tau_J$ for gas mixtures via the height of the plateau. The width of the latter, which is infinite in impact approximation ($\tau_c = 0$), is actually limited by the value of $\kappa = \tau_c/\tau_J \ll 1$. It exists in the interval $1/\tau_J < \omega < 1/\tau_c$. The very far wings ($\omega \gg 1/\tau_c$) may be described only by non-Markovian theory of rotational relaxation taking into account the finite collision time $1/\tau_c$.

2.5 Finite collision time

The behaviour of orientational correlation functions near $t = 0$ carries information on both free rotation and interparticle interaction during collisions. In the impact approximation this information is lost. As far as collisions are considered as instantaneous, impact Eq. (2.48) holds, and all derivatives of exponential $K_J(t)$ have a break at $t = 0$. However,

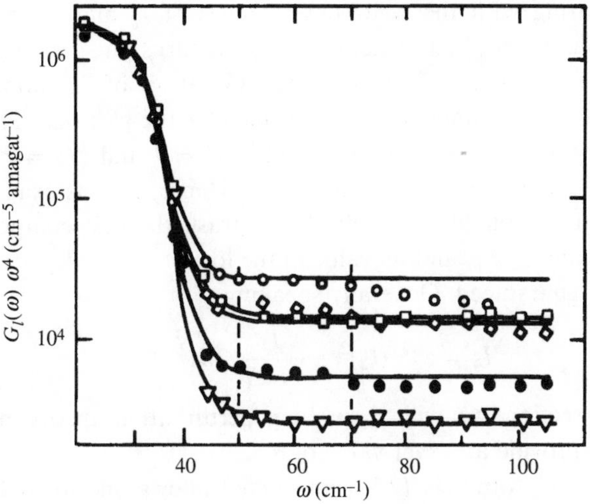

Fig. 2.4. The asymptotic behaviour of the IR spectrum beyond the edge of the absorption branch for CO_2 dissolved in different gases: (○) xenon; (◇) argon; (□) nitrogen; (•) neon; (▽) helium. The points are experimental data, the curves were calculated in [105] according to the quantum J-diffusion model and two vertical broken lines determine the region in which Eq. (2.58) is valid.

there is no break if $\tau_c \neq 0$. If the correlation function of the angular momentum is described by non-Markovian Eq. (1.99), then

$$\dot{K}_J(0) = 0 \quad \text{and} \quad \ddot{K}_J(0) = -\langle M^2 \rangle. \tag{2.63}$$

This information is sufficient to define in a general way the short-time behaviour of

$$K_\ell(t) = 1 - I_2 \frac{t^2}{2!} + I_4 \frac{t^4}{4!} + \cdots . \tag{2.64}$$

It is universally expressed in terms of the spectrum moments

$$I_n = \int \omega^n \, G_\ell(\omega) \, \mathrm{d}\omega = \mathrm{Re} \left((-\mathrm{i})^n \left. \frac{\mathrm{d}^n K_\ell}{\mathrm{d}t^n} \right|_{t=0} \right) .$$

Beginning with the fourth moment, all of them are sensitive to the strength of the interaction during collisions.

In fact, by differentiating Eq. (2.24) with account of (2.63), we find

$$I_2 = \ell\,(\ell + 1)\,kT/I\,, \tag{2.65a}$$

$$I_4 = (I_2)^2 - \frac{\ell(\ell+1)}{r I^2}\,\ddot{K}_J(0) = \left(\frac{\ell(\ell+1)kT}{I}\right)^2 + \frac{\ell(\ell+1)}{r I^2}\,\langle \boldsymbol{M}^2 \rangle\,. \tag{2.65b}$$

To illustrate the accuracy of the perturbation theory these results are worth comparing with the well-known values of I_2 and I_4 for $\ell = 1$ rigorously found from first principles in [8]. It turns out that the second moment in Eq. (2.65a) is exact. The evaluation of I_4, however, is inaccurate: its first component is half as large as the true one. The cause of this discrepancy is easily revealed. Since $\boldsymbol{M} = \dot{\boldsymbol{J}}$ and $\langle \dot{\boldsymbol{J}} \rangle = \boldsymbol{J}/\tau_J$, the second component in $I_4 \tau_J^4$ is linear in $\mathscr{H}_\ell$. Hence, it is as exact in this order as perturbation theory itself. In contrast, the first component in $I_4 \tau_J^4$ is quadratic in $\mathscr{H}_\ell$, and its value in the lowest order of perturbation theory is not guaranteed. Generally speaking

$$I_4 = \alpha_\ell^{(r)} \, (I_2)^2 + \frac{I_2}{rIkT} \langle \boldsymbol{M}^2 \rangle . \tag{2.66}$$

In the next section we will show how perturbation theory must be developed to provide an exact value of $\alpha_\ell^{(r)}$.

Comparison of formulae (2.51) and (2.64) allows one to understand the limits and advantages of the impact approximation in the theory of orientational relaxation. The results agree solely in second order with respect to time. Everything else is different. In the impact theory the expansion involves odd powers of time, though, strictly speaking, the latter should not appear. Furthermore the coefficient $\tilde{I}_4/\tau_J^4$ defined in (2.61) differs from the fourth spectral moment I_4 both in value and in sign. Moreover, in the impact approximation all spectral moments higher than the second one are infinite. This is due to the non-analytical nature of K_J and K_ℓ in the impact approximation. In reality, of course, all of them exist and the lowest two are usually utilized to find from Eq. (2.66) either the dispersion of the torque $\langle \boldsymbol{M}^2 \rangle$ or related R_0 defined in Eq. (1.82):

$$R_0 = \frac{\langle \boldsymbol{M}^2 \rangle}{rIkT} = \frac{I_4}{I_2} - \alpha_\ell^{(r)} I_2 . \tag{2.67}$$

The moments are determined experimentally as for nitrogen in [5, 106] using Rayleigh and Raman depolarized scattering spectra.

The mutual correspondence of non-Markovian and Markovian (impact) approximations becomes clear, if the second derivative of $K_\ell(t)$ is considered. It varies differently within three time intervals with the following bounds: $\tau_c \ll \tau_J \ll \Gamma_\ell^{-1}$ (Fig. 2.5). Orientational relaxation occurs in times Γ_ℓ^{-1}. The gap near zero has a scale of τ_J. A parabolic vertex of extent τ_c and curvature $I_4 > 0$ is inscribed into its acute end. The narrower the vertex, the larger is its curvature, thus, in the impact approximation ($\tau_c = 0$) it is equal to ∞. In reality $\tau_c \neq 0$, and the

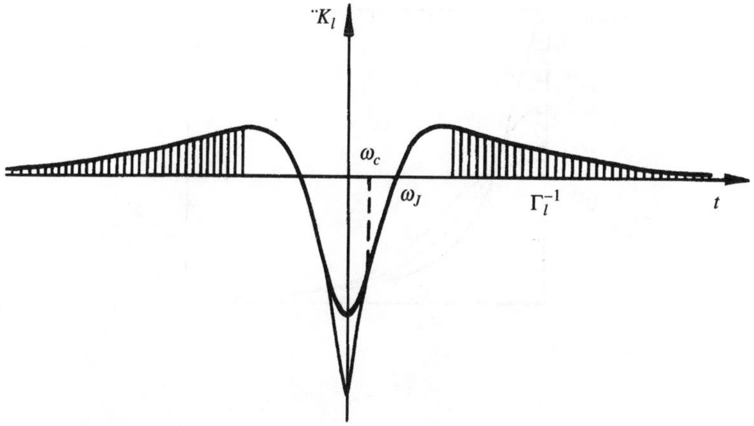

Fig. 2.5. Time evolution of the second derivative of the orientational correlation function. The long-time asymptotics described by impact theory is shadowed.

vertex of the gap is smoothed, while $\tilde{I}_3$ and $\tilde{I}_4$ are actually defined by the quantity $\mathrm{d}^n K_\ell/\mathrm{d}t^n|_{0+\tau_c}$, i.e., by a large scale form of the gap.

Without resorting to the impact approximation, perturbation theory is able to describe in the lowest order in $\mathscr{H}_\ell$ both the dynamics of free rotation and its distortion produced by collisions. An additional advantage of the integral version of the theory is the simplicity of the relation following from Eq. (2.24) for the Laplace transforms of orientational and angular momentum correlation functions [107]:

$$\frac{\ell(\ell+1)}{r I^2}\, \tilde{K}_J\,(\mathrm{i}\omega) = \tilde{K}_\ell^{-1}\,(\mathrm{i}\omega) - \mathrm{i}\omega\,, \tag{2.68}$$

where

$$\tilde{K}_\ell(p) \quad = \quad \int_0^\infty \mathrm{e}^{-pt} K_\ell(t)\,\mathrm{d}t \tag{2.69a}$$

$$\tilde{K}_J(p) \quad = \quad \int_0^\infty \mathrm{e}^{-pt} K_J(t)\,\mathrm{d}t \tag{2.69b}$$

are taken at $p = \mathrm{i}\omega$. This important relation is extended to symmetric rotators in [108]. It was used several times to calculate the orientational correlation function via angular momentum correlation functions found from this or that model [109, 110].

The same was done recently for their spectral properties [111, 112] using Eq. (2.68) to express the spectrum of orientational relaxation (2.13)

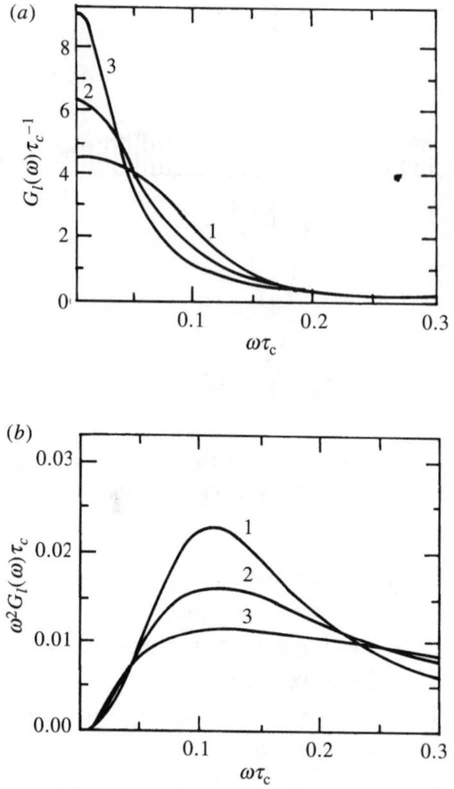

Fig. 2.6. IR spectra for $\omega > 0$ (*a*) and FIR spectra (*b*) at $\bar{\omega}^2\tau_c^2 = 0.01$ and small values of κ^2, viz. 0.02 (curve 1), 0.04 (curve 2), 0.08 (curve 3) [111].

via components of Eq. (1.91):

$$G_\ell(\omega) = \frac{g(\omega)\bar{\omega}^2}{(\pi\bar{\omega}^2 g(\omega))^2 + [\omega + \pi\bar{\omega}^2 f(\omega)]^2}\,. \tag{2.70}$$

The model of angular momentum relaxation defined by Eq. (1.92) or Eq. (1.93) was chosen as an example and used in Eq. (2.70) to calculate IR and FIR spectra of orientational relaxation [111, 112].

Substituting Eq. (1.92) into Eq. (2.70), we obtain spectra proper to the gas phase at $\kappa \ll 1$. These spectra for $\ell = 1$ are depicted in Fig. 2.6. There ω for the IR spectra is measured from the vibrational frequency and only the right-hand half of the symmetrical band is shown. The spectra deviate only slightly from those arising in the J-diffusion model. It is hardly possible to see that their wings are not Lorentzian. The effect

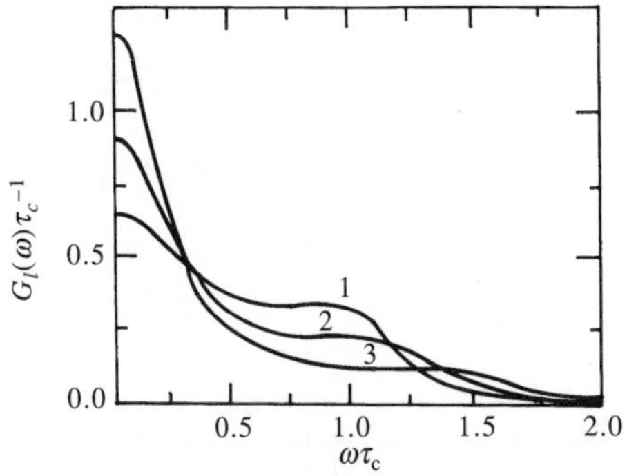

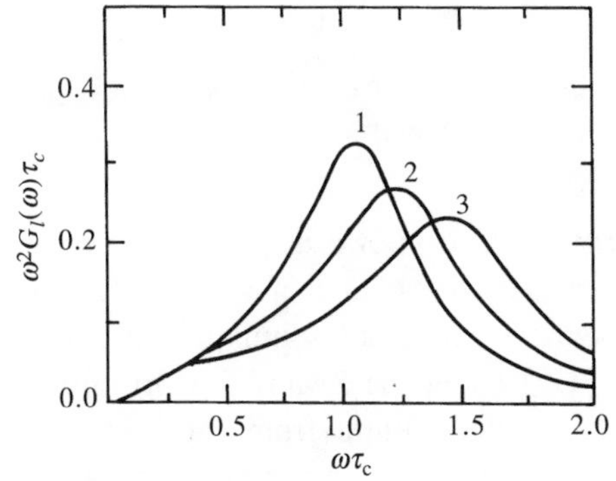

Fig. 2.7. IR spectra for $\omega > 0$ (*a*) and FIR spectra (*b*) at $\bar{\omega}^2\tau_c^2 = 0.5$ and larger values of κ^2, viz. 1 (curve 1), 2 (curve 2), 4 (curve 3) [111].

is better expressed in FIR spectra. They resemble those obtained from the impact Rocard formula that includes inertial effects [90]. The values

$$\mathcal{H}_1 = 2(\tau_J^*)^2 = \bar{\omega}^2\tau_c^2/\kappa^2$$

are 1/2 for curve 1, 1/4 for curve 2 and 1/8 for curve 3. The less it is the narrower is the IR spectrum and the flatter is the Debye plateau corresponding to a rotational Brownian motion theory. The role of the finite collision time at such small κ is negligible.

It becomes crucial for much larger values of κ that correspond to liquids. Although the results displayed in Fig. 2.7 are related to the same $\mathcal{H}_\ell$ as in Fig. 2.6, additional maxima appear on the wings of IR spectra, which correspond to the maxima of $g(\omega)$ shown in Fig. 1.9. It looks like a manifestation of molecular libration during collisions, and changes qualitatively the shape of the wings as compared with the intermediate impact asymptotics shown in Fig. 2.4. In the FIR

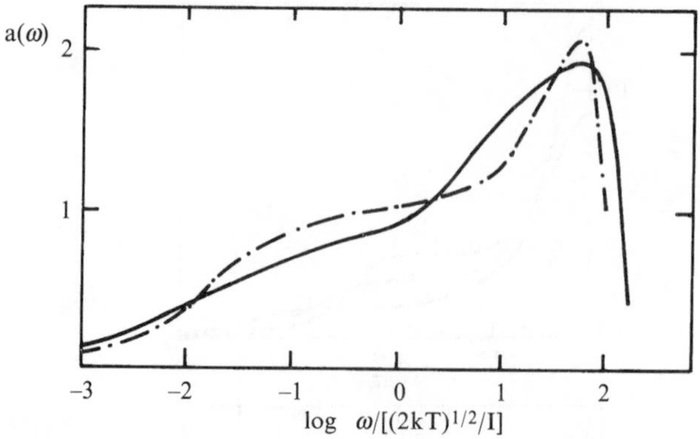

Fig. 2.8. The absorption coefficient of $CHBr_3$ in CCl_4 as a function of dimensionless frequency [113]: (—•—) theory; 2. (—) experiment.

spectra super-absorption (also known as Poley absorption) arises at the librational frequencies. As in the case of $g(\omega)$, the maxima are shifted to higher frequencies as τ_c increases. Similar results were obtained by Guillot and Bratos [113] using the Zwanzig–Mori approach with three phenomenological parameters found from the best fit to experiment. They reached at least qualitative agreement between theoretical and experimental results [114] obtained at 25 °C on approximately spherical bromoform molecules constituting a 20% solution in carbon tetrachloride (Fig. 2.8). It is all that could be expected since the influence of induced electric moments has been neglected. As is evident, the absorption coefficient first reaches the Debye plateau and only then exhibits the super-absorption maximum related to molecular libration. It was shown in [111] that all fitting parameters used in [113] may be related to the physically meaningful times which turn out to be

$$\tau_D = 6.19 \times 10^{-12}\,\text{s}, \ \tau_J = 0.1 \times 10^{-12}\,\text{s}, \ \tau_c = 0.2 \times 10^{-12}\,\text{s}.$$

These values establish the following parametrization of the problem

$$\mathscr{H}_1 = \frac{\tau_J}{\tau_D} = 0.017, \ \kappa = \frac{\tau_c}{\tau_J} = 2.$$

It clearly shows that perturbation theory is valid and the Mori approach is identical to it in this limit. However, the impact approximation is unacceptable since the collision time τ_c exceeds τ_J. This is a fact of principal importance in understanding Poley absorption and related phenomena. At such a condition, the libration maxima appear in spectra of angular momentum relaxation shown in Fig. 1.9 (curves c and d). This is enough to find the same maxima in IR and FIR spectra shown in Fig. 2.7.

This is a rather general conclusion independent of the model of rotational relaxation. It is quite clear from Eq. (2.70) and Eq. (2.16) that the high-frequency asymptotic behaviour of both spectra is determined by the shape of $g(\omega)$:

$$G_\ell(\omega) = \frac{g(\omega)\bar{\omega}^2}{\omega^2}, \quad \frac{a(\omega)}{A(\omega)} = \pi\bar{\omega}^2 g(\omega) \quad \text{at} \quad \omega \gg \bar{\omega}\,. \tag{2.71}$$

Hence, the problem is reduced to whether $g(\omega)$ has its maximum on the wings or not. Any model able to demonstrate that such a maximum exists for some reason can explain the Poley absorption as well. An example was given recently [77] in the frame of a modified impact theory, which considers instantaneous collisions as a non-Poissonian random process [76]. Under definite conditions discussed at the end of Chapter 1 the negative loop in $K_J(t)$ behaviour at long times is obtained, which is reflected by a maximum in its spectrum. Insofar as this maximum appears in $g(\omega)$, it is exhibited in IR and FIR spectra as well. Other reasons for their appearance are not excluded. Complex formation, changing hindered rotation of diatomic species to libration, is one of the most reasonable.

Eq. (2.68) may also be used to solve the inverse problem. The recovery of $g(\omega)$ from experimentally obtained optical spectra may prompt the origin of the maximum. To find $g(\omega)$, it is necessary to determine from the correlation function K_ℓ not only G_ℓ but also

$$\Psi_\ell = \operatorname{Im} \tilde{K}_\ell(\mathrm{i}\omega)/\pi = \int_0^\infty K_\ell(t)\sin\omega t \; \mathrm{d}t/\pi\,. \tag{2.72}$$

Knowing G_ℓ and Ψ_ℓ, we obtain a result of no less importance than Eq. (2.70):

$$\frac{\pi^2\ell(\ell+1)kT}{\ell}\, g(\omega) = \frac{G_\ell(\omega)}{G_\ell^2(\omega) + \Psi_\ell^2(\omega)}\,. \tag{2.73}$$

This is a direct generalization of the Hubbard relation (2.27) to the case $\omega \neq 0$. It is actually an algorithm for extraction of a wide spectral component which forms the pedestal. Bi-Lorentzian spectrum (2.54) may serve as an example of the above algorithm realization. Using its correlation function (2.53) in (2.72), we find Ψ_ℓ in addition to G_ℓ.

Substituting both of them into Eq. (2.73) we obtain to the lowest order on $\mathscr{H}_\ell$:

$$g(\omega) = \frac{\tau_J}{\pi\left(1+\omega^2\tau_J^2\right)}\,. \tag{2.74}$$

If the same procedure is applied to real IR or FIR spectra then the deviation from the Lorentzian shape of the spectrum (2.74) may be found in the wings. These are expected to be pronounced in the liquid phase.

It has been suggested [115, 116] to solve the inverse problem using the simplified relation which is actually a high-frequency limit of Eq. (2.73). This relation can be found, if we take into account that

$$G_\ell^2 + \Psi_\ell^2 = \frac{1}{\pi^2}|\tilde{K}_\ell(\mathrm{i}\omega)|^2 = \frac{1}{\pi^2}\left|\int_0^\infty K_\ell(t)\mathrm{e}^{-\mathrm{i}\omega t}\,\mathrm{d}t\right|^2 \propto \frac{1}{(\pi\omega)^2}$$

at $\omega \to \infty$. By substitution of this result into Eq. (2.73) the linear relation between spectra of perturbation and that of response may be found:

$$\frac{\ell(\ell+1)kT}{rI^2}\,g(\omega) = \omega^2 G_\ell(\omega)\,. \tag{2.75}$$

This very result was obtained by two-sided Fourier transformation of the equation

$$\ddot{K}_\ell(t) = \frac{\ell(\ell+1)}{rI^2}\,K_J(t)\,, \tag{2.76}$$

which was derived by too rough decoupling of perturbation $\boldsymbol{J}(t)$ and response $d_q^\ell(t)$ in [115, 116]. The accuracy of this approximation is ascertained by comparing Eq. (2.76) with its exact analogue obtained by differentiation of Eq. (2.24):

$$\ddot{K}_\ell = -\frac{\ell(\ell+1)}{rI^2}\left[K_J - \int_0^t K_J(t')\dot{K}_\ell(t-t')\,\mathrm{d}t'\right].$$

If the second term on the right-hand side of the equation is omitted, the latter is transformed into Eq. (2.76). As the omission is possible only for $t \ll \tau_J$, Fourier transformation of the reduced equation holds for $\omega\tau_J \gg 1$ only. Consequently, the equality (2.75) is of asymptotic character, and may not be utilized to find full $g(\omega)$ or its Fourier-transform $K_J(t)$ at any times. When it was nevertheless used in [117], the rotational correlation function turned out to be alternating in sign. The oscillatory behaviour of $K_J(t)$ occured not only in a compressed gas, but also at normal pressure, when $K_J(t)$ should vanish monotonically, if not exponentially. The origin of these non-physical oscillations is easily ascertained if we take into account that, in the impact approximation, the spectral wings are described by Eq. (2.58). As long as this expression is used in Eq. (2.75) only true high-frequency behaviour of the impact contour (2.74) is recovered: $g(\omega) \approx 1/\pi\,\tau_J\,\omega^2$ at $\omega\tau_J \gg 1$. However, if the whole contour $G_\ell(\omega)$ is used in the relation (2.75) then a non-physical parabolic hole appears in the centre of the spectrum. The multiplier ω^2 turns $g(0)$ to 0 and creates a maximum at artificial frequency that separates the hole from the wings. This artefact is displayed in harmonic modulation of $K_J(t)$ with this very frequency.

Despite such a fault, the attempt to utilize the experimental data undertaken in [117] proves that the inverse problem is solvable. It is even simpler to do starting from the differential kinetic equation (2.25) whose integral is [118, 119]

$$\ln K_\ell(t) = -\frac{\ell(\ell+1)}{2I_2}\int_0^t (t-s)K_J(s)\,\mathrm{d}s\,. \qquad (2.77)$$

The long-time behaviour of $K_\ell(t)$ is obviously exponential, and the rate is proportional to the diffusion coefficient defined in Eq. (2.42). Therefore the expression

$$D(t) = \int_0^t (1-s/t)K_J(s)\,\mathrm{d}s/rI^2$$

is sometimes considered as its time-dependent analogue instead of that defined in Eq. (2.40) [60]. Of course, both expressions tend to the same diffusion coefficient at $t \to \infty$. Differentiating Eq. (2.77) twice with respect to time, we find that [118]

$$K_J(t) = -\frac{2I_2}{\ell(\ell+1)}\frac{\mathrm{d}^2\ln K_\ell(t)}{\mathrm{d}t^2}\,. \qquad (2.78)$$

Owing to this relation the true shape of $K_J(t)$ may be recovered from $K_\ell(t)$ found from the optical spectra or MD simulations.

2.6 Cumulant expansion

In this section we consider how to express the response of a system to noise employing a method of cumulant expansions [38]. The averaging of the dynamical equation (2.19) performed by this technique is a rigorous continuation of the iteration procedure (2.20)–(2.22). It enables one to get the higher order corrections to what was found with the simplest perturbation theory. Following Zatsepin [108], let us expound the above technique for a density of the conditional probability which is the average over all realizations of the random perturbation:

$$P(g, g_0, t) = \langle \delta\,[g - \Omega(t)] \rangle.$$

Here g and g_0 are a set of angular variables, which define a molecular orientation at instants of time 0 and t, respectively and Ω is the orientation at instant t which was g_0 at $t = 0$. By difference of arguments we mean the difference of turns. In the molecular frame (MS), where the axes are oriented along the main axes of the inertia tensor, $\omega_i = J_i/I_i$. Thus the analogue of Eq. (2.19) is the equation

$$\frac{\mathrm{d}}{\mathrm{d}t}\delta\,[g - \Omega\,(t)] = \mathrm{i}\omega\,\hat{\boldsymbol{L}}\,[g - \Omega(t)]\,, \tag{2.79}$$

and the initial condition for $P(g, g_0, t)$ is

$$P\,(g, g_0, 0) \;=\; \delta\,(g - g_0). \tag{2.80}$$

A simple repetition of the iteration procedure (2.20)–(2.22) results in divergence of higher order solutions. However, a perturbation theory series may be summed up so that all unbound diagrams are taken into account, just as is usually done for derivation of the Dyson equation [120]. As a result P satisfies the integral-differential equation

$$\frac{\partial}{\partial t}P(g, g_0, t) = \int_0^t \hat{M}(t - t')P(g, g_0, t')\,\mathrm{d}t' \tag{2.81}$$

with a kernel ('mass-operator') $\hat{M}(t - t')$ determined by a diagrammatic expansion of cumulants:

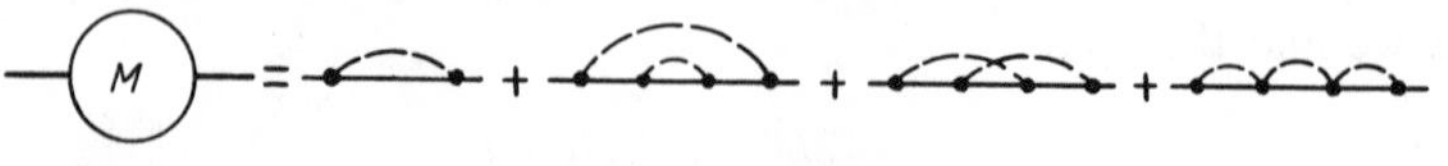

(2.82)

The diagrams are composed of the angular velocity components by the following rules. Function $\eta(t)$ ($\eta = 1$ for $t \geq 0$ and $\eta = 0$ for $t < 0$) corresponds to a solid line and operator $\mathrm{i}\omega\hat{\boldsymbol{L}}\delta(t_1 - t_2)$ to a point with two ends. A dashed line denotes angular velocity components taken at different instants of time and averaged in one and the same cumulant. All cumulants with an odd number of arguments are equal to zero. Analytical expressions corresponding to (2.82) are given in Appendix 1.

Eigenfunctions of the operator $\hat{M}$ are obviously proportional to Wigner D-functions. Therefore a general solution (2.81) with initial condition (2.80) has to be represented as

$$P(g, g_0, t) = \sum_{\ell=0}^{\infty} \sum_{m,m',n} K^{\ell}_{mm'}(t)\,D^{\ell}_{mn}(g)\,D^{\ell\,*}_{m'n}(g_0)\,\frac{2\ell + 1}{8\pi^2}, \tag{2.83}$$

where

$$K^{\ell}_{mm'} = \sum_n \langle D^{\ell}_{mn}\,[g(t)]\;D^{\ell\,*}_{m'n}\,[g(0)]\rangle \tag{2.84}$$

are correlation functions $D^{\ell}_{mm'}[g(t)]$. For spherical and linear rotators, when $m = m' = 0$, they are identical to $K_{\ell}(t)$ introduced earlier

$$K^{\ell}_{00} = \langle D^{\ell}_{00}\,(g - g_0)\rangle = \langle P_{\ell}\;\{\cos\,[\boldsymbol{u}(t)\cdot\boldsymbol{u}(0)]\}\rangle = K_{\ell}. \tag{2.85}$$

In fact, expansion (2.83) is the generalization of Eq. (2.36). It is useful when symmetric rotators are considered. By substituting this result into Eq. (2.81), we get the following equation for correlation functions (instead of (2.24))

$$\frac{\partial}{\partial t}K^{\ell}_{mm'} = \sum_n \int_0^t M_{mn}(t-t')K^{\ell}_{nm'}(t')\,\mathrm{d}t'. \tag{2.86}$$

This equation is exact, and its kernel is expanded into the series (2.82). The decoupling procedure resulting in (2.24) is equivalent to retention of the first term in this series. Using Eq. (2.86), one may develop a consistent perturbation theory which will take into consideration higher orders of $\mathscr{H}_{\ell}$.

The Laplace transform $\tilde{K}^{\ell}_{mm'}$ found from Eq. (2.86), takes the form

$$\tilde{K}^{\ell}_{mm'} = \left[\mathrm{i}\omega - \hat{M}\,(\mathrm{i}\omega)\right]^{-1} K^{\ell}_{mm'}\,(0). \tag{2.87}$$

Its poles are determined to any order of $\mathscr{H}_{\ell}$ by expansion of $\hat{M}$. However, even in the lowest order in $\mathscr{H}_{\ell}$, the inverse Laplace transformation, which restores the time kinetics of $K^{\ell}_{mm'}$, keeps all powers to $\mathscr{H}_{\ell}t/\tau_J$. This is why the theory expounded in the preceding section described the long-time kinetics of the process, while the conventional time-dependent perturbation theory of Dirac [121] holds only in a short time interval after interaction has been switched on. By keeping terms of higher order in $\mathscr{H}_{\ell}$, we describe the whole time evolution to a better accuracy.

To prove this let us make more precise the short-time behaviour of the orientational relaxation, estimating it in the next order of $\mathscr{H}_{\ell}$. The estimate of I_4 given in (2.65b) involves terms of first and second order in $\mathscr{H}_{\ell}$ but the accuracy of the latter was not guaranteed by the simplest perturbation theory. The exact value of I_4 presented in Eq. (2.66) involves numerical coefficient $\alpha^{(r)}_{\ell}$, which is correct only in the next level of approximation. The latter keeps in Eq. (2.86) the terms quadratic to $\mathscr{H}_{\ell}$ emerging from the expansion of $\hat{M}(\mathscr{H}_{\ell})$. Taking into account this correction calculated in Appendix 2, one may readily reproduce the exact

Table 2.2. *The coefficients $\alpha_\ell^{(r)}$ of Eq. (2.66).*

ℓ \ r	2	3
1	2	5/2
2	8/3	17/6

short-time asymptotic behaviour of the correlation functions for linear molecules ($r = 2$), which has been obtained by Gordon [8]:

$$K_1 = 1 - \frac{kT}{I}t^2 + \left[\frac{\langle M^2\rangle}{24I^2} + \frac{1}{3}\left(\frac{kT}{I}\right)^2\right]t^4$$

$$K_2 = 1 - \frac{3kT}{I}t^2 + \left[\frac{\langle M^2\rangle}{8I^2} + 4\left(\frac{kT}{I}\right)^2\right]t^4 \qquad (2.88)$$

Coincidence with the exact expression is caused by the fact that terms of higher order in $\mathcal{H}_\ell$ do not contribute to the second and fourth moments.

Correspondingly, for spherical rotators ($r = 3$) we have

$$K_1 = 1 - \frac{kT}{I}t^2 + \left[\frac{\langle M^2\rangle}{36I^2} + \frac{5}{12}\left(\frac{kT}{I}\right)^2\right]t^4$$

$$K_2 = 1 - \frac{3kT}{I}t^2 + \left[\frac{\langle M^2\rangle}{12I^2} + \frac{17}{4}\left(\frac{kT}{I}\right)^2\right]t^4 \qquad (2.89)$$

just as that obtained in [122]. The values of $\alpha_\ell^{(r)}$ extracted from comparison of these formulae with Eq. (2.64) and Eq. (2.66) are listed in Table 2.2.

Now we refer to the analysis of a functional relationship between the times of orientational and rotational (angular momentum) relaxation that are $\tau_{\theta,\ell}$ and τ_J, respectively. To lowest order in $\mathcal{H}_\ell$, this relationship is given by the Hubbard relation (2.28). It is universal in the sense that it does not depend on the mechanisms of rotational relaxation. However, this relation does not hold when $\tau_{\theta,\ell}$ is calculated to higher order in $\mathcal{H}_\ell$. Corrections to the Hubbard relation are expressed in terms of higher correlation moments of $\omega_i(t)$ whose dependence on τ_J is specific for different mechanisms. Let us demonstrate this, taking the impact theory as an example. In principle it distinguishes correlated behaviour of the angular momentum from non-correlated behaviour but in the rotational diffusion limit the difference in $\tau_{\theta,\ell}$ dependence on τ_J appears only in the second order in $\mathscr{H}_\ell$. Let us find it by Laplace transformation of Eq. (A2.1) from Appendix 2. Taking into account that $\tau_{\theta,\ell} = \tilde{K}_\ell(0)$ where $\tilde{K}_\ell$ is defined in Eq. (2.69a), we get

$$\tau_{\theta,\ell} = [p + \tilde{\mu}(p)]^{-1}_{p\to 0} = \left(\int_0^\infty \mu(t)\,\mathrm{d}t \right)^{-1} . \tag{2.90}$$

The fourth-order correlators obtained in (1.55) were used to determine the integrals in (A2.3). This enables us to calculate the integral in (2.90) and ascertain the relationship between the dimensionless relaxation times $\tau^*_{\theta,\ell} = \tau_{\theta,\ell}(kT/I)^{\frac{1}{2}}$ and $\tau^*_J = \tau_J(kT/I)^{\frac{1}{2}}$ [123]:

$$\tau_{\theta,\ell} = \frac{1}{\ell(\ell+1)\,\tau^*_J} \left[1 + A^{(r)}_\ell \left(\tau^*_J\right)^2 \right], \tag{2.91}$$

where

$$A^{(r)}_\ell = \frac{2}{1+\gamma} \left[\ell\,(\ell+1) - a_r\right] \tag{2.92}$$

and

$$a_r = \begin{cases} 1/2 & \text{for spherical rotators } (r=3) \\ 1 & \text{for linear rotators } (r=2). \end{cases}$$

It is seen that the result obtained is sensitive to both the molecular symmetry and the strength of collision γ, which is a quantitative measure of the degree of correlation. However, the latter affects only correction to the Hubbard relation which appears in the second order in $(\tau^*_J)^2 \ll 1$. Formula (2.91) predicts a linear dependence of product $\tau^*_{\theta,\ell}\tau^*_J$ on $(\tau^*_J)^2$ for any γ, but the slope of the lines differs by a factor of two, being minimal for γ=1 and maximal for γ=0. In principle, it is possible to calculate corrections of the higher orders in $(\tau^*_J)^2$ and introduce them into (2.91). In practice, however, this does not extend the application range of the results due to a poor convergence of the perturbation theory series.

In Fig. 2.9 the theory is compared with some experimental data. In these coordinates the Hubbard relation corresponds to a horizontal straight line with intercept 1/6, but for any finite τ_J deviations from 1/6 are observed. The angle formed by the straight lines with γ=0 and γ=1 defines the acceptable range of slopes. If experimental points are within the angle, the impact description may turn out to be acceptable. The above method for testing experimental data is considerably different from the conventional one. As a rule, the Hubbard relation is checked

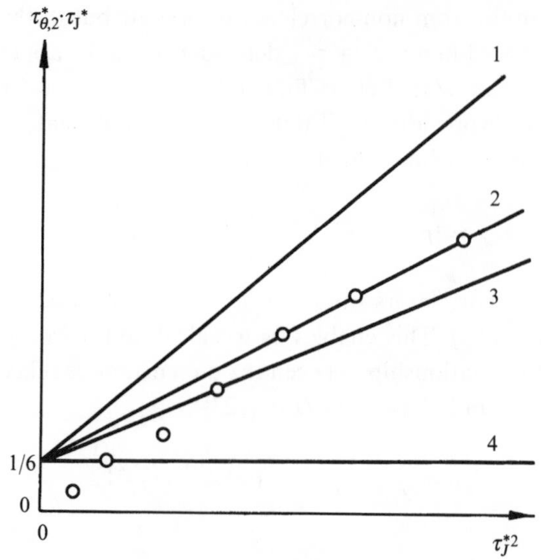

Fig. 2.9. Comparison of the experiment for ClO_3F [94] with theoretical predictions of $\tau_{\Theta,2}(\tau_J)$ dependence. The Hubbard relation is presented by horizontal line (4), the lines with non-zero slopes obtained correspond to the strong collision case $\gamma = 0$ (1), the weak collision case $\gamma = 1$ (3) and the intermediate case $\gamma = 0.6$ (2).

in the coordinates of Fig. 2.10. Here it is presented by a straight line in a dense medium region. The experimental points for a number of heavy molecules approach this line, though in logarithmic coordinates it is hard to see whether the deviations are considerable.

The Hubbard relation is indifferent not only to the model of collision but to molecular reorientation mechanism as well. In particular, it holds for a jump mechanism of reorientation as shown in Fig. 1.22, provided that rotation over the barrier proceeds within a finite time τ_J^0. To be convinced of this, let us take the rate of jump reorientation as it was given in [11], namely

$$1/\tau_{\theta,\ell} = \ell\,(\ell+1)\,\langle\theta^2\rangle/\tau, \tag{2.93}$$

where $\langle\theta^2\rangle = 4D_\theta\tau_J^0$. In view of Eq. (2.42) the mean square angular displacement per jump may be represented as

$$\langle\theta^2\rangle = 4\,\langle\omega^2\rangle\left(\tau_J^0\right)^2/r\,. \tag{2.94}$$

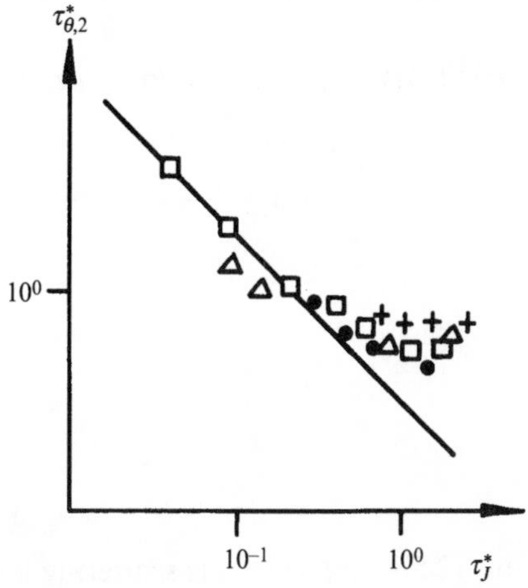

Fig. 2.10. Verification of the Hubbard relation (straight line) in log plot. (□) CCl_3F [67]; (•) ClO_3F [94]; (△) CF_4 [124]; (+) SF_6 [125]).

The rate of activated jumps is given by the Arrhénius formula

$$\frac{1}{\tau} \approx \frac{1}{\tau_J^0}\, \mathrm{e}^{-U_0/kT}, \tag{2.95}$$

where U_0 is the height of a libration barrier. Formulae (2.94) and (2.95) reveal the meaning of phenomenological parameters of the jump model and determine the relationship between orientational relaxation time and the time of above-barrier rotational (angular momentum) relaxation τ_J^0:

$$\frac{1}{\tau_{\theta,\ell}} \simeq \frac{\ell\,(\ell+1)\,\langle\omega^2\rangle}{r}\,\left(\tau_J^0\,\mathrm{e}^{-U_0/kT}\right). \tag{2.96}$$

As one would expect, the rate of orientational relaxation in the jump model is activated, and the higher the libration barrier U_0, the lower the rate. However, the Hubbard relation obtained as a result of Eq. (1.123) used in Eq. (2.96) does not involve this characteristic parameter of the solid-like model:

$$\frac{1}{\tau_{\theta,\ell}} \simeq \langle\omega^2\rangle\ell\,(\ell+1)\,\tau_J/r = kT\,\ell\,(\ell+1)\,\tau_J/I.$$

It is just the same as in gas models. Thus, the Hubbard relation itself allows one neither to choose any definite gas model, nor to distinguish it from a libration one.

3

Transformation of isotropic scattering spectra

As is seen from relations (2.5)–(2.8), isotropic scattering is independent of orientational relaxation. Since the isotropic component of the polarization tensor is invariant to a molecule's reorientation, the corresponding correlation function K_0 describes purely vibrational relaxation. This invariance does not mean however that vibrational relaxation is completely insensitive to angular momentum relaxation. Interaction between vibrations and a molecule's rotation determines the rotational structure of the isotropic scattering spectra observed in highly rarefied gases. The heavier the molecule, the smaller is the constant α_e of the Q-branch rotational structure. In fact this thin structure is easily resolved only in hydrogen and deuterium. The isotropic Raman spectrum of most other gases is usually unresolved even at rather low pressure and when describing its shape at higher densities one may consider J a classical (continuous) variable.

Within the framework of the impact theory $J(t)$ is a purely discontinuous Markovian process. The same is valid for the corresponding frequency, or 'rotational component', which changes its position in the spectrum after each collision. This phenomenon, known as spectral diffusion or rotational frequency exchange, is accompanied by adiabatic dephasing of the vibrational transition caused by these same collisions. Both processes contribute to observed spectral transformation with increasing collision frequency, however they have opposite effects. While frequency exchange leads to collisional narrowing, dephasing results in the spectrum-broadening. If dephasing is weak and the collision frequency is small, the tendency for the spectrum to narrow prevails. However, at higher gas densities the spectrum may become so narrow that dephasing is able to compete with the residual rotational broadening, and eventually the situation changes to the opposite case of purely vibrational broadening.

The initial 'static spectrum' and the well-narrowed one may be described quite generally without specifying any model for a molecule's interaction with a medium. To trace the spectral transformation between these limits we developed a quasi-classical impact theory for alternative models of strong and weak collisions. It was generalized later for collisions of arbitrary strength and applied to both linear and spherical tops. Since the dephasing of diatomics is usually very weak, 'collisional narrowing' in nitrogen is rather pronounced. It will be studied first in the impact approximation valid at relatively low and moderate densities where dephasing is negligible. These results will be sewed together with those obtained by perturbation theory at moderate and higher densities. Since both theories overlap in the rapid modulation limit, the whole density range is covered this way, from gas phase to cryogenic liquid. In conclusion the theoretical results for linear and spherical molecules are compared with experiments done in both phases and above critical point.

3.1 Static contour

The origin of the rotational structure of the isotropic Q-branch ($\Delta v = 0$, $\Delta j = 0$) is connected with the dependence of the vibrational transition frequency shift on rotational quantum number j [121, 126]

$$\Omega = \alpha_e\, j\,(j+1) \approx \alpha J^2, \tag{3.1}$$

where $\alpha = \alpha_e/\hbar^2$. The last expression is used for a quasi-classical description of the Q-branch shape which is possible when the rotational structure is unresolved. The quasi-classical theory holds at relatively high pressures when the collisional broadening of rotational components exceeds their splitting. The broadening is mainly determined by the rate of collisions which change J^2 and may be roughly estimated as $1/\tau_E$. Thus the validity criterion of the classical theory is given by the inequality

$$\frac{1}{\tau_E} \gg \alpha_e\langle j(j+1) - (j-1)j\rangle = 2\alpha_e\langle j\rangle = 2\hbar\alpha\langle J\rangle. \tag{3.2}$$

As soon as condition (3.2) is satisfied the 'static contour' $G_0^{st}(\omega)$ arises, which is already insensitive to broadening of individual rotational components, but has not yet transformed as a whole. The static contour is almost indifferent to further increase in gas density until

$$\alpha\langle J^2\rangle \gg 1/\tau_E. \tag{3.3}$$

The shape of the static contour is readily calculated from Eq. (2.13) with the correlation function

$$K_0^{st} = \int_0^\infty \varphi_B(J) \exp\left(\mathrm{i}\alpha J^2 t\right) \,\mathrm{d}t. \tag{3.4}$$

For instance, in the case of linear rotators ($r = 2$), we find from Eq. (3.4) and (1.1) that

$$K_0^{st} = \frac{1}{1 - \mathrm{i}\omega_Q t}. \tag{3.5}$$

The corresponding static contour has the exponential form [127]

$$G_0^{st}(\omega) = \begin{cases} 0 & \text{at } \omega/\omega_Q < 0 \\ (1/\omega_Q) \exp\left(-\omega/\omega_Q\right) & \text{at } \omega/\omega_Q \geq 0, \end{cases} \tag{3.6}$$

where

$$\omega_Q = \int_0^\infty \omega G_0(\omega) \,\mathrm{d}\omega = 2\alpha d = 2\alpha_e I kT/\hbar^2.$$

As discussed in [91], the shape of a static spectrum determines significantly the spectral transformation as frequency exchange increases. In particular, spectral narrowing will take place only if the second moment of the spectrum is finite. In our case

$$\langle(\omega - \omega_Q)^2\rangle = \int_0^\infty (\omega - \omega_Q)^2 G_0(\omega) \,\mathrm{d}\omega = \omega_Q^2. \tag{3.7}$$

Consequently, when the gas density increases, the increase of the frequency exchange rate causes the narrowing of the isotropic scattering spectrum.

The phenomenon discussed has analogues in magnetic resonance ('exchange narrowing' [128]) and atomic spectroscopy ('Dicke effect' [129]). In both cases, however, the initial static contour is symmetric and Gaussian in form, which considerably simplifies the description of its collapse. From this point of view, isotropic scattering is a special example of a spectral transformation which is complicated by the spectrum asymmetry. In all cases the necessary condition for the narrowing is the absence or weakness of phase breaking, which often accompanies frequency exchange.

3.2 Perturbation theory

In order to describe the shape of the Q-branch after its collapse, it is sufficient to use the stochastic perturbation theory expounded in the preceding chapter. The dynamical equation

$$\frac{dK}{dt} = i\alpha J^2(t)K + i\Delta\Omega(t)K \tag{3.8}$$

determines the time evolution of

$$K(t) = \bar{\alpha}(0)\bar{\alpha}(t)/\bar{\alpha}^2 . \tag{3.9}$$

This is a random molecule's response to both random perturbations $J(t)$ and $\Delta\Omega(t)$, where $\Delta\Omega(t)$ represents the shift in the vibrational frequency during collisions if in gas.

When collisional effects are negligible the second component in (3.8) vanishes, and J becomes a constant. The solution obtained in this case is used in Eq. (3.4). If collisions occur they change J and frustrate the vibrational phase simultaneously. Nevertheless, the processes are usually considered to be statistically independent:

$$\langle J^2(t)\,\Delta\Omega(t')\rangle = \langle J^2(t)\rangle\;\langle\Delta\Omega(t')\rangle . \tag{3.10}$$

Physically the independence reflects the fact that dephasing is performed by weak long-range interactions, and rotational relaxation results mainly from short-range, repulsive forces. In other words the rotational state is changed solely when the distance between molecules becomes rather short, while the phase is frustrated in all cases and the contribution of frontal collisions is not so significant.

In using perturbation theory, first we have to 'centre' random processes by subtracting their equilibrium values $\omega_Q = \alpha\langle J^2\rangle$ and $\Delta = \langle\Delta\Omega\rangle$. As a result Eq. (3.8) is reduced to

$$\frac{dK}{dt} = i(\omega_Q + \Delta)K + i(\alpha\,\delta J^2 + \delta\Omega)K, \tag{3.11}$$

where $\delta J^2 = J^2(t) - \langle J^2\rangle$, $\delta\Omega = \Delta\Omega(t) - \Delta$ are 'centred' functions. In view of Eq. (3.10), we have

$$\langle\delta J^2(t)\;\delta\Omega(t')\rangle = \langle\delta J^2(t)\rangle = \langle\delta\Omega(t')\rangle = 0. \tag{3.12}$$

Assuming that both perturbations are weak, let us find the molecule's response

$$K_0 = \langle K\rangle ,$$

averaged over all realizations of the random perturbations. Stated in such a way, the problem is identical to that considered in the preceding chapter. The kinetic equation, which is similar to Eq. (2.22), may be immediately written as

$$\dot{K}_0 = \mathrm{i}\left(\omega_Q + \Delta\right) K_0 - 4\alpha^2 I^2 \int_0^t \tilde{K}_E\left(t - t'\right) K_0(t')\,\mathrm{d}t' - \int_0^t \langle \delta\Omega(t)\ \delta\Omega(t')\rangle K_0(t')\,\mathrm{d}t', \tag{3.13}$$

where $\tilde{K}_E = K_E(t) - K_E(\infty)$ and $K_E(t - t') = \langle J^2(t) J^2(t')\rangle / 4I^2$. Though the exact functional form of both kernels in dense media is unknown, we may use the Markovian simplification of Eq. (3.13), which is valid for $t \gg \tau_E, \tau_c$

$$\dot{K}_0 = \mathrm{i}\left(\omega_Q + \Delta\right) K_0 - \left[4\alpha^2 I^2 \tilde{K}_E(0)\tau_E + \langle \Delta\Omega^2\rangle\,\tau_c\right] K_0\,. \tag{3.14}$$

This treatment describes only the long-time exponential behaviour such as

$$K_0(t) = \exp\left[\mathrm{i}\left(\omega_Q + \Delta\right) t - \left(w_0 + \gamma_{\mathrm{dp}}\right) t\right], \tag{3.15}$$

where

$$w_0 = 4\alpha^2 d^2 \tau_E = \omega_Q^2 \tau_E, \tag{3.16}$$

$$\gamma_{\mathrm{dp}} = \langle \delta\Omega^2\rangle\ \tau_c. \tag{3.17}$$

Here τ_E, τ_c are the correlation times of rotational and vibrational frequency shifts. The isotropic scattering spectrum corresponding to Eq. (3.15) is the Lorentzian line of width $\Delta\omega_{1/2} = w_0 + \gamma_{\mathrm{dp}}$. Its maximum is shifted from the vibrational transition frequency by the quantity ω_Q due to the collapse of rotational structure and by the quantity Δ due to the displacement of the vibrational levels in a medium.

According to the validity criteria of perturbation theory, the above results hold when

$$w_0\tau_E = \left(\omega_Q \tau_E\right)^2 \ll 1, \tag{3.18}$$

$$\gamma_{\mathrm{dp}}\tau_c = \langle \delta\Omega^2 \tau_c^2\rangle \ll 1. \tag{3.19}$$

Condition (3.19) is usually satisfied in processes of vibrational dephasing [41, 130, 131, 132]. Because of this condition the dephasing is weak and the effect of rotational structure narrowing is pronounced. A much more important constraint is imposed by inequality (3.18). It shows that perturbation theory must be applied to a rather dense medium and even then only the central part of the spectrum (at $|\Delta\omega| \ll 1/\tau_E,\ 1/\tau_c$) is Lorentzian.

Formulae (3.15)–(3.17) are quite general because the method is indifferent to how the perturbation changes in time. It may be a sequence of collisions or random continuous noise. Thus, the results are valid when the gas is condensed into a liquid. Motional narrowing of rotational structure progresses with increase of density as long as

$$\gamma_{\text{dp}} \ll w_0 . \tag{3.20}$$

Only after the above inequality is inverted, is the narrowing replaced by a broadening.

To verify that γ_{dp} and w_0 change with density in the opposite way, one should return to the impact theory where $\langle \delta\Omega^2 \rangle = \langle \delta\Omega_c^2 \rangle \tau_c / \tau_0$ according to Eq. (1.81) and hence

$$\gamma_{dp} = \langle \delta\Omega_c^2 \tau_c^2 \rangle / \tau_0 = n\sigma_{\text{dp}} \langle v \rangle . \tag{3.21}$$

In view of Eq. (3.19) the phase shift performed by a single collision is small, therefore

$$\sigma_{\text{dp}} \ll \sigma_0 , \tag{3.22}$$

where σ_{dp} is the dephasing cross-section and σ_0 is a kinetic cross-section. As seen from Eq. (3.21), the width of the line determined by dephasing monotonically increases with increase in the density of gas n. In contrast according to Eq. (3.16) and Eq. (1.57), the residual rotational width of the collapsed spectrum is given by

$$w_0 = \omega_Q^2 \tau_J / (1 + \gamma) . \tag{3.23}$$

According to (1.124) this width monotonically decreases as n increases. Using linear approximation (1.124) in Eq. (3.21) one can relate γ_{dp} to τ_J,

$$\gamma_{\text{dp}} = \frac{\sigma_{\text{dp}}}{\sigma_J} \frac{1}{\tau_J} . \tag{3.24}$$

Hence the dephasing rate is directly proportional to the rotational relaxation rate $1/\tau_J$, while w_0 in (3.23) is inversely proportional to it.

One may hope that the results presented in Eq. (3.23) and Eq. (3.24) remain valid beyond the framework of impact theory. As is seen from Chapter 1, linear in density Eq. (3.21) and Eq. (1.124) become invalid in highly dense media. However, it is unlikely that their relative efficiency will be considerably changed. Thus the direct proportion (3.24) may be retained even in the case where $1/\tau_J$ increases nonlinearly with increase in density (see Fig. 1.23). Since it is easier to measure τ_J in the liquid than τ_E, it is of some importance to express the isotropic spectrum width as a function of τ_J.

In the conclusion of the present chapter we show how comparison of NMR and Raman scattering data allows one to test formulae (3.23) and (3.24) and extract information about the relative effectiveness of dephasing and rotational relaxation. In particular, spectral broadening in nitrogen caused by dephasing is so small that it may be ignored in a relatively rarefied gas when spectrum collapse proceeds. This is just what we are going to do in the next sections devoted to the impact theory of the isotropic Raman spectrum transformation.

3.3 The impact theory of frequency exchange

Now our main objective is to consider gradual transformation of the highly asymmetric static contour of the Q-branch into the Lorentzian line centred on frequency ω_Q and progressively narrowing with increasing density. It is clear from the above discussion that such a transformation is caused by a random change of $J(t)$ which results in migration of the corresponding frequency over the static contour. In the impact theory J is a Markovian, purely discontinuous perturbation, which obeys the Feller equation (1.3). In order to describe the system's response to this perturbation, one should refer to sudden modulation theory (see Chapter 4 of [20] and [24]). According to the theory

$$K_0 = \int_0^\infty K(t, J)\,\mathrm{d}J \tag{3.25}$$

is resolved into spectral components $K(t, J)$ which satisfy the following kinetic equation:

$$\dot{K} = \mathrm{i}\alpha J^2 K - \frac{1}{\tau_0}K + \frac{1}{\tau_0}\int_0^\infty f(J', J)K(t, J')\,\mathrm{d}J'\,. \tag{3.26}$$

The first component on the right-hand side controls dynamic development of the response in the same way as in Eq. (3.8), and the other two control spectral exchange due to collisions. Solution of Eq. (3.26) should satisfy the initial condition

$$K(0, J) = \varphi_B(J) = \exp\left(-J^2/2d\right) J/d \tag{3.27}$$

which accounts for the equilibrium distribution of linear rotators over $J = |\boldsymbol{J}|$.

It is necessary to get insight into the kernel of the integral equation (3.26). Since frequency exchange is initiated by non-adiabatic collisions, it is reasonable to use the Keilson–Storer model. However, before employing kernel (1.16) it should be integrated over the angle

$\beta = \arccos(\boldsymbol{J} \cdot \boldsymbol{J}'/JJ')$, which does not influence the scattered light frequency. The result is

$$f(J', J) = \int_0^\pi f(\boldsymbol{J} - \gamma \boldsymbol{J}')J \, \mathrm{d}\beta$$
$$= \frac{J}{d(1-\gamma^2)} I_0 \left(\frac{\gamma J' J}{(1-\gamma^2)d} \right) \mathrm{e}^{-(J^2+\gamma^2 J'^2)/2d(1-\gamma^2)}, \quad (3.28)$$

where I_0 is the modified Bessel function [37]. The parameter γ determines the magnitude of interaction during collisions. At $\gamma = 0$ interaction is so strong that it restores equilibrium immediately, while as $\gamma \to 1$ the collisions are weak and do not considerably change the rotational energy. Since there is a significant difference between these situations from the mathematical perspective they will be considered separately.

The validity condition of sudden modulation theory is determined by the requirement of a quick change in J^2 in comparison with its conservation time:

$$\tau_c \ll \tau_E \, . \quad (3.29)$$

This requirement is consistent with condition (3.18) if

$$\omega_Q \tau_c \ll 1, \quad (3.30)$$

which is usually the case ($\omega_Q \tau_c \leq 10^{-2}$). Consequently, impact theory and perturbation theory must coincide in the region

$$\omega_Q < 1/\tau_E < 1/\tau_c \, . \quad (3.31)$$

Owing to this fact, they may be sewed together in order to describe quantitatively the Q-branch transformation with density from a gas to the liquid state [133–5].

3.4 Transformation of the contour by strong collisions

In the limit of strong collisions ($\gamma = 0$) we have $f(J', J) = \varphi_B(J)$, and Eq. (3.26) is considerably simplified to

$$\dot{K} - \left(\mathrm{i}\alpha J^2 - 1/\tau_J\right) K = \left(1/\tau_J\right) \varphi_B(J) K_0 \, , \quad (3.32)$$

where $\tau_J = \tau_0$, according to Eq. (1.22). Solving this linear differential equation and averaging the result over J, we obtain the integral equation for K_0

$$K_0(t) = \frac{\exp\left(-t/\tau_J\right)}{1 - \mathrm{i}\omega_Q t} + \frac{1}{\tau_J} \cdot \int_0^t \frac{\exp\left[-(t-t')/\tau_J\right] K_0(t') \, \mathrm{d}t'}{1 - \mathrm{i}\omega_Q (t - t')} \, . \quad (3.33)$$

Fourier transformation of this equation yields the isotropic scattering spectrum [133]

$$G_0(x) = \frac{1}{\pi\omega_Q}\,\mathrm{Im}\left(\frac{\exp(-z)\,\mathrm{Ei}(z)}{1 + \mathrm{i}\Gamma\,\exp(-z)\,\mathrm{Ei}(z)}\right). \tag{3.34}$$

Here $z = x - \mathrm{i}\Gamma$, $x = \omega/\omega_Q$ and $\Gamma = 1/\omega_Q\tau_J$ is a dimensionless parameter proportional to the gas density, and $\mathrm{Ei}(z)$ is an integral exponent [37].

At $\Gamma \to 0$ the spectrum (3.34) is obviously reduced to a static one, given in Eq. (3.6). In the opposite limit

$$\Gamma \gg 1 \tag{3.35}$$

it collapses to the following form

$$G_0(\omega) = \frac{1}{\pi\omega_Q}\,\frac{\Gamma/(x^2+\Gamma^2)}{\left[x - 1 - x/(x^2+\Gamma^2)\right]^2 + \left[\Gamma/(x^2+\Gamma^2)\right]^2}. \tag{3.36}$$

In its central part where $|x - 1| \ll \Gamma$ (or $|\Delta\omega| = |\omega - \omega_Q| \ll 1/\tau_J$) the spectrum has the Lorentzian form. It is described by perturbation theory and has half-width at half-height as calculated in Eq. (3.23) at $\gamma = 0$:

$$\Delta\omega_{1/2} = \omega_Q/\Gamma = \omega_Q^2\,\tau_J\,. \tag{3.37}$$

The intensity at the periphery of the line ($|\Delta\omega| \gg 1/\tau_J$) decreases as $\omega_Q^2/\pi\tau_J\omega^4$ in accordance with the general rule (2.62) [20, 104]. However, the most valuable advantage of general formula (3.34) is its ability to describe continuously the spectral transformation from a static contour to that narrowed by motion (Fig. 3.1). In the process of the spectrum's transformation its maximum is gradually shifted, the asymmetry disappears and it takes the form established by perturbation theory.

3.5 Transformation of the contour by weak collisions

In the case of weak collisions the change in J is so slight that one may proceed from an integral description of the process to a differential one, just as in Eq. (1.23). However, the kernel of the integral equation (3.26) specified in Eq. (3.28) is different from that in the Feller equation. Thus, the standard procedure described in [20] is more complicated and gives different results (see Appendix 3). The final form of the equation obtained in the limit $\gamma \to 1$, $\tau_0 \to 0$ with

$$\frac{1-\gamma}{\tau_0} = \frac{1}{\tau_J} = \mathrm{const} \tag{3.38}$$

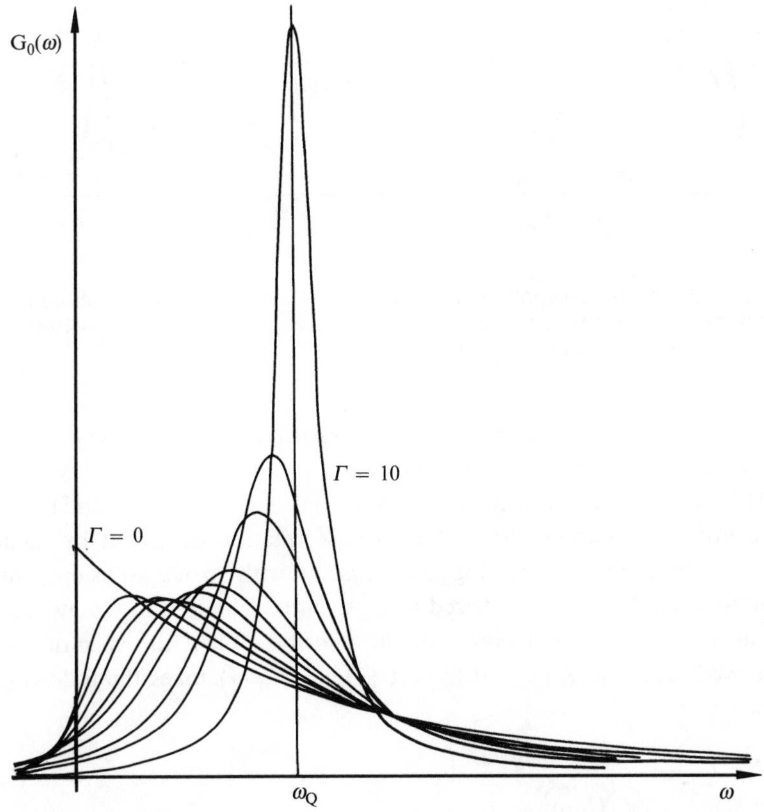

Fig. 3.1. The pressure transformation of the isotropic Raman spectrum in the strong collision limit from $\Gamma = 0$ up to $\Gamma = 10$. The intermediate values of Γ are 0.1; 0.2; 0.3; 0.5; 0.7; 1; 2; 3.

is as follows [134]:

$$\frac{\partial K}{\partial t} = \mathrm{i}\alpha J^2 K + \frac{1}{\tau_J}\left(1 + \frac{d}{J^2}\right) K + \frac{J}{\tau_J}\left(1 - \frac{d}{J^2}\right)\frac{\partial K}{\partial J} + \frac{d}{\tau_J}\frac{\partial^2 K}{\partial J^2}\,. \quad (3.39)$$

This equation may be solved by separation of the variables, which results in formula (A4.12) of Appendix 4 at $\ell = 0$. By Fourier transformation of the latter we find the spectrum

$$G_0 = \frac{1}{\pi\omega_Q}\,\mathrm{Re}\left(\frac{2\,{}_2F_1\left\{p, 1, p+1, \left[(1-\sqrt{B})/(1+\sqrt{B})\right]^2\right\}}{\left(1+\sqrt{B}\right)^2\,\Gamma p}\right), \quad (3.40)$$

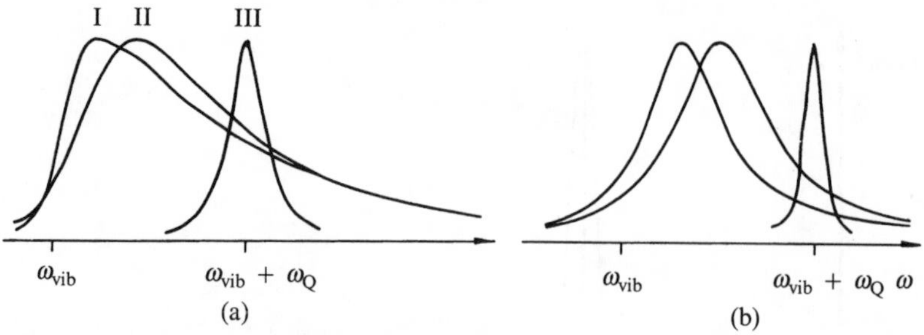

Fig. 3.2. Q-branch transformation with increase of density in strong collision (*a*) and weak collision (*b*) approximation at $\Gamma = 0.1$ (I); $\Gamma = 0.3$ (II); $\Gamma = 10$ (III). All spectra are normalized to 1 at their maxima.

where $B = 1 - 2i/\Gamma$, $p = [ix - (1 - \sqrt{B})\Gamma]/2\sqrt{B}\Gamma)$, and ${}_2F_1$ is the generalized hypergeometric function [37].

Although from a mathematical point of view formulae (3.34) and (3.40) have little in common, the spectral transformation described by them proceeds in a similar way (Fig. 3.2). Just as with strong collisions, the contour is gradually symmetrized and its centre is shifted to the average frequency ω_Q with an increase in the density. When the spectrum is narrowed (at $\Gamma \gg 1$), its central part $\left(|\Delta\omega| \ll 1/\tau_J\right)$ takes the following form:

$$G_0 = \frac{\omega_Q^2\,\tau_J}{2\pi\left[(\Delta\omega)^2 + \left(\omega_Q^2\,\tau_J/2\right)^2\right]}, \qquad \Delta\omega = \omega - \omega_Q\,. \tag{3.41}$$

The width of this Lorentzian line is half as large as that found in (3.37). This, however, is not a surprise because the perturbation theory equation (3.23) predicted exactly this difference in the width of the line narrowed by strong and weak collisions. This is the maximal difference expected within the framework of impact theory when the Keilson–Storer kernel is used and $0 < \gamma < 1$.

It should be noted that the same method for calculation of isotropic scattering spectra is applied to spherical molecules as well. The only difference between linear and spherical molecules is the shape of the static spectrum, while its collapse proceeds in a qualitatively similar way.

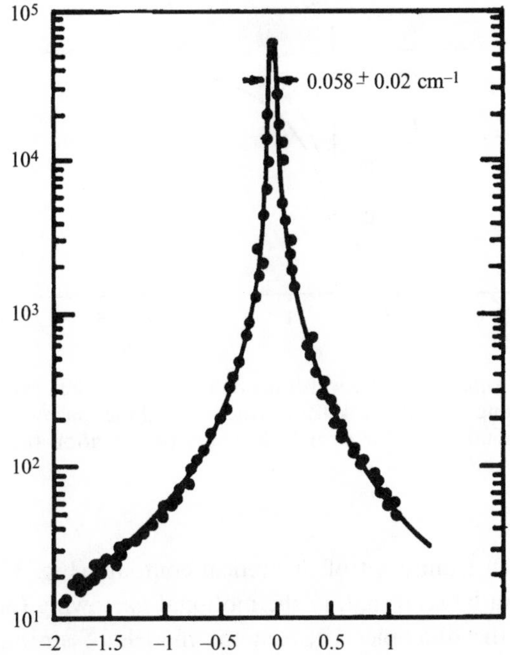

Fig. 3.3. Lorentzian line shape (solid line) and experimental CARS data (points) of liquid nitrogen ($T = 77$ K) from [136].

3.6 Comparison with experiment: linear molecules

The quasi-classical description of the Q-branch becomes valid as soon as its rotational structure is washed out. There is no doubt that at this point its contour is close to a static one, and, consequently, asymmetric to a large extent. It is also established [136] that after narrowing of the contour its shape in the liquid is Lorentzian even in the far wings where the intensity is four orders less than in the centre (see Fig. 3.3). In this case it is more convenient to compare observed contours with calculated ones by their characteristic parameters. These are the half width at half height $\Delta\omega_{1/2}$ and the shift of the spectrum maximum $\omega_{max} - \omega_v = \delta\omega + \Delta$, which is usually assumed to be a sum of the rotational shift $\delta\omega$ that is nonlinear in density and a linear shift of larger scale Δ determined by vibrational dephasing.

In the case of the isotropic spectrum it is useful to consider the functional dependence of $\Delta\omega_{1/2}$ and $\delta\omega$ on τ_E, since it is rotational energy relaxation that causes frequency modulation in this spectrum. Such a

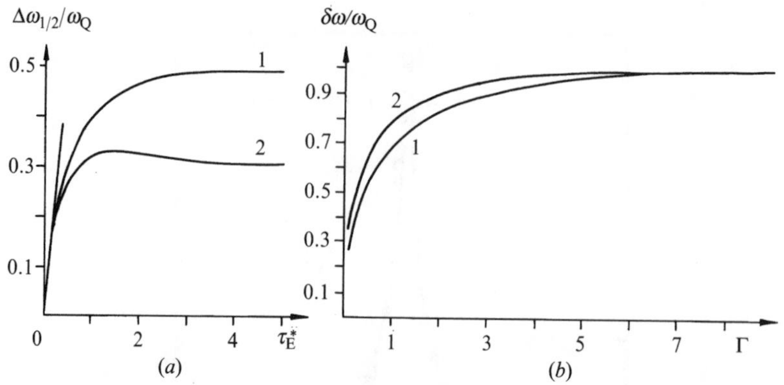

Fig. 3.4. The isotropic Q-branch width (*a*) and rotational shift (*b*) calculated in the models of strong (1) and weak (2) collisions as a function of $\tau_E^* = \omega_Q \tau_E$ and $\Gamma = 1/\tau_E^*$ correspondingly. The straight lines are perturbation theory estimates of spectral width and shift.

dependence taken from a set of theoretical contours (Fig. 3.2) is shown in Fig. 3.4. It can be seen that, in the motional narrowing (perturbation theory) region, the difference between the models of strong and weak collisions vanishes. This is connected with the universal character of the perturbation method, which expresses results in terms of correlation time τ_E irrespective of whether the perturbation is correlated or not, or of impact or continuous type.

Broadening caused by strong collisions may be distinguished from that caused by weak collisions only in a relatively rarefied medium beyond the perturbation theory region. When collisions are weak, narrowing is preceded by weak broadening of the spectrum, while in the case of strong collisions no broadening is observed and the line monotonically narrows with increase in $1/\tau_E$. It should be noted that, in the opposite limit of $\tau_E \to \infty$, the same width inherent in a static contour is reproduced in both situations. However, data extrapolation to this limit (Fig. 3.4a), is senseless. When $\tau_E^* = \omega_Q \tau_E > 10$ the rotational structure of the Q-band is resolved and the quasi-classical theory is not applicable.

In both models the rotational shift of the line $\delta\omega$ is the same either in the static limit, where it is equal to zero, or in the case of extreme narrowing where it reaches its maximum value ω_Q. A slight difference in its dependence on τ_E is observed in the intermediate region only. The experimentally observed density dependence of the shift shown in Fig. 3.5 is in qualitative agreement with theory.

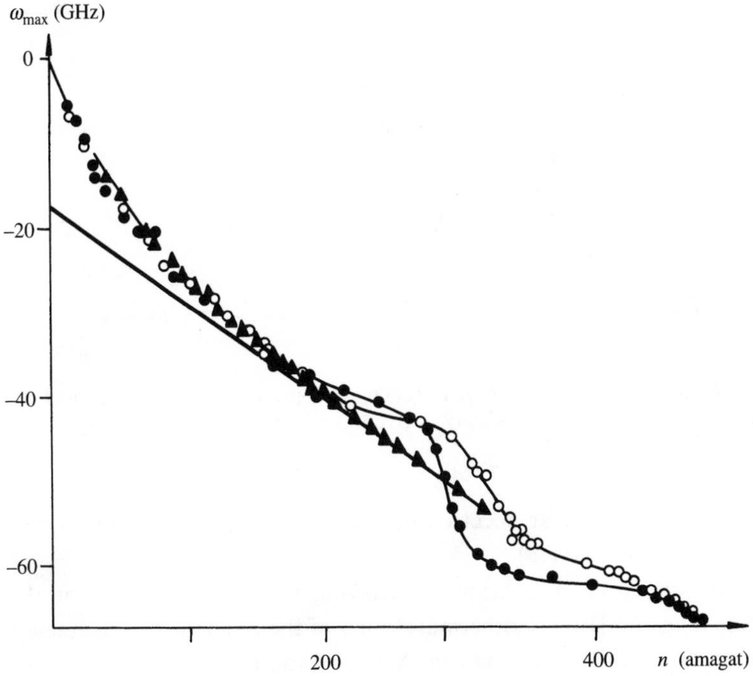

Fig. 3.5. The shift of oxygen Q-branch [137]. The contribution of vibrational dephasing Δ that is linear in density is shown as a straight line. The initial deviation of this line reproduces the theoretical behaviour of $\delta(n)$ shown in Fig. 4.4(*b*). The experiment was performed at a few temperatures higher than the critical one 7.87 K (▲), 0.95 K (○) and 0.12 K (•).

Though the difference between the models of strong and weak collisions manifests itself in a limited range of densities, it is much more explicitly pronounced at the collapse of the Q-branch than at the analogous transformation of the Doppler contour [20, 13]. The cause is in the initial asymmetry of the isotropic scattering spectrum which allows the broadening effect of spectral exchange to manifest itself at a quasi-static stage of the process. The distinctions are even more pronounced if nonlinear methods are employed. If any rotational component is saturated by stimulated Raman scattering, then one can judge the strength of collisions carrying out spectral exchange by the shape of the hole in the Q-branch spectrum [138, 139] and by the kinetics of its burning and recovering [140, 141]. However, in the subsequent discussion we confine

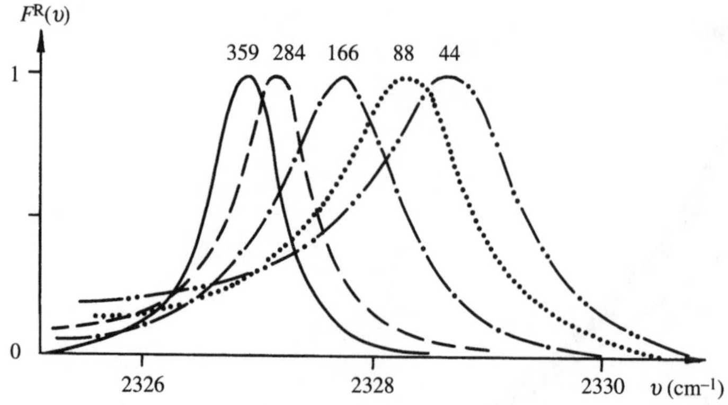

Fig. 3.6. Density transformation of nitrogen isotropic Raman spectrum normalized to a maximum [89] (gas density is given in amagat).

ourselves either to linear spectroscopy data or to coherent anti-Stokes Raman scattering (CARS).

The phenomenon of Q-branch narrowing was demonstrated for the first time long ago [142]. Isothermal studies of the gas density dependence of the spectral width presented in [89] gave the most suitable data for verification of the theory (Fig. 3.6). Cross-sections of all processes remain the same at fixed temperature, and the density dependence of the effect appears in pure form. According to [89], the maximum is particularly pronounced in this dependence at 22 amagat. The same maximum is also seen in Fig. 3.4(*a*) on the theoretical curve corresponding to weak collisions. As far as strong collisions are concerned, the initial broadening of the spectrum with increase in density is not observed. Hence, real collisions appear to be weak. It is even possible to estimate the degree of weakness having a solution at arbitrary γ (how to do this will be demonstrated further in this chapter).

In the pioneering work the same information was extracted from the extremum position assuming it is independent of γ [143]. This is actually the case when isotropic scattering is studied by the CARS spectroscopy method [134]. The characteristic feature of the method is that it measures $|\tilde{K}_0(i\omega)|^2$ not the real part of $K_0(i\omega)$, as conventional Raman scattering does. This is insignificant for symmetric Lorentzian contours, but not for the asymmetric spectra observed in rarefied gas. These CARS spectra are different from Raman ones both in shape and width until the spectrum collapses and its asymmetry disappears. In particular, it turns out that

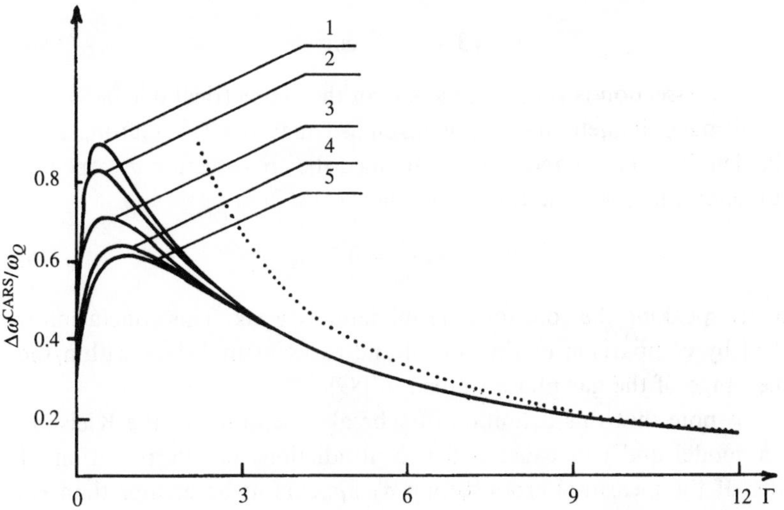

Fig. 3.7. The CARS line width dependence on $\Gamma = 1/\omega_Q\tau_E$ at different collision strengths [144]: (1) $\gamma = 0$ (strong collisions); (2) $\gamma = 0.4$; (3) $\gamma = 0.7$; (4) $\gamma = 0.9$; (5) $\gamma = 1$ (weak collisions). The dots denote the perturbation theory result: $\Delta\omega_{1/2}^{CARS} = 2\omega_Q/\Gamma$.

the width of CARS spectra passes through a maximum with increase in density, and the position of the maximum is insensitive to the strength of collisions [134]. The solution for an arbitrary correlation adduced in Section 7 affirms that the position of the maximum hardly depends on γ (see Fig. 3.7). Under this condition the fitting performed for the first time appeared to be very successful [143]. Therefore it is reproduced here.

Consider the dependence of the spectral width on the dimensionless parameter Γ, which in the framework of impact theory linearly increases with increase in density. Then, according to the theory expounded in the preceding section, the extremum is at the point

$$\Gamma^{max} = 1/\omega_Q\tau_J^{max} = T_{rot}/\alpha_e T\tau_J^{max} = 0.3\,, \tag{3.42}$$

where $T_{rot} = \hbar^2/2Ik$. Since for nitrogen α_e=0.02 cm^{-1} [89], the value

$$1/\tau_J^{max} = n_{max}\sigma_J\left(16kT/\pi m\right)^{1/2} = 1.2 \times 10^{11}\ s^{-1} \tag{3.43}$$

corresponds to the maximum for the dependence of $\Delta\omega_{1/2}$ on density obtained in [89] at $T = 300$ K. As already mentioned, the maximum has actually been observed at $n_{max} = 22$ amagat. Its position theoretically calculated in Eq. (3.43) coincides with the actual one, if

$$\sigma_J = 3 \times 10^{-15}\ \mathrm{cm}^2\,. \tag{3.44}$$

Such a cross-section is somewhat less than the geometrical one $\sigma_0 = \pi d_0^2$, if the nitrogen diameter $d_0 = 2\rho$ is taken as equal to 3.7 Å. Getting back to the definition (1.22) and having obtained the cross-section σ_J, one can try to determine γ via the following relation:

$$\gamma = 1 - \frac{\sigma_J}{\sigma_0} = 0.3\,. \tag{3.45}$$

Roughly speaking the collisions are moderately weak. This conclusion is verified by comparison of theoretical and experimental data within the whole range of the gas phase studied in [89].

Let us note that this definition of γ breaks the limits of the Kielson–Storer model and can cause a few contradictions in interpretation of results. If the measured cross-section σ_J appears to be greater than σ_0, then, according to (3.45), the sought γ does not exist. To be exact, this assertion is valid relative to the cross-section of the rotational energy relaxation $\sigma_E = (1-\gamma^2)\sigma_0$, since γ^2 is always positive. As to σ_J, taking into account the domain of negative values of γ, corresponding to the anticorrelated case (see Chapter 2), formula (3.45) fails to define γ when $\sigma_J > 2\sigma_0$.

The other cause of trouble in interpretation is far less trivial. The weak collision limit of the Kielson–Storer model implies that $\gamma \to 1$, $\tau_0 = (n\sigma_0 v)^{-1} \to 0$ but puts no restriction for $\sigma_J = (1/nv) \lim\, (1-\gamma)/\tau_0$ compared with σ_0. Therefore, the cross-section resulting from the fitting may be nearly equal to the geometrical one. According to Eq. (3.45) this is an indication that collisions are strong although $\gamma \approx 1$ is imposing them to be weak. The only possible way to eliminate this contradiction is not to use relation (3.45) for definition of γ. As a matter of fact, this relation contains the cross-section σ_0 that corresponds to no physical relaxation channel. As the Kielson–Storer model is self-consistent, none of the solutions at any γ contain σ_0 as such, but only in the combination $\sigma_0(1-\gamma^2) = \sigma_E$. Contrarily, parameter γ may appear on its own. So, no contradiction appears, if fitting is performed over the σ_E and γ variables. Only at $\gamma \ll 1$ may an estimation of Eq. (3.45) type be used without complications.

The cross-section of J-changing collisions found in Eq. (3.44) is sufficient for conversion of any gas density value to the quantity

$$\Gamma = 1/\omega_Q\tau_J = \sigma_J \left(T_{\mathrm{rot}}/\alpha_e\right) \left(16k/\pi T m\right)^{1/2} n\,. \tag{3.46}$$

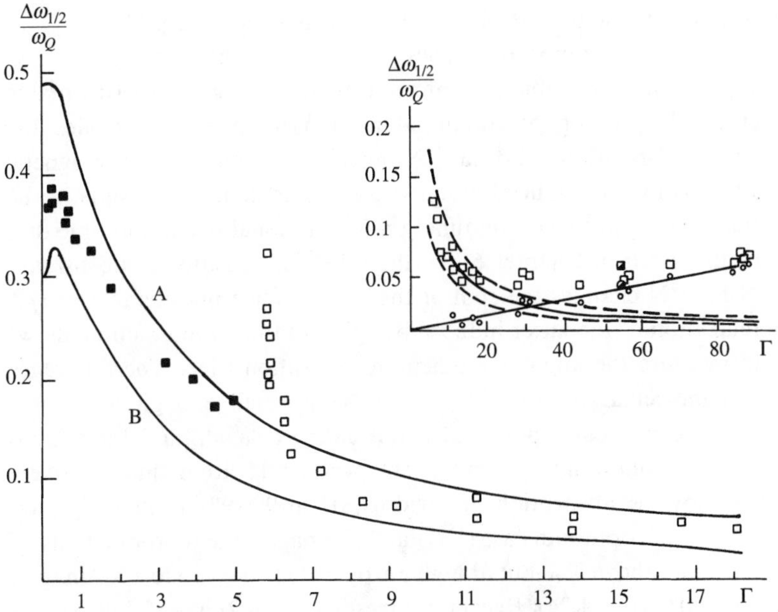

Fig. 3.8. The Q-branch Raman width alteration with condensation of nitrogen. The theoretical results for the strong (A) and weak (B) collision limits are shown together with experimental data for gaseous [89] (■) and liquid nitrogen [145] (□) (point ◪ is taken from the CARS experiment of [136]). The broken curves in the inset are A and B limits whereas the intermediate solid curve presents the rotational contribution to line width at $\gamma = 0.3$. The straight line estimates the contribution of vibrational dephasing [143], and the circles around it are the same liquid data but without rotational contribution.

It should be remembered, however, that a linear relationship between Γ and n is only valid within the limits of binary impact theory. Its restrictions have already been discussed in connection with Fig. 1.23, where the straight line drawn through zero corresponds to relation (3.46). The latter is acceptable within the whole region of the gas phase up to nearly the critical point. Therefore we used Eq. (3.46) to plot experimental data in Fig. 3.8. The coincidence of maxima in theoretical and experimental dependence $\Delta\omega_{1/2}(\Gamma)$ is rather good, as it is achieved by choice of cross-section (3.44), which is the only fitting parameter of the theory. Moreover, within the whole range of the gas phase the experimental widths do not fall outside the narrow corridor of possible values established by the theory. The upper curve corresponds to strong collisions and the lower to the weak collision limit. As follows from (3.23), they differ by a factor of 2. The experimental points are located somewhere in the middle, and this is quite consistent with the value of γ estimated in Eq. (3.45).

Since the above comparison relies on the numerical value σ_J obtained in Eq. (3.44), it is desirable to confirm it by measuring τ_J experimentally. Direct NMR probing of rotational motion in nitrogen provides the necessary information. For a ^{14}N nucleus of spin $I = 1$ magnetic relaxation is mainly induced by quadrupole interaction. It results in the dependence of spin-relaxation times on orientational relaxation time that will be discussed in Chapter 6. On the other hand, isotopic substitution of ^{15}N for ^{14}N changes the spin of the NMR active nucleus to $I = 1/2$. The relaxation of the latter induced by spin–rotational interaction allows one to measure the angular momentum relaxation time. The difference between molecular masses $^{15}N_2$ and $^{14}N_2$ is small enough to assume that the intermolecular potential is not changed at all, and there is no difference in rotational dynamics. Transverse relaxation times T_2 were measured by the absorption line width [81] in 99.99% chemically pure $^{15}N_2$ gas. In the optimum case ($^{15}N_2$ at 2–5 amagat) the reproducibility of T_2 values was above 5%, but at high nitrogen densities (up to 100 amagat) it was 10–20% (Fig. 1.24). Recently Jameson *et al.* [82] have fulfilled more accurate measurements of the longitudinal relaxation time T_1 in $^{15}N_2$ gas (pure and dissolved in various buffers), but the results are almost the same. With increasing temperature the J-diffusion rate determined from both T_2 and T_1 measurements significantly decreases. Fig. 1.25 shows that this dependence practically coincides with the inverse proportionality $\sigma_J \propto 1/T$. At 150 K the cross-section becomes 1.5 times higher than that calculated by the van der Waals diameter 3.7 Å. If one considers this effective cross-section to be the real geometrical characteristic of a molecule, then it is impossible to obtain a close-packed liquid structure with spheres of such large size. This fact clearly indicates that binary estimation of collisional frequency is inapplicable at high densities, and the rigid sphere model is appropriate only above critical temperatures. However, at room temperature the estimate given in Eq. (3.44) practically coincides with $\sigma_J \approx 34$ Å^2 found experimentally in [81], which seems to be physically reasonable.

If the J-diffusion model is valid but only the energy relaxation time is known then Eq. (1.57) may be used to find the other:

$$\tau_J/\tau_E = 1 + \gamma = Z_J/Z_E. \tag{3.47}$$

Here $Z_J = \sigma_0/\sigma_J$ and $Z_E = \sigma_0/\sigma_E$ are the numbers of collisions sufficient for recovery of equilibrium in J-space and in energy space respectively.

Taking into consideration Eq. (3.45), we see that

$$Z_J = \frac{1}{1-\gamma} = 1.5, \quad Z_E = \frac{1}{1-\gamma^2} = 1.1\,. \tag{3.48}$$

These quantities are close to those usually observed. For instance, in SF_6 $Z_J = 1.5$ [146], and in NO $Z_E \approx 1$ [147]. It should be noted that the calculation of Z_E provides a lower estimate because it is performed under the assumption that J-diffusion is extended to the whole rotational spectrum. Actually, non-adiabatic transitions are possible solely between relatively low rotational levels (see Chapter 4), whereas higher excited rotational states are separated from each other by an energy spacing considerably exceeding $1/\tau_c$. The perturbation of these levels by collisions is of an adiabatic character, and the greater their energy the longer their lifetimes [148]. Thus, it is reasonable that the experimentally found value of Z_E for nitrogen is within the range $3.6 < Z_E < 5.2$ [149, 150]. As dispersion in the lifetimes of different rotational levels was ignored, when analysing data we should be satisfied with the coincidence of the estimate obtained in Eq. (3.48).

For a system with moderate density, Fig. 3.9 suggests a very efficient method of interpretation of experimental data in coordinates $(\Delta\omega_{1/2}/\omega_Q)$ $(1/\omega_Q\tau_J)$. Really, as the detailed experimental study in [151] demonstrates for CO and N_2 in various buffer gases, simple plotting of the dependence of a half-width of the isotropic bands (in cm^{-1}) on density (in amagat) produces a rather chaotic picture (see Fig. 3.9a). The same data were replotted in [151] in anamorphosing coordinates of Fig. 3.8 with the abscissa determined by NMR measurements of τ_J for the same molecule and under the same conditions as in the Raman experiment. The result is shown in Fig. 3.9(*b*). As can be seen, the theoretically calculated interval between the limiting cases of strong and weak collisions exceeds experimental error in measurements of Q-branch half-widths just slightly. All the systems studied, except CO in CO_2, are within the 'corridor' determined theoretically. For this system the authors of [151] have stated that collapse of rotational structure takes place, because the band shape changes from initially asymmetric to symmetric, though its width remains the same up to ≈ 380 amagat. A simultaneous broadening of the band caused by vibrational dephasing was successfully separated from the rotational contribution under the assumption that collisions are rather strong. The vibrational contribution found as a result turns out to be a linear function of density with a broadening coefficient around 0.45×10^{-2} cm^{-1} $amagat^{-1}$.

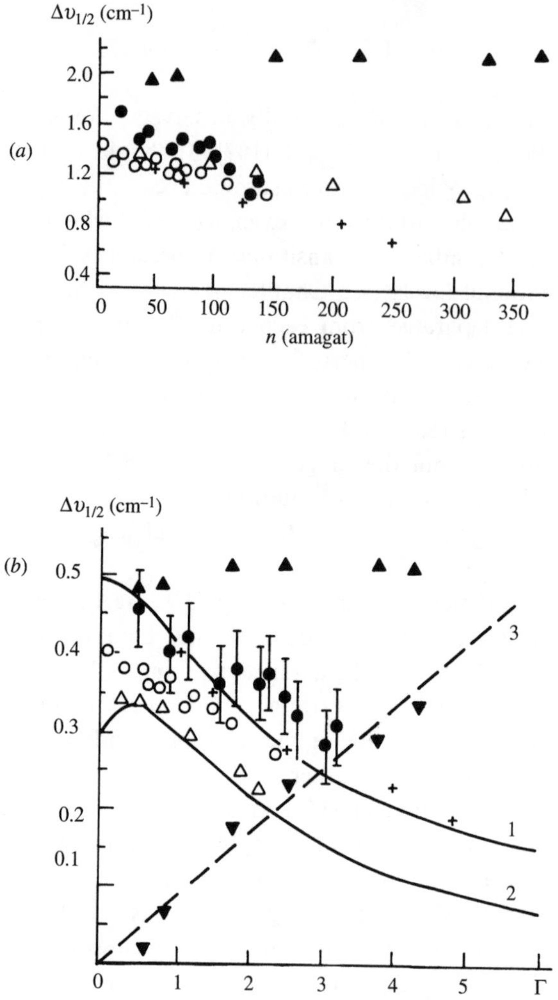

Fig. 3.9. (*a*) Dependence of the experimental half-width of the isotropic Q-branch of N_2 and CO on the density: (•) CO, 295 K; (+) CO in CF_4, 273 K; (▲) CO in CO_2, 323 K; (○) N_2, 295 K; (△) N_2 in CO_2, 323 K. The error in the measurements of half-width is ± 0.2 cm^{-1}. (*b*) The same data as in (*a*) but in relation to measured $\Gamma = 1/\omega_Q \tau_J$. Theoretical curves for strong (curve 1) and weak (curve 2) collision limits are identical to curves A and B in Fig. 3.8. The upside-down triangles near the broken line present the difference between the actual half-width of CO in CO_2 (△) and curve 1. The error in all cases is approximately the same as that indicated for CO.

We proceed now to analysis of the spectral behaviour under further condensation of the medium. Since the isothermal gas experiment is performed within a relatively narrow range of densities, we have to refer to data obtained by other authors [145] which are taken from the co-existence curve in the liquid phase. Each point on this curve corresponds to its own temperature, and all of them are different from those used in the gas experiment. Comparison with such data in universal coordinates of Fig. 3.8 becomes possible owing to the fact that their abscissa Γ is calculated by the quantity τ_J known from NMR experiments [71, 152]. The advantages are as follows: first, points are plotted without any fitting, and, second, there is no arbitrariness in taking account of a non-linear dependence of τ_J on density, unlike situations where it is determined from a model. In particular, use of the rough sphere model, which gives just an approximate description of the actual change in $1/\tau_J(n)$, results in a different distribution of points in Fig. 3.8 and creates the illusion of full agreement between experiment and theory [80]. In the contrast, Fig. 3.8 shows a direct comparison of Raman spectroscopy data ($\Delta\omega_{1/2}$) with nuclear resonance data ($1/\tau_J$). The theory predicts quite definite correspondence between them that throws light upon the contribution of rotational and vibrational broadening of the line. Insofar as the spectrum has already collapsed, this correspondence is expressed by the perturbation theory formula

$$\Delta\omega_{1/2} = w_0 + \gamma_{\mathrm{dh}}\,, \tag{3.49}$$

where w_0 and γ_{dh} were given in Eq. (3.23) and Eq. (3.24).

Comparison of gas and liquid data was performed as stated above and shown in Fig. 3.9. It displays the peculiarity localized within a short range of temperatures near the critical point. When studying the isothermal dependence of $\Delta\omega_{1/2}$ on density in the above-critical region, the same peculiarity is experimentally revealed for oxygen [137] (Fig. 3.10). Its origin is connected with anomalous vibrational dephasing in the vicinity of the critical point. The critical state is known to be characterized by large fluctuations of density, which decay slowly. The fluctuations result in an inhomogeneous broadening that is not effectively averaged and hence produces relatively large (quasi-static) dephasing. If the critical region ($\Delta T < 5$ K) is excluded, narrowing of the line in liquid nitrogen continues as Γ increases, and experimental points do not fall outside the limits established by rotational collapse theory. Only in a highly dense liquid does the width of the line found experimentally exceed the theoretically calculated one. The disagreement should be attributed to a

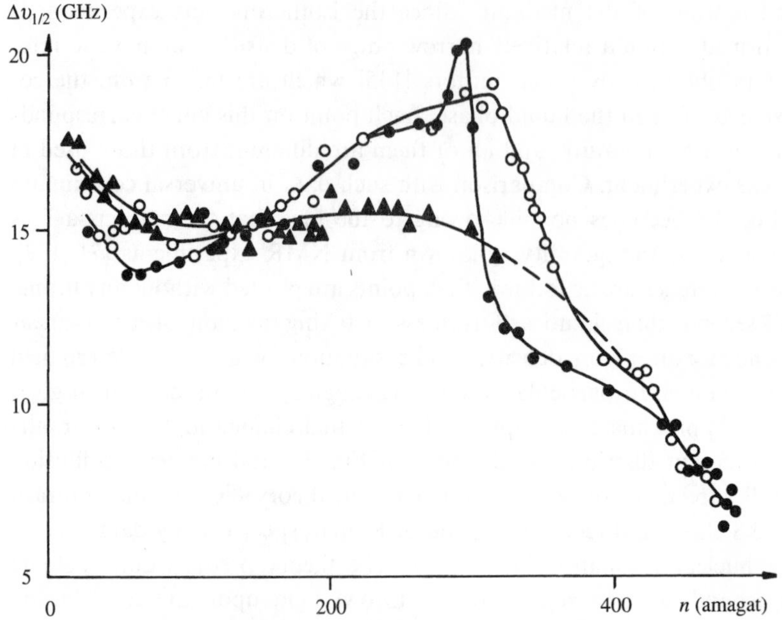

Fig. 3.10. Anomalous broadening of oxygen isotropic Raman spectrum in the vicinity of (but above) the critical point [137]. The notation is the same as in Fig. 3.5.

complementary mechanism of broadening which is related not to rotation but to vibrational dephasing. Broadening may also take place due to decay of excited vibrational states; however, the rate of this process is usually less than that of pure dephasing [153–155].

As mentioned, the shape of the liquid nitrogen spectrum is rigorously Lorentzian (see Fig. 3.3), as it is the Fourier transform of the correlation function (3.15). According to Eq. (3.23) and Eq. (3.24), the width (3.49) is given by

$$\frac{\Delta\omega_{1/2}}{\omega_Q} = \frac{1}{(1+\gamma)\Gamma} + \frac{\sigma_{\mathrm{dp}}}{\sigma_J}\,\Gamma\,.$$

It implies that relaxation times obey relation (3.47) even after gas condensation, although both $1/\tau_E$ and γ_{dp} become nonlinear in density. The contribution of the rotational broadening represented by the first component may be estimated to a rather high accuracy via the value of γ found in (3.45). Subtrncting it from the width observed, we obtain the dephasing contribution which is linear in Γ (see inset in Fig. 3.8). The slope of the straight line $\sigma_{dp}/\sigma_J = 7 \times 10^{-4}$ makes it possible to estimate the dephasing cross-section taking into account the estimate (3.44):

$$\sigma_{\rm dp} = 4.7 \times 10^{-4}\ \sigma_0. \tag{3.50}$$

As expected, it satisfies condition (3.22). In other words, the shift in oscillation phase caused by collisions is very small.

The weakness of adiabatic dephasing is an extremely rare phenomenon in optical spectroscopy. It is peculiar for vibrational but not for electronic transitions in atoms and molecules. When the latter were considered by adiabatic impact theory in the point particle approximation, the collisional phase shift may not be small as it diverges at zero impact parameter [58]. A. I. Burshtein and Yu. I. Naberukhin [156] first mentioned that the finite size of the particles is of principal importance since it restricts the phase, which becomes small at sufficiently large temperature. Within the framework of impact theory the motional narrowing phenomenon arises just above this temperature where it may be as well described by the perturbation theory (with respect to interparticle interaction) [157, 158]. For NMR and ESR spectroscopy, motion becomes fast enough at rather low temperature but for atomic spectra it is unattainable. It turned out later that strong dephasing becomes weak just in the region of vibrational spectroscopy [41, 130–132].

It should be noted that there is a considerable difference between rotational structure narrowing caused by pressure and that caused by motional averaging of an adiabatically broadened spectrum [158, 159]. In the limiting case of fast motion, both of them are described by perturbation theory, thus, both widths in Eq. (3.16) and Eq (3.17) are expressed as a product of the frequency dispersion and the correlation time. However, the dispersion of the rotational structure (3.7) defined by intramolecular interaction is independent of the medium density, while the dispersion of the vibrational frequency shift $\langle \delta\Omega^2 \rangle$ in (3.21) is linear in gas density. In principle, correlation times of the frequency modulation are also different. In the first case, it is the free rotation time τ_E that is reduced as the medium density increases, and in the second case, it is the time of collision $\tau_c \approx \rho/\langle v \rangle$ that remains unchanged. As the density increases, the rotational contribution to the width decreases due to the reduction of τ_E, while the vibrational contribution increases due to the dispersion growth. In nitrogen, they are of comparable magnitude after the initial (static) spectrum has become ten times narrower. At 77 K the rotational relaxation contribution is no less than 20% of the observed Q-branch width. If the rest of the contribution is entirely determined by dephasing, its time $T_{dp} = 274$ ps [143]. We will pursue this discussion at the end of Chapter 5, presenting some recent isothermic data, obtained for compressed gas and liquid.

In summary it must be stressed that Q-branch narrowing by pressure occurs owing to exceptionally weak dephasing in diatomics. Otherwise the phase shift during collisions would inevitably dominate over frequency exchange and provide spectral broadening typical of the optical spectroscopy. The idea that collisions break the phase of monochromatic radiation goes back to Lorentz and to the origin of the Lorentzian line as a result of impact broadening. It has gained such a strong foothold that the discovery of Q-branch 'non-broadening' in a gas (Fig. 3.11) [160, 161] has been an enigma until lately. It seems to be in contradiction with the energy–time uncertainty principle. However, if not a phase but a frequency is broken in collision then reduction of the monochromatic radiation time does not necessarily increase the uncertainty of the energy. Unlike phase shift, frequency shift results in the opposite effect, namely monochromatization increases, as the rate of collision grows. The phenomenon is well known in radiophysics and may be easily observed in magnetic resonance, since spins do not normally experience phase shift. Particle motion modulates only a local magnetic field and related Larmor frequency [9]. The situation is quite opposite in optical spectroscopy because electrical interaction between particles is much more efficient in phase breaking than magnetic interaction. Phase modulation of the radiation may be avoided under special conditions established in [158] that we will discuss in the next chapter. Diatomic Q-branch narrowing is one of the rare exceptions in which these conditions are well fulfilled. Matters become more problematic in the case of polyatomic molecules considered below.

3.7 The Q-branch band shape in the Keilson–Storer model

Recently, the development of CARS higher sensitivity and spectral resolution has made possible detailed studies of line shapes of Raman-active transitions in polyatomic molecules in gases in a wide density range. It covers all stages of rotational structure transformation into a single symmetrical line. In [162] there were investigated the spectra of pure methane and silane and of their mixtures with argon at densities up to 150 amagat at room temperature. Fig. 3.12 represents both a pattern of resolved structure and a broadening of symmetrized Q-branch envelope under high pressures. Highly pronounced shift of the smoothened

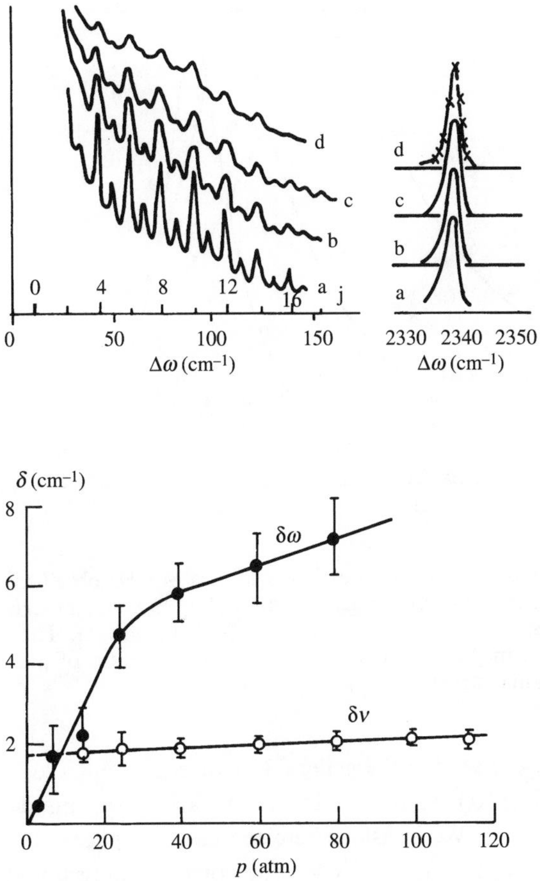

Fig. 3.11. Comparison of side branch broadening with Q-branch 'non-broadening', made for nitrogen in [160] for 27°C and different pressures: 15 atm (curve a), 25 atm (curve b), 40 atm (curve c), 60 atm (curve d). In the lower part $\delta\omega$ is the width of resolved rotational components, $\delta\nu$ is the width of the non-resolved Q-branch, which is primarily isotropic.

contour as a whole is mainly due to vibrational dephasing, which in the case of spherical molecules is much more efficient than for nitrogen and other linear molecules. That is why even under moderate pressures the dephasing contribution to the half-width prevails over the rotational one (Fig. 3.13). The narrowing of rotational structure is completely masked and manifests itself only in nonlinearity of broadening at low pressures until collapse is accomplished.

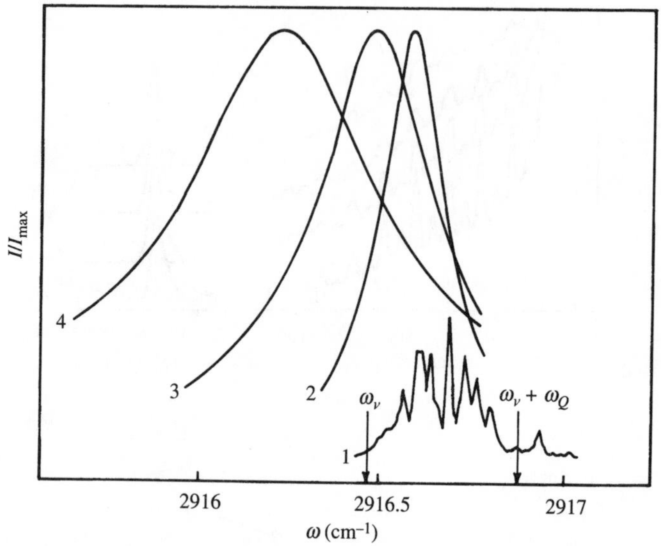

Fig. 3.12. The room temperature CARS spectra of CH_4 obtained in [162] at the following densities: (1) 0.1 amagat of pure CH_4; (2) 5 amagat CH_4; (3) 5 amagat CH_4 + 35 amagat Ar; (4) 5 amagat CH_4 + 85 amagat Ar. The position of the vibration frequency ω_v is indicated as well as the centre of gravity of the Q_{01} branch rotational structure $\omega_v + \omega_Q$.

In the present section the general kinetic equation (3.26) will be solved within the Keilson–Storer model for an arbitrary angular momentum correlation [163]. We consider here the case of spherical molecules (for linear molecules see Appendix 5). The corresponding initial condition is the equilibrium distribution

$$K\ (0, J) = \varphi_B(J) = \frac{2J^2}{(2\pi d^3)^{\frac{1}{2}}} \exp\left(-\frac{J^2}{2d}\right), \tag{3.51}$$

where $d = \hbar kT/(4\pi Bc)$. As the phase space of the angular momentum of spherical tops is three-dimensional, the kernel of the integral part of Eq. (3.26) also changes relative to that of linear molecules. Summing the probabilities of all reorientations of $\boldsymbol{J}'$ into $\boldsymbol{J}$, which cause the same transformation of J' into J, we have

$$f(J', J) = \int f\left(\boldsymbol{J} - \gamma\boldsymbol{J}'\right) J^2 \, \mathrm{d}\Omega_{\boldsymbol{J}}$$
$$= \frac{2}{[2\pi d(1-\gamma^2)]^{\frac{1}{2}}} \frac{J}{\gamma J'} \exp\left(-\frac{J^2 + \gamma^2 J'^2}{2d(1-\gamma^2)}\right) \sinh\left(\frac{\gamma J J'}{d(1-\gamma^2)}\right). \tag{3.52}$$

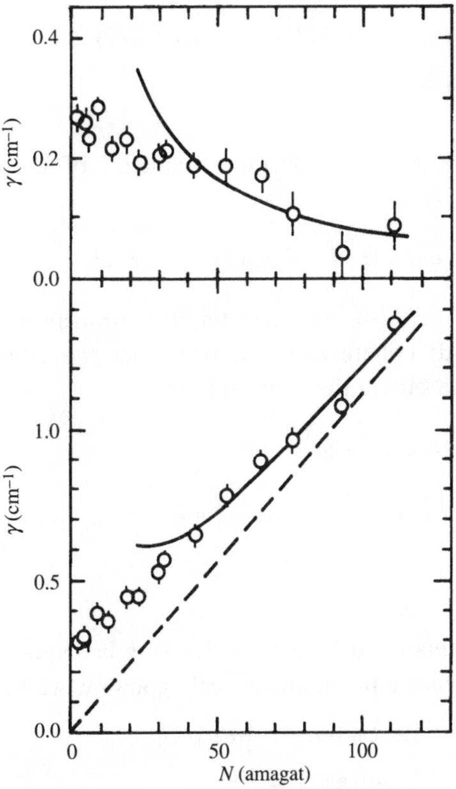

Fig. 3.13. Density-dependence of the Q_{01} branch line width γ of methane (the dashed line is for pure vibrational dephasing, supposed to be linear in density). (○) experimental data (with error bars) [162]; Top part: rotational contribution γ_R and its theoretical estimation in motional narrowing limit [162] (solid line); the points were obtained by subtraction of dephasing contribution γ_{dp} from experimental HWHM $\gamma = \gamma_R + \gamma_{\text{dp}}$.

It is shown in Appendix 6 that the generalized Laguerre polynomials are eigenfunctions of the integral operator (3.26) with kernel (3.52). Let us search for the solution of (3.26) in the form of expansion over these eigenfunctions

$$K(t, J) = \varphi_B(J) \sum_{n=0}^{\infty} a_n(t) L_n^{1/2}\left(J^2/2d\right) . \tag{3.53}$$

Substituting (3.53) into (3.26) for the unknown coefficients of the expansion and taking into account initial conditions and the explicit form of the eigenvalues (Appendix 5), one obtains

$$\begin{aligned}\frac{\mathrm{d}}{\mathrm{d}t}\, a_n(t) &= \frac{2}{3}\mathrm{i}\omega_Q \left[(2n+3/2)\, a_n(t) - (n+3/2)\, a_{n+1}(t) - n a_{n-1}(t)\right] \\ &\quad -\frac{1-\gamma^{2n}}{\tau_0}\cdot a_n(t), \end{aligned} \tag{3.54}$$

where $a_n(t = 0) = \delta_{n,0}$. The average frequency of the Q-branch of spherical molecules is

$$\omega_Q = \alpha_e \int_0^\infty J^2 \varphi_B(J)\, \mathrm{d}J = 3\alpha_e d. \tag{3.55}$$

Proceeding in Eq. (3.54) to the Laplace transform: $a_n(\omega) = (1/\pi)\int_0^\infty \mathrm{e}^{-i\omega t} a_n(t)\, \mathrm{d}t$ (where ω is counted from the vibrational transition frequency), we obtain the recurrent relation

$$\begin{aligned}\mathrm{i}\omega a_n(\omega) &= \pi^{-1}\delta_{n,0} + \frac{2}{3}\mathrm{i}\omega_Q \\ &\quad \times \left[\left(2n+\frac{3}{2}\right) a_n(\omega) - \left(n+\frac{3}{2}\right) a_{n+1}(\omega) - n a_{n-1}(\omega)\right] \\ &\quad - \frac{1-\gamma^{2n}}{\tau_0} a_n(\omega). \end{aligned} \tag{3.56}$$

As the isotropic Raman and CARS spectra may be expressed via $a_0(\omega)$ by virtue of the Laguerre polynomials' orthogonality, we have

$$G_0(\omega) \sim \mathrm{Re}\, a_0(\omega)$$

$$G_0^{\mathrm{CARS}}(\omega) \sim |a_0(\omega)|^2.$$

Solving (3.56) relative to $a_0(\omega)$, we have

$$a_0(\omega) = \frac{1}{\pi}\left[\mathrm{i}(\omega-\omega_Q) + \mathrm{i}\omega_Q/b_0(\omega)\right]^{-1}, \tag{3.57}$$

where $b_n(\omega) \equiv a_n(\omega)/a_{n+1}(\omega)$ satisfies the recurrent relation

$$b_n(\omega) = 2 + \frac{3}{2(n+1)}\left(1 - \frac{\omega}{\omega_Q} + \mathrm{i}\frac{1-\gamma^{2(n+1)}}{\omega_Q\tau_0}\right) - \left(1+\frac{3}{2(n+1)}\right)\frac{1}{b_{n+1}(\omega)}. \tag{3.58}$$

Consequent substitution of (3.58) into (3.57) leads to a continuous fraction representation of the solution

$$a_0(\omega) = \cfrac{1/\pi}{i(\omega-\omega_Q) + \cfrac{i\omega_Q}{2+\frac{3}{2}\left(1-\frac{\omega}{\omega_Q}+i\frac{1}{\omega_Q\tau_E}\right) - \cfrac{5/2}{2+\frac{3}{4}\left(1-\frac{\omega}{\omega_Q}+i\frac{1+\gamma^2}{\omega_Q\tau_E}\right)-\cdots}}}. \tag{3.59}$$

In some cases Eq. (3.59) may be expressed through well-known functions.

(i) The limit of zero pressures (static contour):

$$\lim_{\tau_E\to\infty} |a_0(\omega)|^2 = \left(\frac{3}{\pi\omega_Q}\right)^2 \left\{ \left| 1 - \left(\frac{6\omega}{\omega_Q}\right)^{1/2} \exp\left(\frac{3\omega}{2\omega_Q}\right) \int_0^{\left(\frac{3\omega}{2\omega_Q}\right)^{1/2}} \exp\left(x^2\right) \mathrm{d}x \right|^2 + \frac{3\pi\omega}{2\omega_Q} \exp\left(-\frac{3\omega}{2\omega_Q}\right) \right\}. \quad (3.60)$$

(ii) Non-correlated rotational relaxation ($\gamma = 0$):

$$a_0(\omega) = \frac{1}{\pi} \int_0^\infty \frac{\varphi_B(J)\,\mathrm{d}J}{\mathrm{i}\left(\omega - \alpha_e J^2\right) + 1/\tau_E} \left(1 - \frac{1}{\tau_E} \int_0^\infty \frac{\varphi_B(J)\,\mathrm{d}J}{\mathrm{i}\left(\omega - \alpha_e J^2\right) + 1/\tau_E}\right)^{-1}. \quad (3.61)$$

(iii) Correlated rotational relaxation ($\gamma \to 1$):

$$a_0(\omega) = \frac{1}{\pi} \int_0^\infty \mathrm{d}t\, \mathrm{e}^{-\mathrm{i}\omega t} \exp\left(\frac{3\left(1-\sqrt{c}\right)}{4\tau_E} t\right) \left[\left(1+\sqrt{c}\right)^2 - \left(1-\sqrt{c}\right)^2 \exp\left(-\frac{\sqrt{c}}{\tau_E} t\right)\right]^{-3/2}, \quad (3.62)$$

where $c = 1 - 8\mathrm{i}\alpha_e d\tau_E$.

In order to calculate (3.59) in a general case, let us consider the asymptotic behaviour of $b_n(\omega)$ at large n. Assuming that the limit exists in (3.59), we find ($|\gamma| \neq 1$)

$$\lim_{n\to\infty} b_n(\omega) = b_\infty(\omega) = 2 - \frac{1}{b_\infty(\omega)} = 1. \quad (3.63)$$

Owing to the impossibility of permutation of the limits $N \to \infty$ and $\gamma \to 1$, hereafter the magnitude of the fraction at $\gamma = 1$ is found by continuity. With finite $N \gg 1$, let us find b_N in the form $b_N = 1 + \delta_N$. From (3.59) we have

$$\delta_N + \delta_N\,\delta_{N+1} = \delta_{N+1} + \frac{3Z_N}{2(N+1)}\left(1 + \delta_{N+1}\right) - \frac{3}{2(N+1)} \quad (3.64)$$

where

$$Z_N = 1 - \frac{\omega}{\omega_Q} + \mathrm{i}\frac{1}{\omega_Q \tau_E} \sum_{m=0}^{N} \gamma^{2m}.$$

The inequalities

$$|\delta_N| \ll 1; \left| \frac{\mathrm{d}}{\mathrm{d}N} (1/\delta_N) \right| \ll 1, \tag{3.65}$$

being valid for the asymptotics over N, the following expression holds:

$$\delta_N \approx \pm \left(\frac{3}{2N} (Z_N - 1) \right)^{\frac{1}{2}}. \tag{3.66}$$

In Eq. (3.66) the sign '+' is chosen to provide the decay in time of the spectrum correlation function. When the approximate solution (3.66) is used for the back iterations in Eq. (3.58) from $b_N = 1 + \delta_N$ up to b_0 and subsequent calculation of $a_0(\omega)$ the error does not accumulate. This was proved by comparison of approximate numerical calculations of limiting cases 2 and 3 with exact formulae (3.61) and (3.62).

3.8 Comparison with experiment: spherical molecules

With formulae (3.58), (3.59) and (3.66) Q-branch contours are calculated for CARS spectra of spherical rotators at various pressures and for various magnitudes of parameter γ (Fig. 3.14). For comparison with experimental data, obtained in [162], the characteristic parameters of the spectra were extracted from these contours: half-widths and shifts of the maximum subject to the density. They are plotted in Fig. 3.15 and Fig. 3.16. The corresponding experimental dependences for methane were plotted by one-parameter fitting. As a result, the cross-section for rotational energy relaxation σ_E is found:

$$1/\tau_E = n\sigma_E \langle v \rangle. \tag{3.67}$$

The obtained magnitude of $\sigma_E \approx 24$ Å^2 differs somewhat from that found by experimentalists themselves because the experimental data (shown in the inset in Fig. 3.15) were fitted in [162] with the perturbation theory formula

$$w_0 = \frac{2}{3} \omega_Q^2 \tau_E. \tag{3.68}$$

Fig. 3.15 demonstrates that in this pressure domain the half-width of the spectrum still differs essentially from its perturbation theory estimate

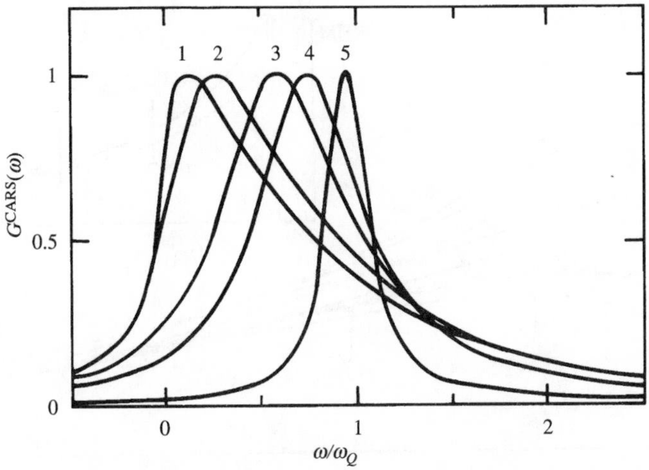

Fig. 3.14. Pressure-induced narrowing of the CARS spectrum of a spherical top at $\gamma = 0.5$. The intensity is plotted in arbitrary units, the frequency in units of ω_Q, pressure in units $\Gamma = 1/\omega_Q\tau_E$: (1) $\Gamma = 0.05$, (2) $\Gamma = 0.1$, (3) $\Gamma = 0.5$, (4) $\Gamma = 1$, (5) $\Gamma = 5$.

(3.68), though $\Delta\omega_{1/2}$ is already indifferent to the model of rotational relaxation.

Vibrational broadening in [162] was taken into account under the conventional assumption that contributions of vibrational dephasing and rotational relaxation to contour width are additive as in Eq. (3.49). This approximation provides the largest error at low densities, when the contour is significantly asymmetric and the perturbation theory does not work. In the frame of impact theory these relaxation processes may be separated more correctly under assumption of their statistical independence. Inclusion of dephasing causes appearance of a factor

$$\beta = \langle \exp(\mathrm{i}\varphi) \rangle_{\mathrm{coll}} \tag{3.69}$$

before the integral part of Eq. (3.26) [164]:

$$\begin{aligned} \dot{K} - \mathrm{i}\alpha J^2 K &= -\frac{1}{\tau_0}K + \frac{\beta}{\tau_0}\int_0^\infty f(J',J)K(t,J')\,\mathrm{d}J' \qquad (3.70) \\ &= -\frac{1}{\tau_{\mathrm{dp}}}K - \frac{\beta}{\tau_0}\left[K - \int_0^\infty f(J',J)K(t,J')\,\mathrm{d}J'\right], \end{aligned}$$

$$1/\tau_{\mathrm{dp}} = (1-\beta)/\tau_0 . \tag{3.71}$$

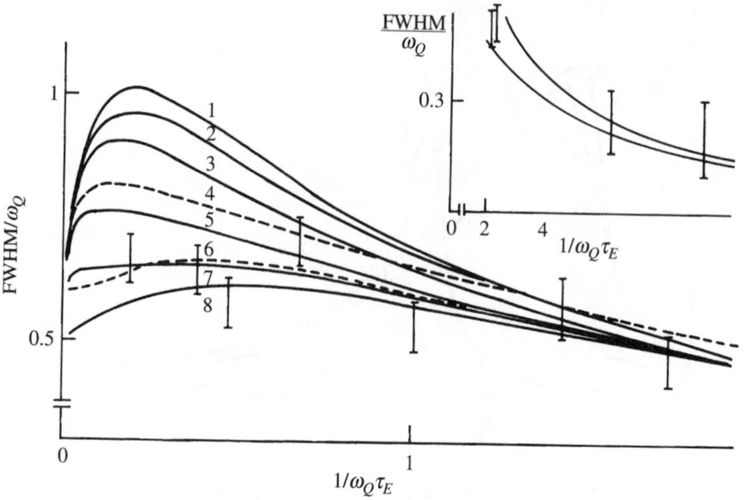

Fig. 3.15. The CARS spectrum rotational width versus methane density for various values of parameter γ: (1) $\gamma = 0$, (2) $\gamma = 0.3$, (3) $\gamma = 0.5$, (4) $\gamma = 0.7$, (5) $\gamma = 0.75$, (6) $\gamma = 0.9$, (7) $\gamma = 0.95$, (8) $\gamma = 1$. Curves (4) and (6) are obtained by subtraction of the dephasing contribution from the line width calculated taking account of vibrational broadening. The other dependences are found assuming purely rotational broadening (vibrational relaxation neglected).

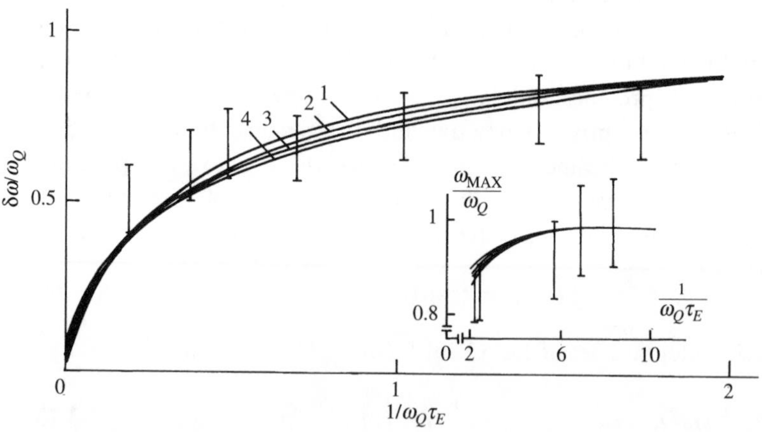

Fig. 3.16. The density-dependence of the frequency shift of the Q-branch maximum. The γ values for the curves are in the notation of Fig. 3.15. When plotting the experimental data, the cross-section found in the fitting of the density dependence of the width was employed (Fig. 3.15).

By a simple redefinition

$$\beta/\tau_E = 1/\tau_e$$

the problem reduces to a trivial convolution of a calculated set of contours (depending on parameters τ_e) with a Lorentzian contour with half-width $\gamma_{\rm dh} = 1/\tau_{\rm dp}$.

The contribution of dephasing may be easily extracted, if the experiment provides high enough pressures. When the contribution of rotational broadening (3.68) is relatively small, $\Delta\omega_{1/2} \approx 1/\tau_{\rm dp}$ is linear over density and the slope is $\sigma_{dp}\langle v\rangle$. On the other hand the 'true' relaxation time of rotational energy $\tau_E = \tau_0/(1-\gamma^2)$ may be expressed via τ_e and $\tau_{\rm dp}$ found experimentally as follows:

$$\tau_E = \tau_e \left[1 + \left(1-\gamma^2\right)\tau_e/\tau_{\rm dp}\right]^{-1} . \tag{3.72}$$

In the general case parameters τ_e and $\tau_{\rm dp}$ should be determined by means of self-consistent two-parameter fitting.

In [162] experiments on methane provided a linear pressure dependence of the contour width. This made it possible to find the dephasing cross-section and to discriminate between contributions of rotational and vibrational relaxation to the contour width. This was done under the above-mentioned simplifying assumption that they are additive. (Let us note that processing of experimental data on linear molecules was always performed under this assumption.) The points found by this method are shown in Fig. 3.15, curves (4) and (6).

We have verified the applicability of the assumption concerning the additivity of the widths for methane [163]. The vibrational contribution, known in advance, was subtracted not from the experimental contour width, but from that calculated for given magnitudes of σ_E and $\sigma_{\rm dp}$ with the following formulae

$$\begin{aligned}
d_0(\omega) &= \left[\mathrm{i}\left(\omega-\omega_Q\right)+\mathrm{i}\frac{\omega_Q}{b_0(\omega)}+\frac{1}{\tau_{\rm dp}}\right]^{-1}, \\
b_n(\omega) &= 2+\frac{3Z_n'}{2(n+1)}-\left[1+\frac{3}{2(n+1)}\right]\frac{1}{b_{n+1}(\omega)}\ ; \quad 0 \le n \le N, \\
b_N(\omega) &= 1+\left(\frac{3}{2N}\left(Z_N'-1\right)\right)^{\frac{1}{2}},
\end{aligned} \tag{3.73}$$

$$Z_n' = 1 - \frac{\omega}{\omega_Q} + \frac{\mathrm{i}}{\omega_Q \tau_{\mathrm{dp}}} + \frac{\mathrm{i}}{\omega_Q \tau_e} \sum_{m=0}^{n} \gamma^{2m},$$

where τ_e is determined from τ_E by inversion of (3.72). The corresponding curve in Fig. 3.15 shows that when dephasing is relatively small and rotational relaxation is rather close to the weak collision limit, the simple additive procedure can be used to process data.

For silane, another system studied in [162], the rigorous two-parameter fitting (3.72) and (3.73) is necessary. Rough estimations yield the cross-sections of the rotational energy relaxation $\sigma_E \approx 100$ Å^2, $\gamma \approx 1$. Let us compare certain values of the cross-sections of rotational energy relaxation σ_E and angular momentum relaxation σ_J. The latter quantity may be found from T_1 measurements of NMR relaxation induced by spin–rotational interaction. For this purpose we have obtained a correct relation between times T_1 and τ_J, which holds for arbitrary correlation of rotational relaxation of spherical molecules [165]. With this relation, one may use the density-dependence of time T_1 in methane [166] and silane [167] to show that the rotational relaxation of these molecules is close to a correlated one. Good agreement exists in this case as well as in the case of optical spectra for the following magnitudes of the fitting parameter σ_J : $\sigma_J^{CH_4} \approx 16$ Å^2; $\sigma_J^{SiH_4} \approx 57$ Å^2. In the Keilson–Storer model σ_E and σ_J are connected by the equality

$$\sigma_E = (1+\gamma)\,\sigma_J\,. \tag{3.74}$$

As the process of rotational relaxation is close to a correlated one ($\gamma \approx 1$) for both gases, according to (3.74) the σ_E cross-section is twice as large as σ_J. This result agrees with experiment and it appears that quasi-classical impact theory may be applied to description of rotational relaxation in moderately dense gases.

Let us note that cross-sections σ_E, obtained from optical data and verified by NMR measurements, exceed the corresponding cross-sections found by acoustic methods ($\sigma_E^{CH_4} \approx 3$ Å^2) [168]. A similar disagreement between the data obtained by NMR and molecular spectroscopy with the acoustic data exists for linear molecules as well.

4

Quantum theory of spectral collapse

The quasi-classical theory of spectral shape is justified for sufficiently high pressures, when the rotational structure is not resolved. For isotropic Raman spectra the corresponding criterion is given by inequality (3.2). At lower pressures the well-resolved rotational components are related to the quantum number j of quantized angular momentum. At very low pressure each of the components may be considered separately and its broadening is qualitatively the same as of any other isolated line in molecular or atomic spectroscopy.

At the beginning, line shape theory concentrated on calculation of the width and shift of an isolated line broadened by collisions considered as instantaneous. This approach, known as 'impact theory', which originated with the pioneering work of Lorentz and Weisskopf, was initially purely adiabatic. The assumed adiabaticity of collisions excluded in principle any interference between spectral lines in the frame of impact theory [58]. The situation changed with enhanced study of Stark multiplets of atoms in plasmas [169]. The Stark sublevels were so weakly split in a weak electrical field of ions that a condition similar to (1.7) was met ($\Delta E\tau_c \ll 1$) and a non-adiabatic generalization of impact theory became necessary. Transitions between Stark sublevels as an effective mechanism of their broadening were first taken into account by Kolb [169]. Subsequently nonadiabatic theory was employed to describe overlapping Stark multiplets [170–172]. It was mentioned that a qualitatively new feature arises when collisions are non-adiabatic: collisionally induced interference between components of the Stark structure causes spectral collapse. However, narrowing of the collapsed spectrum in a gas phase was only shown to exist in 1967 in the work of Burshtein and Naberukhin [158]. These authors established the conditions under which interference is not accompanied by dephasing. This is an extreme case when spectral exchange results in collisional narrowing of the whole spectrum. The Q-branch transformation was soon pointed out as a clear example of such a narrowing effect [173]. Another example was found independently by Lightman and Ben-Reuven [174] who applied the Fano theory [175] to FIR spectra of NH_3.

We will show below when and how the line interference and its special case, 'spectral exchange', appear in spectral doublets considered as an example of the simplest system. It will be done in the frame of conventional 'impact theory' as well as in its modern 'non-Markovian' generalization. Subsequently we will concentrate on the impact theory of rotational structure broadening and collapse with special attention to the shape of a narrowed Q-branch.

4.1 Spectral exchange in impact theory

In quantum theory as well as in classical theory, linear absorption of light at frequency ω is described by a spectral function

$$I(\omega) = \mathrm{Re}\frac{1}{\pi}\int_0^\infty \exp\,(-\mathrm{i}\omega t)\; K(t)\; \mathrm{d}t\,. \tag{4.1}$$

The correlation function

$$K(t) = \mathrm{Sp}\;[\rho \mathbf{d}\bar{\mathbf{d}}(t)] \tag{4.2}$$

is defined as a product of operators $\mathbf{d}$ acting either in j–m space or some other Liouville space. The product is averaged over the equilibrium density matrix ρ. In optical transitions the upper state is usually empty and the sublevels of the lower state are equally populated in the high-temperature limit. In this limit or in the case of non-degenerate ground state one can put for simplicity $\rho = 1$ for lower sublevels and 0 for upper states.

The stochastic problem is to describe properly the time evolution of the Heisenberg operator $\bar{\mathbf{d}}(t)$ averaged over all the realizations of collisional process in the interval (0,t). The averaging, performed in the impact theory, results in the phenomenological kinetic equation [170, 158]

$$\dot{\bar{\mathbf{d}}} = \mathrm{i}[\mathbf{H_0}, \bar{\mathbf{d}}] - \hat{\mathbf{\Gamma}}\bar{\mathbf{d}}, \tag{4.3}$$

where $\mathbf{H_0}$ is an unperturbed Hamiltonian. The result of instantaneous collision is presented by a 'super-operator' $\hat{\mathbf{\Gamma}}$. In the Liouville space where $\bar{\mathbf{d}}$ is a vector $\bar{d}_{lm}$ the collisional operator is

$$\mathbf{\Gamma}_{ik,lm} = \langle \delta_{il}\delta_{mk} - S^*_{il}S_{km} \rangle\,. \tag{4.4}$$

It is calculated in the **S**-matrix formalism and averaged over impact distances b and velocities v with Maxwellian distribution $f(v)$

$$\langle \ldots \rangle = n \int_0^\infty 2\pi b \, \mathrm{d}b \int_0^\infty v f(v) \, \mathrm{d}^3 v (\ldots). \tag{4.5}$$

Now the microscopic problem arises as to how to calculate

$$\mathbf{S} = \mathbf{T} \exp \left(\int_{-\infty}^{+\infty} \mathrm{e}^{-\mathrm{i}\mathbf{H}_o t} \mathbf{V}(t) \mathrm{e}^{\mathrm{i}\mathbf{H}_o t} \, \mathrm{d}t \right), \tag{4.6}$$

where $\mathbf{V}(t)$ is an interparticle interaction potential and **T** is the chronological operator. Two evident simplifications are possible.

(i) One may put $\mathbf{H}_o \tau_c = 0$ for purely non-adiabatic theory.
(ii) Perturbation theory with respect to $\mathbf{V}\tau_c \ll 1$ may be employed.

At the beginning we will use both to simplify the problem, as much as it was done in the pioneering works [158, 176].

From (4.1)–(4.4) one obtains

$$I(\omega) = \mathrm{Re} \frac{1}{\pi} \sum_i \rho_i d_{ik}^* d_{lm} G_{ik,lm}^{-1}, \tag{4.7}$$

where

$$\mathbf{G}_{ik,lm} = \mathrm{i}(\omega - \omega_{ik}) \delta_{il} \delta_{km} + \mathbf{\Gamma}_{ik,lm}. \tag{4.8}$$

One of the simplest examples of line interference is impact broadening of H atom Lα Stark structure, observed in plasmas [176] (Fig. 4.1.(*a*)). For a degenerate ground state the impact operator is linear in the **S**-matrix:

$$\hat{\mathbf{\Gamma}} = \langle \mathbf{I} - \mathbf{S} \rangle.$$

As charge–dipole interaction between the electron and the atom is small, the perturbation theory expansion may be used to estimate $\hat{\mathbf{\Gamma}}$. The odd terms of this expansion disappear after averaging over impact parameters due to isotropy of collisions. In the second order approximation only those elements of $\hat{\mathbf{P}}$ that are bilinear in **V** are non-zero. Straightforward calculation showed [176] that all components of the Stark structure are broadened but only those for which $m = 0$ interfere with each other:

$$\hat{\mathbf{G}} = \begin{pmatrix} \mathrm{i}(\omega - \omega_{10}) + 2\gamma & 0 & 0 & -\gamma \\ 0 & \mathrm{i}(\omega - \omega_{20}) + \gamma & 0 & 0 \\ 0 & 0 & \mathrm{i}(\omega - \omega_{30}) + \gamma & 0 \\ -\gamma & 0 & 0 & \mathrm{i}(\omega - \omega_{40}) + 2\gamma \end{pmatrix}, \tag{4.9}$$

where $\gamma = Cn$ and C is a function of temperature and the transition moment. Two lines (2–0 and 3–0) which belong to a degenerate doublet, do not interfere at all. The off-diagonal element $\mathbf{P}_{02,03}$ binding similar lines in some phenomenological three-level system (0,2,3) was initially considered to be non-zero [170–172]. In reality $\mathbf{\Gamma}_{02,03} = -\langle S_{23}\rangle$ is averaged to zero when the interaction is spherically isotropic ($\langle \mathbf{V}\rangle = 0$):

$$\langle S_{23}\rangle = -\mathrm{i}\int \langle V_{23}(t)\rangle \exp\left(+\mathrm{i}\omega_{23}t\right)\,\mathrm{d}t = 0.$$

As a consequence these two lines are broadened independently of each other as well as of the rest and according to Eq. (4.7) they have the usual impact shape

$$I_k = \frac{|d_{k0}|^2}{\pi}\frac{\gamma}{(\omega-\omega_{k0})^2+\gamma^2}, \qquad k = 2,3\,.$$

Broadening of the interfering lines (0–1 and 0–4) is qualitatively different. This doublet undergoes spectral collapse that we will discuss later, together with the next example.

This is a simplified model of the rotational structure of a linear molecule's vibrational spectra (Fig. 4.1(*b*)). There are only two lower rotational sublevels ($j = 0, 1$) in each vibrational state (ground state and first excited state). Two spectral lines are permitted by selection rules for the free rotator absorption spectrum: one for the P-branch (ω_{32}) and one for the R-branch (ω_{41}). Two other transitions belong to the Q-branch, which is forbidden in absorption spectra of linear molecules but may be obtained by Raman scattering. The principal difference, with regard to the preceding example, is the rotational degeneracy of both optically connected vibrational states. This degeneracy is removed by rotational–vibrational interaction which changes differently the rotational splitting of lower and upper vibrational states. Since the splitting is small enough, the interparticle interaction $\mathbf{V}$ induces non-adiabatic transitions between rotational sublevels with comparable cross-sections in both vibrational states. However, not only line broadening but also interference in each pair of lines is induced by non-adiabatic collisions. It was shown in [158] that

$$\hat{\mathbf{G}} = \begin{pmatrix} \mathrm{i}(\omega-\omega_{31})+\gamma_1 & -\beta_1 & 0 & 0 \\ -\beta_1^* & \mathrm{i}(\omega-\omega_{42})+\gamma_1^* & 0 & 0 \\ 0 & 0 & \mathrm{i}(\omega-\omega_{32})+\gamma_2^* & \beta_2^* \\ 0 & 0 & \beta_2 & \mathrm{i}(\omega-\omega_{41})+\gamma_2 \end{pmatrix}, \tag{4.10}$$

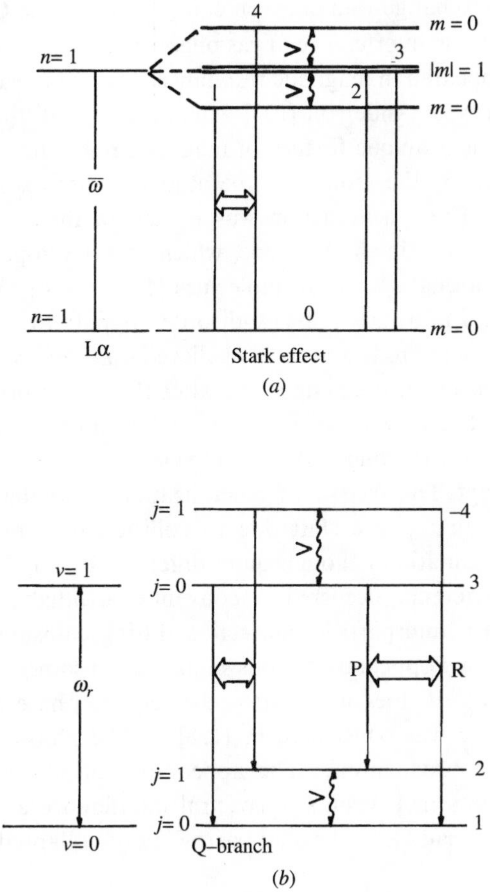

Fig. 4.1. The $L\alpha$ line of the H atom and its structure in the constant electric field (*a*) and the rotational structure of the vibrational transition (*b*). Wavy arrows show collision-induced transitions, thick horizontal arrows indicate the optical transitions that mutually interfere.

where

$$\gamma_1 = 1 - \langle S_{11}S_{33}^* \rangle,\ \gamma_2 = 1 - \langle S_{11}S_{33} \rangle, \beta_1 = \langle S_{12}S_{34}^* \rangle, \beta_2 = \langle S_{12}S_{34} \rangle. \quad (4.11)$$

As can be seen, the interference in each pair of lines does not disappear even in the lower order of perturbation theory when off-diagonal elements of the S-matrix are linear in V_{ik}. In the doublet, which represents the absorption spectrum, β_1 is the quantum equivalent of Gordon's classical

P–R exchange rate. In the other doublet β_2 is the quantum estimation of the frequency exchange rate between components of the Q-branch.

Discovery of line interference in gas phase spectroscopy occurred much later than it appeared in magnetic resonance. The latter was first found in the famous work of Anderson [104] who stressed that 'the phenomenon of 'narrowing' is a unique feature of magnetic resonance line breadths... which is caused by the motion of the atoms themselves in gases...' and other phases. This phenomenon was missed in the impact theory of spectral lines – even in the best one, which was developed by the same author for rotational spectra of molecules [177]. This probably resulted from the crudely understood 'phase-disorder' hypothesis imposed on the basis of the H-theorem. It was widely believed that each collision destroys any phase relations in the system and after that the process starts from the beginning. The Lorentz idea that any collision breaks completely the phase of radiation ('strong collision' approximation) was in accordance with this concept. The Weisskopf generalization of adiabatic broadening theory implies that phase shift due to collision may be small ('weak collision approximation') though quite different for any individual line. Hence no interference occurred in conventional adiabatic theory. As a matter of fact, interparticle interaction during collisions changes the phase of connected optical transitions similarly and sometimes identically. In the latter case it does not destroy the relative phase of the lines at all and therefore was considered in [158] as 'the phase memory effect in spectral line broadening'. The appearance of off-diagonal, 'phase memory' elements in $\hat{\mathbf{\Gamma}}$ results in spectral interference and collapse. In extreme cases it leads to collisional narrowing of collapsed spectra.

4.2 Collapse of the spectral doublet

The above-mentioned examples are particular cases of the following **G**-matrix:

$$\hat{\mathbf{G}} = \begin{pmatrix} \mathrm{i}(\omega - \bar{\omega} + \Delta) + \Gamma & -\beta \\ -\beta^* & \mathrm{i}(\omega - \bar{\omega} - \Delta) + \Gamma \end{pmatrix}, \tag{4.12}$$

where $\bar{\omega}$ is the average frequency of the doublet and 2Δ is the splitting of its components. When the components of the doublet are equally allowed and both transition moments are d the spectrum of the doublet in the high-temperature limit is [176]

$$I(\omega) = \begin{cases} \dfrac{|d|^2}{\pi}\left[\dfrac{\Gamma+\beta\Omega^{-1}(\omega-\bar{\omega}+\Omega)}{(\omega-\bar{\omega}+\Omega)^2+\Gamma^2}+\dfrac{\Gamma-\beta\Omega^{-1}(\omega-\bar{\omega}-\Omega)}{(\omega-\bar{\omega}-\Omega)^2+\Gamma^2}\right] & \beta<\Delta \\ \dfrac{|d|^2}{\pi}\left[\left(1+\dfrac{\beta}{\Omega'}\right)\dfrac{\Gamma-\Omega'}{(\omega-\bar{\omega})^2+(\Gamma-\Omega')^2}\right. & \\ \left.+\left(1-\dfrac{\beta}{\Omega'}\right)\dfrac{\Gamma+\Omega'}{(\omega-\bar{\omega})^2+(\Gamma+\Omega')^2}\right] & \beta>\Delta \end{cases} \tag{4.13}$$

where $\Omega = (\Delta^2 - |\beta|^2)^{\frac{1}{2}}$ and $\Omega' = -i\Omega$. Impact broadening of the doublet is assumed to be non-adiabatic, i.e.,

$$\Delta \ll 1/\tau_c, \tag{4.14}$$

but the ratio Δ/β is free to be large or small and correspondingly Ω may be either real or imaginary.

The value of Ω is crucial. The formulae (4.13) show that for real Ω the spectrum consists of two split components of equal width and intensities which are asymmetric to an extent of the ratio $\beta/\Delta \propto n$. When this parameter is small, they are slightly shifted relative to each other from their initial positions (Fig. 4.2(*a*)). The shift $\delta \propto \beta^2/\Delta$ is quadratic in buffer gas density n. Collapse occurs when Ω becomes 0 with increase of density. After that Ω becomes imaginary and the two lines merge at the centre of gravity of the spectrum $\bar{\omega}$, i.e. their shifts vanish (Fig. 4.2(*b*)). Now, also, the doublet components differ from each other not only in their widths but also in their intensities, which are of opposite signs. It is noticeable that, with further increase in gas density, the broader line disappears altogether and the whole spectrum intensity transfers to the narrower one (Fig. 4.3(*a*)). The density dependence of their widths is also different as shown in Fig. 4.3(*b*). Generally speaking the width of the thin component after collapse first decreases and then starts to increase with density as $\Gamma - \beta$:

$$\Delta\omega_{1/2} = \Gamma - \Omega' = \Gamma - (\beta^2 - \Delta^2)^{\frac{1}{2}} \approx (\Gamma - \beta) + \frac{\Delta^2}{2\beta} \quad \text{at } \Delta \ll \beta. \tag{4.15}$$

However, for the $L\alpha$ example, where $\Gamma = 2\beta$, the minimum is not pronounced. In fact for $\Gamma/\beta \geq 2$ the minimum is not obvious and the spectrum of such doublets looks like it is continuously broadened with increase of the collision rate proportional to n. In contrast, for $1 \leq \Gamma/\beta < 2$ the minimum is much deeper. In the extreme case

$$\Gamma = \beta \tag{4.16}$$

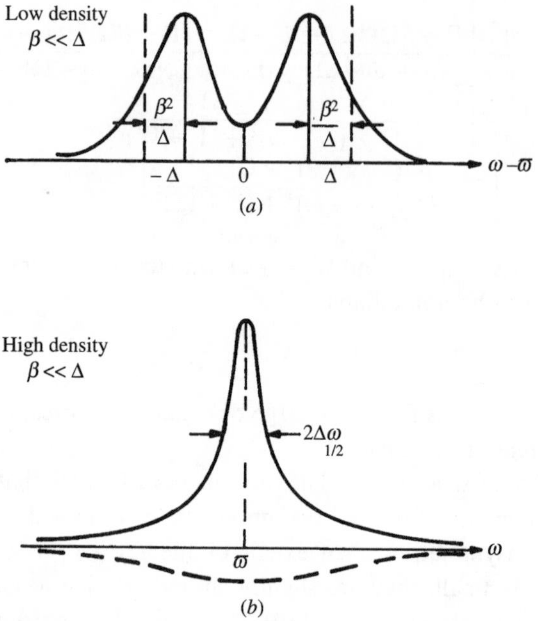

Fig. 4.2. The structure of the doublet: (*a*) before collapse and (*b*) after (the broken line is the broad component of the collapsed doublet, that has negative intensity).

the spectral width monotonically decreases with n and approaches 0 as $n \to \infty$ (Fig. 4.3(*b*)). This effect is the line narrowing produced by collisions. The 'motional averaging and narrowing' of the spectral structure is well known from the pioneering work of McConnell [9]. Condition (4.16) is considered to be natural in NMR spectroscopy. As a matter of fact it is not met in our first example ($L\alpha$) and is unlikely to be met in atomic spectroscopy at all. In the second example we must assume [158]

$$S_{11} = S_{33}, \; S_{12} = S_{34} \tag{4.17}$$

to obtain the 'spectral exchange' arrangement

$$\gamma_1 = 1 - \langle |S_{11}|^2 \rangle = W = \langle |S_{12}|^2 \rangle = \beta_1 \, . \tag{4.18}$$

In this manner the necessary condition (4.16) of the line narrowing effect is satisfied only for the doublet that belongs to the Q-branch. This result reflects the fact that a rotational phase shift does not affect Q-branch

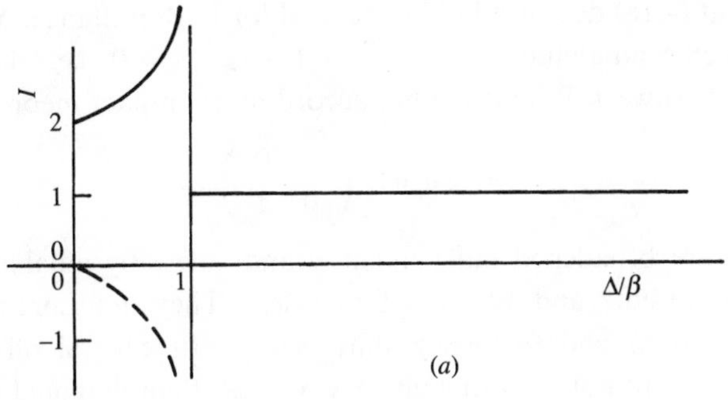

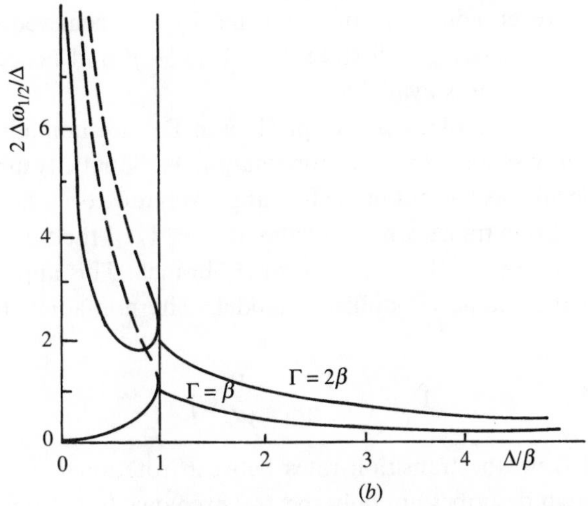

Fig. 4.3. The change of the integral intensities of the components (*a*) and their widths (*b*) with $\beta \propto n$ (the broken lines belong to the broader component of the collapsed doublet).

broadening at all. The Q-components are exchanged by collisions of arbitrary adiabaticity with the universal rate W which is actually the rate of transitions between rotational sublevels.

The situation is rather different for the other doublet which models P–R exchange. Rates

$$\gamma_2 = 1 - \langle S_{11}^2 \rangle \langle S_{12}^2 \rangle = \beta_2$$

are not the same and differ from W since

$$S_{11}^2 = |S_{11}|^2 e^{i\alpha},$$

where α is a rotational phase shift per collision. It is due to this phase relation that (4.18) does not hold in general for P–R exchange. Moreover, in the extreme adiabatic limit $|S_{12}|^2 = 1 - |S_{11}|^2 = 0$, i.e. there is no interference between P- and R-lines according to impact theory:

$$\hat{\mathbf{\Gamma}} \to \langle 1 - e^{i\alpha} \rangle \begin{pmatrix} 1 & 0 \\ 0 & 1 \end{pmatrix} .$$

Both lines are broadened independently and solely by adiabatic phase shift as in Lorentz and Weisskopf theories. They are Lorentzians of width $\langle 1 - \cos\alpha \rangle$ and frequency shift $\langle \sin\alpha \rangle$. In general off-diagonal elements of $\hat{\mathbf{\Gamma}}$ are not zero though they are less than diagonal elements. Consequently, the spectrum may collapse even in the adiabatic case when $\Delta \gg 1/\tau_c$. However, adiabatic collapse is hardly ever achieved in the gas phase where $1/\tau_c \gg 1/\tau_o > \beta$ since $\Delta \gg 1/\tau_c \gg \beta$ and hence only the resolved doublet limit is available.

Fortunately most molecules, except H_2 and D_2, are non-adiabatically broadened. Only small corrections for rotational adiabaticity are required for such molecules as N_2, but in the first approximation even these may be neglected. In this extreme, which is valid at $\Delta \ll 1/\tau_c$, the **S**-matrices are real and therefore $\gamma_2 = W = \beta_2$ as in the Q-branch. This approximation is similar to the classical J-diffusion model. The non-adiabatic impact operator

$$\hat{\mathbf{\Gamma}} = \begin{pmatrix} W & -W \\ -W & W \end{pmatrix} \tag{4.19}$$

is composed from the transition rates between rotational sublevels $j = 0, 1$. Therefore it describes not only spectral exchange but also population relaxation of these levels: $N_j(t)$. Introducing a population vector $\langle\langle \mathbf{N}| = (N_0, N_1)$ one can write the master equation for the relaxation as

$$\langle\langle \dot{\mathbf{N}}| = -\langle\langle \mathbf{N}|\hat{\mathbf{\Gamma}} . \tag{4.20}$$

However, the transition rates down and up are equal, as in Eq. (4.19), only in the high-temperature limit. In general the master equations are

$$\dot{N}_0 = -N_0 W_{01} + N_1 W_{10}, \; \dot{N}_1 = +N_0 W_{01} - N_1 W_{10} . \tag{4.21}$$

Hence we have

$$\hat{\mathbf{\Gamma}} = \begin{pmatrix} W_{01} & -W_{01} \\ -W_{10} & W_{10} \end{pmatrix} \tag{4.22}$$

The conservation law for the total number of particles $\dot{N} = (\dot{N}_0 + \dot{N}_1) = 0$ is reflected by the property

$$\hat{\mathbf{\Gamma}}|\mathbf{I}\rangle\rangle = 0 \tag{4.23}$$

that is known as 'the sum over any line of $\hat{\Gamma}$ is 0'. This property is common for operator (4.22) and its high-temperature analogue (4.19). What is actually different is the property

$$\langle\langle\rho|\hat{\Gamma} = 0\,. \tag{4.24}$$

where $\langle\langle\rho| = (\rho_0, \rho_1)$. This reflects the demands of the detailed balance principle

$$\frac{W_{01}}{W_{10}} = \frac{\rho_1}{\rho_0} = \mathrm{e}^{-\hbar\omega_{10}/kT}, \tag{4.25}$$

which must be satisfied at any finite temperature. The semiclassical calculation of **S**-matrices, which ignores the influence of rotational transition on recoil (translational motion of colliding particles), brings us to high-temperature result (4.19) as in [176] and many other works. Its use in Eq. (4.20) leads to equalization of populations, which is not the equilibrium distribution at any finite temperature. Even more difficulties appear if the semiclassical **S**-matrix is employed for the full rotational spectrum: the relaxation times of rotational energy and momentum diverge. To avoid such difficulties one has to correct the quasi-classical $\hat{\Gamma}$ *a posteriori* in order to satisfy the demands of the detailed balance principle [61].

In the phenomenological theories of spectral exchange in NMR as well as in recently proposed 'fitting laws' for the Q-branch collisional operator both of the conditions (4.23) and (4.24) are fulfilled from the beginning. This arrangement results in inevitable narrowing of the collapsed line. However, even for the Q-branch relations (4.17) are not obligatory. If the adiabatic phase shifts of vibrational states arising from collision are not equal, though small, then $S_{11} \neq S_{33}$ and roughly speaking

$$\gamma_1 = W + \gamma_{dp},\ \beta_1 = W\,,$$

where γ_{dp} is a vibrational dephasing rate. Using these relations in Eq. (4.15) one finds a result similar to that obtained in Eq. (3.49) and Eq. (3.16):

$$\Delta\omega_{1/2} = \frac{\Delta^2}{2W} + \gamma_{\mathrm{dp}}\,,$$

where Δ^2 is the frequency dispersion and $2W = 1/\tau_E$. Thus narrowing of the Q-branch has to be replaced by broadening at very fast collision rates.

4.3 Non-Markovian binary theory

If the splitting of the doublet is so large that inequality (4.14) changes sign, then the broadening becomes significantly adiabatic and the transfer rate W exponentially small. As was mentioned above, only the interference between the components of the resolved doublet is worthy of discussion in this limit. However, this limit may not be treated within impact theory. The latter is only valid for $t \gg \tau_c$ and hence unable to describe the periphery of the resonances at $|\omega \pm \Delta| \gg 1/\tau_c$. To describe properly the far wings of the doublet as well as its centre, one must use non-Markovian binary theory, which does not consider collisions to be instantaneous events. At least two versions of such a theory are possible, differential and integral [178, 179]. Though different, they are equivalent at very low pressures. We begin with the integral theory [180].

The necessary generalization of the description is achieved by using instead of Eq. (4.3) the integral kinetic equation

$$\dot{\bar{\mathbf{d}}} = \mathrm{i}[\mathbf{H_0}, \bar{\mathbf{d}}] + \mathrm{i}\int_{-\infty}^{+\infty} \mathrm{d}s\ [\langle \mathbf{V}(s)\rangle, \bar{\mathbf{d}}] - \int_0^t \mathrm{d}\tau\ \hat{\mathbf{M}}(\tau) e^{\mathrm{i}\mathbf{H_0}\tau} \bar{\mathbf{d}}(t-\tau) e^{-\mathrm{i}\mathbf{H_0}\tau} \tag{4.26}$$

with a kernel

$$\mathbf{M}_{ik,lm}(\tau) = \left\langle \int_{-\infty}^{+\infty} \mathrm{d}s\ \mathrm{e}^{\mathrm{i}(\omega_{il}+\omega_{mk})s} \left(\frac{\partial^2}{\partial s\, \partial s'} \mathbf{U}_{il}(s, s')\mathbf{U}^{+}_{mk}(s, s') \right)_{s'=s-\tau} \right\rangle, \tag{4.27}$$

where all terms averaged according to Eq. (4.5) are linear in n. In this equation the evolution operator $\mathbf{U}(t, t') = \mathbf{U}(0, t, t')$ where

$$\mathbf{U}(s, t, t') = \mathbf{T}\ \exp\left(\mathrm{i}\int_{t'}^{t} \mathrm{e}^{-\mathrm{i}\mathbf{H_0}\tau} \mathbf{V}(s+\tau) \mathrm{e}^{+\mathrm{i}\mathbf{H_0}\tau}\ \mathrm{d}\tau \right).$$

Here s is the time between the beginning of observation ($t = 0$) and the instant of closest approach where the intermolecular interaction $\mathbf{V}$ reaches its maximum. The operator $\mathbf{U}$ rules the alteration of the operator

$$\tilde{\mathbf{d}} = \mathrm{e}^{-\mathrm{i}\mathbf{H_0}t} \mathbf{d}(t) \mathrm{e}^{\mathrm{i}\mathbf{H_0}t} \tag{4.28}$$

during collision. The evolution operator $\mathbf{U}$ obeys the corresponding Schrödinger equation (in the interaction representation) for the pair particle problem. In contrast to the impact theory, $\hat{\mathbf{M}}(\tau)$ contains information on the evolution of the system over times comparable to and less than τ.

From Eq. (4.1) and Eq. (4.2) we find, omitting the bars over $\mathbf{d}$ for simplicity, that

$$I(\omega) = \mathrm{Re\ Sp}\ [\rho \mathbf{d}\mathbf{d}(\omega)]/\pi, \tag{4.29}$$

where $\mathbf{d}(\omega)$ is the Fourier transform of $\mathbf{d}(t)$. If the latter is found from Eq. (4.26) and used in Eq. (4.29) one can easily reproduce Eq. (4.7) with

$$\mathbf{G}_{ik,lm} = \mathrm{i}(\omega - \omega_{ik} - \Delta\dot{\omega}_{ik})\delta_{il}\delta_{km} + \mathbf{R}_{ik,lm}(\omega), \tag{4.30}$$

where

$$\Delta\omega_{ik} = \int_{\infty}^{\infty} \mathrm{d}s \langle V_{ii}(s) - V_{kk}(s)\rangle, \tag{4.31a}$$

$$\mathbf{R}_{ik,lm}(\omega) = \int_0^{\infty} \mathrm{d}\tau\ \mathbf{M}_{ik,lm}(\tau)\ \exp\left[\mathrm{i}(\omega_{ik} - \omega)\tau\right]. \tag{4.31b}$$

The evolution of the spectrum is thus reduced to inversion of the matrix $\mathbf{G}_{ik,lm}(\omega)$ and its convolution according to the recipe given in Eq. (4.7).

In non-Markovian theory, the off-diagonal elements of $\hat{\mathbf{G}}$ cannot be looked upon as transfer rates between two or more discrete eigenfrequencies of the system as they are functions of the continuous variable ω. The transfer rates concept is only acceptable in the Markovian limit of the theory ($t \gg \tau_c$) when ω-dependence is eliminated. To obtain this limit, we must first pass to differential formulation of non-Markovian theory and after that let $t \to \infty$. In the literature there is complete unanimity on how the transition from integral to differential formalism can be carried out correctly. According to Eq. (4.28) the integrand in Eq. (4.26) may be written as

$$\mathrm{e}^{\mathrm{i}\mathbf{H_o}\tau}\mathbf{d}(t-\tau)\mathrm{e}^{-\mathrm{i}\mathbf{H_o}\tau} = \mathrm{e}^{\mathrm{i}\mathbf{H_o}t}\tilde{\mathbf{d}}(t-\tau)\mathrm{e}^{-\mathrm{i}\mathbf{H_o}t}. \tag{4.32}$$

It is considered that the transition from $\mathbf{d}$ to $\tilde{\mathbf{d}}$ completely removes the fast motion with eigenfrequencies of $\mathbf{H}_o$ peculiar to the former, leaving behind only slow decay with relaxation time equal to or greater than τ_o. Since the other cofactor in the integrand of Eq. (4.26), $\hat{\mathbf{M}}(\tau)$, decays much more rapidly (with time $\tau_c \ll \tau_o$) one may use the substitution

$$\tilde{\mathbf{d}}(t-\tau) \to \tilde{\mathbf{d}}(t), \tag{4.33}$$

that transforms (4.32) into $\mathbf{d}(t)$. The latter may be taken out of the integral transforming the integral Eq (4.26) into differential:

$$\dot{\mathbf{d}} = \mathrm{i}[\mathbf{H_o}, \mathbf{d}] + \mathrm{i}\int_{-\infty}^{+\infty} \mathrm{d}s\ [\langle \mathbf{V}(s)\rangle, \mathbf{d}] - \int_0^t \hat{\mathbf{M}}(\tau)\ \mathrm{d}\tau\ \mathbf{d}(t). \tag{4.34}$$

It is once again the non-Markovian equation in a sense that the relaxation rates are time-dependent. They become constant for the times which are long enough to extend the integration over t to ∞. This leads to the basic Markovian equation

$$\dot{\mathbf{d}} = \mathrm{i}[\mathbf{H_0}, \mathbf{d}] + \mathrm{i}\int_{-\infty}^{+\infty} \mathrm{d}s\ [\langle \mathbf{V}(s)\rangle, \mathbf{d}] - \hat{\gamma}\mathbf{d}\,. \tag{4.35}$$

The off-diagonal elements of the Markovian relaxation operator

$$\hat{\gamma} = \int_0^\infty \hat{\mathbf{M}}(\tau)\ \mathrm{d}\tau \tag{4.36}$$

create interference. They do not disappear even in the adiabatic limit, unlike the situation in impact theory. This is a principal difference between the two approaches.

Using Eq. (4.35) in Eq. (4.29) we can confirm the general result of Eq. (4.7) but with

$$\mathbf{G}_{ik,lm} = i(\omega - \omega_{ik} - \Delta\omega_{ik})\delta_{il}\delta_{km} + \gamma_{ik,lm} \tag{4.37}$$

This is the Markovian equivalent of the impact result of Eq. (4.8). It can be seen that the impact and Markovian theories are identical if

$$\hat{\mathbf{\Gamma}}_{ik,lm} = \hat{\gamma}_{ik,lm} - \Delta\omega_{ik}\ \delta_{il}\delta_{km}\,.$$

However, this equality does not hold in general. It holds if the factor $\mathrm{e}^{\mathrm{i}(\omega_{il}+\omega_{mk})s}$ in Eq. (4.27) is omitted as if it were 1. The integrals over s and τ are then trivial and the evolution matrices reduce to **S**-matrices (4.6). This limit is justified for non-adiabatic collisions [181], which occur at $(\omega_{il} + \omega_{mk})\tau_c \ll 1$.

The situation is more complicated in the adiabatic limit when this inequality is reversed. According to Eq. (4.36) and Eq. (4.4) the off-diagonal parts of Γ and γ are different. To elucidate this difference and explore its consequences we shall examine the spectra of the four-level system passing from non-adiabatic to adiabatic broadening.

4.4 Line interference in stochastic perturbation theory

Let us reconsider the four-level system shown in Fig. 4.1(b), which has two doublets in the spectrum split by 2Δ and 2ϵ (Fig 4.4.(a)). Since diagonal elements of $\mathbf{G}_{ik,lm}$ are the same in impact and Markovian theories we assume that $V_{ii} = 0$ without any restriction of generality. This is actually the case for any electric multipolar interaction and hence $\Delta\omega_{ik} = 0$. The non-zero elements of the perturbation

$$V_{12}(t) = V_{34} = V(t) \tag{4.38}$$

are supposed to be real and small enough that standard stochastic perturbation theory may be used to calculate

$$\hat{\mathbf{R}}(\omega) = R(\omega)\begin{pmatrix} 1 & -1 \\ -1 & 1 \end{pmatrix}. \tag{4.39}$$

Here

$$R(\omega) = \overline{V^2}[F(\omega - \epsilon) + F(\omega + \epsilon)] \tag{4.40}$$

is proportional to the dispersion of interparticle interaction $\overline{V^2}$ and to spectral densities $F(\omega) = \int_0^\infty K(\tau)\ \exp\ (-i\omega\tau)\ d\tau$ of the normalized correlation function $K(\tau) = \overline{V(t)V(t+\tau)}/\overline{V^2(t)}$. As an example of the latter we choose the function

$$K(\tau) = \frac{(2\tau_c/\pi)^2}{\tau^2 + (2\tau_c/\pi)^2}\,. \tag{4.41}$$

It corresponds to the long-range interaction $(1/r^2)$ [20]. The Fourier transform of Eq. (4.41) is

$$F(\omega) = \tau_c\left\{e^{-|x|} + \frac{i}{\pi}[e^x\ \mathrm{Ei}\ (-x) - e^{-x}\mathrm{Ei}^*\ (x)]\right\}, \tag{4.42}$$

where $x = 2\omega\tau_c/\pi$ and Ei and Ei* are the integral exponent functions [182]. The real and imaginary parts of this expression are shown as functions of frequency in Fig. 4.4(b). The shape of the spectrum (4.30) depends via (4.40) on the value of $F(\omega)$ at any frequency, not only at resonances.

Much simpler is the situation in the Markovian limit. The spectrum is governed by the ω-independent operator

$$\gamma_{ik,lm} = \mathbf{R}_{ik,lm}(\omega_{ik})$$

which is for our system

$$\hat{\gamma} = \Gamma\begin{pmatrix} 1 & -1 \\ -1 & 1 \end{pmatrix} + i\delta\begin{pmatrix} 1 & 1 \\ -1 & -1 \end{pmatrix}. \tag{4.43}$$

Its real part is determined by

$$\Gamma = \overline{V^2}[\mathrm{Re}\ F(\epsilon - \Delta) + \mathrm{Re}\ F(\epsilon + \Delta)] = \frac{W_{21} + W_{43}}{2}\,, \tag{4.44}$$

where

$$W_{21} = \overline{V^2}\int_{-\infty}^{+\infty} K(\tau)e^{-i\omega_{21}\tau}\ d\tau \quad W_{43} = \overline{V^2}\int_{-\infty}^{+\infty} K(\tau)e^{-i\omega_{43}\tau}\ d\tau \tag{4.45}$$

are the rates of transitions between lower and upper sublevels induced

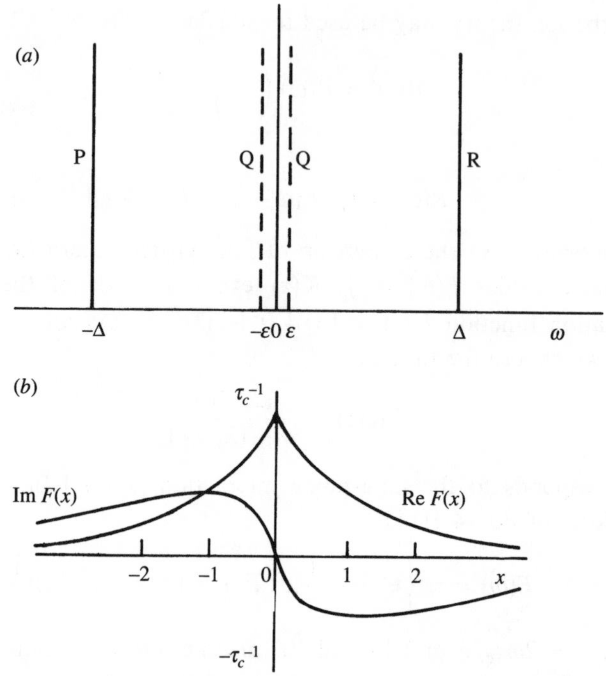

Fig. 4.4. Positions of spectral lines (*a*) and real and imaginary parts of the function $F(x)$ (*b*).

by collisions (or any other weak perturbation). The imaginary part is given by

$$\delta = \bar{V}^2[\mathrm{Im}\ F(\epsilon - \Delta) - \mathrm{Im}\ F(\epsilon + \Delta)], \tag{4.46}$$

where

$$\mathrm{Im}\ F(\omega) = \begin{cases} (4/\pi^2)\omega\tau_c^2 \ln|\omega\tau_c| & \omega\tau_c \ll 1 \\ -1/\omega & \omega\tau_c \gg 1 \end{cases} \tag{4.47}$$

and the high-frequency asymptotic behaviour of Im F is independent of the choice of $K(\tau)$.

As shown in Fig. 4.4a, splitting of the Q-branch doublet is usually so small that it can be neglected when calculating $\hat{\gamma}$. Consequently the additional condition

$$2\epsilon = \omega_{12} - \omega_{34} = 0 \tag{4.48}$$

is added to the above assumed equality $V_{12}(t) = V_{34} = V(t)$. As a result we obtain

$$\delta = 0 \tag{4.49a}$$

$$W_{21} = W_{43} = W = \overline{V^2} \int_{-\infty}^{+\infty} K(\tau) \mathrm{e}^{-\mathrm{i}\,\Delta\tau}\, \mathrm{d}\tau , \tag{4.49b}$$

which are the same conditions as in Eq. (4.17). For the Q-branch $\hat{\gamma} = \hat{\Gamma}$ defined in Eq. (4.19) and leads via spectral exchange to subsequent narrowing of the collapsed line. The non-adiabaticity of collisions is not obligatory as the Massey parameter $\Delta\tau_c$ is still free. The rate of transitions

$$W = 2\overline{V^2}\tau_c \mathrm{e}^{-2\,\Delta\tau_c/\pi} \tag{4.50}$$

may be exponentially smaller than its non-adiabatic estimation $2\overline{V^2}\tau_c$ and more easily satisfies the validity condition of perturbation theory $(W\tau_c \ll 1)$.

This treatment does not bring any qualitative changes in the broadening and Q-branch collapse, however, it does affect the interference of P and R lines that are not degenerate. The larger their splitting 2Δ the greater is the Massey parameter and the adiabaticity of collisions may become quite significant. To describe this case we have to use Eq. (4.40) with Δ substituted for ϵ. Using after that (4.39) and (4.30) in Eq. (4.29) we obtain the spectrum of the doublet in the general form

$$I(\omega) = \frac{2R'(\omega)\Delta^2}{\pi\{[\Delta^2 - \omega^2]^2 - 2\omega R'(\omega)]^2 + 4\omega^2 R'^2(\omega)\}} , \tag{4.51}$$

where R' and R'' are the real and imaginary parts of R. As Δ is inversely proportional to the moment of inertia, P–R exchange in heavy molecules is purely non-adiabatic and therefore

$$R'(\omega) \to R'(0) = W, \quad R'(\omega) \to R'(0) = 0, \quad \omega\tau_c \sim \Delta\tau_c \ll 1. \tag{4.52}$$

With these simplifications the general formula (4.50) reduces to the well-known result of the Markovian (as well as impact) theory of spectral exchange in the doublet [9, 20]

$$I(\omega) = \frac{2W\Delta^2}{\pi\{[\Delta^2 - \omega^2]^2 + 4\omega^2 W^2\}} , \quad \omega \ll 1/\tau_c . \tag{4.53}$$

Spectral exchange is considered here as sudden frequency modulation with switching time $\tau_c = 0$. As a result the spectral periphery is described by a universal power law $I(\omega) = 1/\omega^4$ as in Eq. (2.58) and Eq. (2.62)

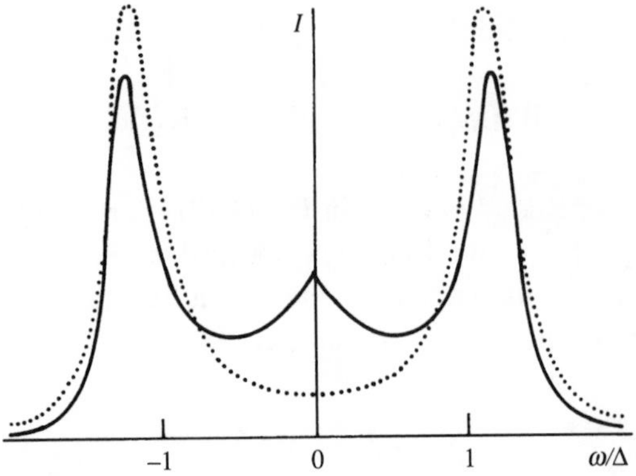

Fig. 4.5. The broadening of the P–R doublet ($\Delta\tau_c = \pi/2$, $\overline{V^2}\tau_c^2 = \pi/8$) in the integral non-Markovian theory (solid line) and in the Markovian approximation (dotted line).

and the central part undergoes collapse and motional narrowing with increase of W as described above.

A qualitatively new result appears in the opposite limit when broadening is adiabatic ($\Delta\tau_c \gg 1$). In this case only the resolved spectrum is worthy of discussion but in the vicinity of P and R lines the broadening is almost the same as before (Fig. 4.5). The significant changes are on the periphery of both resonances. The far wings of the spectrum do not follow the power law but instead, for $\omega \gg 1/\tau_c$

$$I(\omega) \propto \omega^{-4} e^{-\omega\tau_c}$$

is cut off exponentially on account of the finite time of collisions. Even more dramatic events occur at the very middle of the spectrum, which is also the periphery for both lines. Since vibrational–rotational coupling is neglected together with $\epsilon = 0$, the Q-lines are degenerate and forbidden for a free rotator. However, rotation during collision is not free and this ban is removed on a time scale τ_c. As a result a small Q-branch appears and the spectrum looks like a triplet (Fig. 4.5). To find its shape one has to use $F(\omega)$ from Eq. (4.42) to define $R'(\omega) + iR'(\omega) = 2\overline{V^2}F(\omega)$ and insert the results into Eq. (4.51). Following this procedure we find that the peak in the centre corresponding to the Q-branch duplicates the shape of

$$\mathrm{Re}\ F(\omega) = \tau_c e^{-2\omega\tau_c/\pi}.$$

It decreases exponentially on either side and the intensity at the centre is

$$I(0) = \frac{2\overline{V^2}\tau_c}{\pi\Delta^2}. \tag{4.54}$$

In a gas $\overline{V^2} \propto V_c^2\tau_c/\tau_o$ where V_c^2 is the maximum strength of the interparticle interaction on collision. Hence $I(0)$ as well as $\overline{V^2} \propto 1/\tau_o \propto n$ increases with buffer gas density. The integral intensity of the forbidden line is

$$\frac{I(0)}{\tau_c} \sim \frac{V_c^2}{\Delta^2}\frac{\tau_c}{\tau_o},$$

but at any reasonable density it is smaller than that for P- and R-lines. The relative intensity is a product of two small parameters: the adiabatic perturbation theory parameter $V_c^2/\Delta^2 \ll V_c^2\tau_c^2 \ll 1$ and the binary parameter $\tau_c/\tau_o \ll 1$. The latter must restrict the Q-branch intensity even if perturbation becomes strong. In the binary theory, this spectral component is viewed as a collision-induced line, which exists only during the time of collision. The transformation of the IR spectrum of HCl in He and Ar to a triplet with increase of pressure is of this kind. In liquids the intensity of the induced Q-branch is also restricted if stochastic perturbation theory is used to find it as in Chapter 7. It seems unlikely that this feature can reach the intensities of the side-branches. If it nevertheless happens, this is clear evidence of complex formation, which results in full rotation stoppage. In the spectra of complexes the Q-branch dominates and the complex concentration increases with n. It actually occurs with HCl in other mixtures and solutions. This picture was proved experimentally for cryogenic solutions [183].

In conclusion, it has been emphasized that line interference in non-Markovian theories does not disappear in the adiabatic limit as it does in impact theory. Of course it is a small effect since the off-diagonal terms of $\hat{\mathbf{R}}$ are less important for large splittings. The 'adiabatic interference' is also presented in the Markovian theory (see Eq. (4.43)) but this does not mean that Markovian theory is preferable to the impact one. Markovian description of interference between permitted and forbidden lines, if it occurs, leads to a meaningless negative intensity at the peak of the forbidden line [180]. Nothing like this appears in impact theory, nor in integral versions of non-Markovian theory. This and other difficulties of Markovian theory are considered in [180] as a disadvantage of differential non-Markovian theory as a whole, not only its long-time Markovian limit. Its imperfection originates from the unjustified transformation (4.33). The latter implies that $\tilde{\mathbf{d}}(t)$ varies more slowly than $\hat{\mathbf{M}}(t)$ decays, which is not the case when line interference occurs. It may be shown, *a posteriori*, that $\tilde{\mathbf{d}}(t)$ found by Markovian theory contains terms that oscillate with the doublet splitting frequency 2Δ whereas $\hat{\mathbf{M}}(t)$ decays with relaxation time τ_c. Consequently, it is just in the adiabatic limit that one errs in passing from the integral to the differential version of non-Markovian theory at $\Delta\tau_c \gg 1$. Thus we come to following recommendations.

(i) Integral non-Markovian theory must be used for better description of the spectral lines periphery especially in the adiabatic limit.
(ii) Impact theory is the preferred choice for simplified description of spectra, especially in the non-adiabatic limit.

The latter will be used below to describe broadening of resolved isotropic Raman spectra as well as their collapse and subsequent pressure-narrowing.

4.5 Impact theory of rotational spectra

In quantum theory as in classical theory the isotropic Raman spectrum is expressed in terms of the average value of the polarizibility tensor $\alpha^{(0)} = (1/3)\,\mathrm{Sp}\,\hat{\alpha}$ randomly changing in time due to collisions:

$$I(\omega) = \mathrm{Re}\frac{1}{\pi}\int_0^\infty \exp(-i\omega t)\,\mathrm{Sp}\,[\rho\alpha^{(0)}(t)\alpha^{(0)}]\,\mathrm{d}t. \tag{4.55}$$

The only difference is that $\alpha^{(0)}$ is now an operator acting in $|jm\rangle$ space of angular momentum eigenfunctions. This space consists of an infinite number of states, unlike those discussed above which had only four. This complication may be partly avoided if one takes into account that the scalar product in Eq. (4.55) does not depend on the projection index m. From spherical isotropy of space, Eq. (4.55) may be expressed via reduced matrix elements $\langle j\|\alpha^{(0)}\|j'\rangle$ as follows

$$I(\omega) = \mathrm{Re}\frac{1}{\pi}\int_0^\infty \exp(-i\omega t)\sum_{j_i j_f}\rho_{j_i}\langle \boldsymbol{j}_i\|\alpha^{(0)}(t)\|j_f\rangle\langle j_f\|\alpha^{(o)}\|j_i\rangle\,\mathrm{d}t, \tag{4.56}$$

where i and f are vibrational quantum numbers. Similar formulae exist for anisotropic Raman and IR spectra. The correlation functions are different but the time evolution of any averaged dynamical operator is the same, as it obeys the impact kinetic equation shown below.

Owing to space isotropy and isotropy of collisions in free space, only reduced elements of any vectorial quantity

$$\langle j_i m_i | A_q^{(\kappa)} | j_f m_f \rangle = (-1)^{j-m} \begin{pmatrix} j_i & \kappa & j_f \\ -m_i & q & m_f \end{pmatrix} (2j_i+1)^{1/2} \langle j_i \| A^{(\kappa)} \| j_f \rangle$$

are time-dependent. The closed kinetic equation for reduced elements, given by impact theory, is [184, 61]:

$$\frac{\mathrm{d}}{\mathrm{d}t} \langle j_i \| A^{(\kappa)} \| j_f \rangle = \mathrm{i} \langle j_i \| [\mathbf{H_0}, A^{(\kappa)}] \| j_f \rangle - \sum_{j_i' j_f'} \langle j_i' \| A^{(\kappa)} \| j_f' \rangle \mathbf{\Gamma}^{(\kappa)}_{j_i' j_f', j_i j_f}, \tag{4.57}$$

where

$$\mathbf{\Gamma}^{(\kappa)}_{j_i j_f, j_i' j_f'} = \frac{(2j_i'+1)^{1/2}}{(2j_i+1)^{1/2}}$$
$$\left\langle \frac{\pi}{k^2} \sum_{ll' q \{m\}} (-1)^{j_i - m_i + j_i' - m_i'} \begin{pmatrix} j_i & \kappa & j_f \\ -m_i & q & m_f \end{pmatrix} \begin{pmatrix} j_i' & \kappa & j_f' \\ -m_i' & q & m_f' \end{pmatrix} \right. \tag{4.58}$$
$$\left. \times (\delta_{ii'} \delta_{ff'} \delta_{ll'} \delta_{mm'} - \langle l'm', j_i' m_i' | S_i | lm, j_i m_i \rangle^* \langle l'm', j_f' m_f' | S_f | lm, j_f m_f \rangle) \right\rangle .$$

Here $\hbar k$ is a translational moment of a pair of reduced mass μ. Averaging over kinetic energy $E_k = (\hbar k)^2/2\mu$ is included in the operation

$$\langle \ldots \rangle = \frac{n\bar{v}}{(kT)^2} \int (\ldots) E_k \, \exp(-E_k/kT) \, \mathrm{d}E_k .$$

The elements of **S**-matrices are determined in the basis of orbital angular momentum l and rotational moments j_i, j_f of vibrational states i, f and their projections (m, m_i, m_f). Both **S**-matrices in Eq. (4.58) have to be calculated for the same energy E_k of colliding particles.

The half-width (at half-height) and the shift of any vibrational–rotational line in the resolved spectrum is determined by the real and imaginary parts of the related diagonal element $\mathbf{\Gamma}^{(\kappa)}_{j_i j_f, j_i j_f}$. For linear molecules the blocks of the impact operator at $\kappa = 0, 2$ correspond to Raman scattering and that at $\kappa = 1$ to IR absorption. The off-diagonal elements in each block $\mathbf{\Gamma}^{(\kappa)}_{j_i j_f, j_i' j_f'}$ perform interference between corresponding spectral lines (optical transitions $j_i \to j_f$ and $j_i' \to j_f'$). At the same time, relaxation of rotational populations of vibrational state i is governed by the operator $\mathbf{\Gamma}^{(o)}_{j_i j_i, j_i' j_i'}$. The block $\hat{\mathbf{\Gamma}}^{(1)}$, which governs the relaxation of the angular momentum **J**, gives rise not only to transitions between rotational sublevels but also to m-relaxation within each of them, caused by momentum reorientation from collisions. Even when transitions between rotational states are negligible due to adiabaticity of collisions, $\mathbf{J}$ is still relaxing due to 'm-diffusion'. In other words, in the adiabatic limit $\hat{\mathbf{\Gamma}}^{(o)} = 0$ but $\mathbf{\Gamma}^{(1)}_{j_ij_i,j_i'j_i'} = \Gamma_j \delta_{j_ij_i'}$.

To obtain a more detailed idea of the impact operator, it is customary to employ a 'semiclassical calculation', assuming that the orbital angular momentum of colliding particles may be considered unchanged despite transitions between rotational states. In such a case scattering occurs in a collision plane determined by impact parameter b and initial velocity v. As a result

$$\langle l'm', j_i'm_i'|S_i|lm, j_im_i\rangle \approx \delta_{ll'}\delta_{mm'}\langle j_i'm_i'|S_i(b,v)|j_im_i\rangle \,. \tag{4.59}$$

Employing this simplification (known as the 'external field approximation' – EFA) in Eq. (4.58) and replacing summation over l with integration over the impact parameters, we obtain

$$\tilde{\mathbf{\Gamma}}^{(\kappa)}_{j_ij_f,j_i'j_f'} = \frac{(2j_i'+1)^{1/2}}{(2j_i+1)^{1/2}}$$
$$\left\langle \sum_{q\{m\}} (-1)^{j_i-m_i+j_i'-m_i'} \begin{pmatrix} j_i & \kappa & j_f \\ -m_i & q & m_f \end{pmatrix} \begin{pmatrix} j_i' & \kappa & j_f' \\ -m_i' & q & m_f' \end{pmatrix} \right. \tag{4.60}$$
$$\left. \times (\delta_{ii'}\delta_{ff'} - \langle j_i'm_i'|S_i|j_im_i\rangle^* \langle j_f'm_f'|S_f|j_fm_f\rangle) \right\rangle .$$

The tilde over operator $\mathbf{\Gamma}$ here and below indicates that the operator is calculated in the EFA, as was done in [185, 186]. This treatment ignores the influence of rotational transitions, caused by the anisotropic part of the interaction, on relative translational motion of colliding particles. Therefore $\tilde{\mathbf{\Gamma}}^{(\kappa)}_{j_ij_f,j_i'j_f'}$ differs slightly from the true operator $\mathbf{\Gamma}^{(\kappa)}_{j_ij_f,j_i'j_f'}$. What the difference is and how it may be removed *a posteriori* will be discussed in the next chapter.

The selection rules for isotropic Raman spectra $j_i = j_f = j$ greatly simplify the formalism. The frequency matrix has only diagonal elements

$$\langle j_f|\omega|j_i\rangle = \omega_{fi} - \alpha_e j(j+1) = \Omega_j \,, \tag{4.61}$$

where Ω_j is the frequency of the jth rotational Q-branch component near vibrational frequency ω_{fi}. As we concentrate on the lowest vibrational transition later, let $f = 1, i = 0$ and $\omega_{fi} = \omega_{10}$. Since $\mathbf{A}^{(0)} = \alpha^{(0)}$ only diagonal elements $\langle j|\alpha^{(0)}|j\rangle = \alpha_j$ are of interest. In the Liouville space of quantum transitions these elements compose the vector $\boldsymbol{\alpha}$ and obey the impact equation

$$\frac{\mathrm{d}}{\mathrm{d}t}\alpha_j = \mathrm{i}\Omega_j\alpha_j - \sum_{j'}\alpha_{j'}\boldsymbol{\Gamma}^{(0)}_{(01)j'j} \tag{4.62}$$

The relaxation operator $\hat{\boldsymbol{\Gamma}}^{(0)}_{(01)}$ carries out rotational broadening as well as vibrational dephasing between 0 and 1 states.

If the interaction $\mathbf{V}$ does not depend significantly on vibrational quantum numbers as in Eq. (4.38) then dephasing is negligible and

$$\hat{\boldsymbol{\Gamma}}^{(0)}_{(01)} = \hat{\boldsymbol{\Gamma}}^{(0)}_{(00)}\,. \tag{4.63}$$

That means the frequency exchange operator in the Q-branch coincides with that ruling relaxation of rotational populations $N_j = \sum_m N_{jm}$ in the ground state:

$$\frac{\mathrm{d}}{\mathrm{d}t}N_j = -\sum_{j'}N_{j'}\boldsymbol{\Gamma}^{(0)}_{j'j}\,. \tag{4.64}$$

Thus this operator must satisfy the necessary demands of the particle conservation law and detailed balance principle as was the case for its simpler analogue in Eq. (4.23) and Eq. (4.24). They are

$$\hat{\boldsymbol{\Gamma}}^{(0)}|I\rangle\rangle = 0\,, \tag{4.65}$$

$$\langle\langle\rho|\hat{\boldsymbol{\Gamma}}^{(0)} = 0. \tag{4.66}$$

Here $\langle\langle\rho|_j = (2j+1)\rho_j$ and ρ_j is the equilibrium population of the state $|jm\rangle$:

$$\rho_j = \exp\,[-\beta j(j+1)]/Z\,, \tag{4.67}$$

where $\beta = B/kT \quad (B = \hbar^2/2I)$. At $t \to \infty, \quad N_j \to (2j+1)\rho_j$.

4.6 Collapse of isotropic Raman spectra

Now we are well equipped to approach the main problem of spectral collapse theory. Presenting Eq. (4.62) in the form

$$\langle\langle\dot{\alpha}(t)| = \langle\langle\alpha(t)|(\mathrm{i}\hat{\boldsymbol{\Omega}} - \hat{\boldsymbol{\Gamma}}^{(0)}) \tag{4.68}$$

we must keep in mind that $\langle\langle\alpha(0)| = \langle\langle\rho\alpha|$ and the spectrum (4.55) has to be rewritten as

$$I(\omega) = \mathrm{Re}\frac{1}{\pi}\int_0^\infty \exp\,(-\mathrm{i}\omega t)\langle\langle\alpha(t)|I\rangle\rangle\langle\langle I|\alpha\rangle\rangle\;\mathrm{d}t\,. \tag{4.69}$$

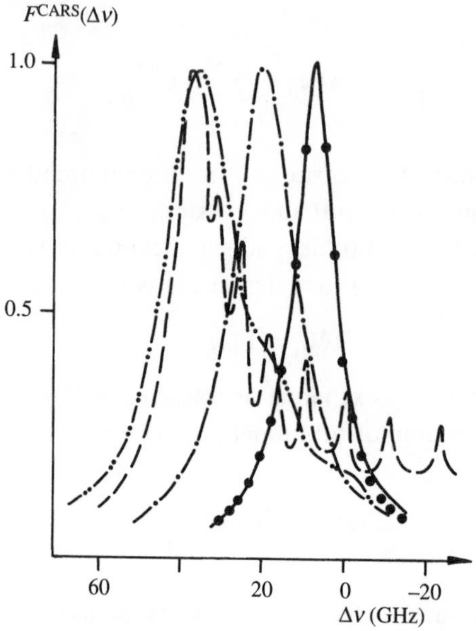

Fig. 4.6. The CARS spectrum collapse of acetylene Q-branch with increase of gas density [187]: (– – –) 0.55 atm of C_2H_2; (—··—) 1.7 atm of C_2H_2; (—·—) 1.7 atm of C_2H_2 + 40 atm of N_2; (—) 1.7 atm of C_2H_2 +120 atm of N_2. All contours are normalized to a maximum value, and the narrowest one has a Lorentzian shape.

The quantum theory must describe not only the shape of a resolved rotational structure of the Q-branch but its transformation with increase of pressure to a collapsed and well-narrowed spectrum as well. A good example of such a transformation is shown in Fig. 4.6. The limiting cases of very low and very high pressures are relatively easy to treat as they relate to slow modulation and fast modulation limits of frequency exchange.

In the slow modulation limit the splittings of spectral components are much larger than the rate of line interference and the off-diagonal part of the impact operator may be simply omitted to a first approximation:

$$\Gamma_{j'j} \approx \Gamma_{jj}\delta_{j'j}.$$

This so called 'secular simplification' of the problem reduces it to the conventional situation of independently broadened rotational lines. Comparing Eq. (4.68) with Eq. (4.20) we can see that secular simplification is possible due to the presence of the frequency operator $\hat{\boldsymbol{\Omega}}$ in Eq. (4.68).

For

$$\Omega_j - \Omega_{j'} \gg \Gamma^{(0)}_{j'j}$$

off-diagonal elements of $\hat{\Gamma}^0$ may be either completely neglected or taken into account as second-order corrections with respect to gas density. Nothing like this can be done with the impact operator in Eq. (4.20) as there is no $\hat{\mathbf{\Omega}}$.

To approach the alternative limit we have to introduce a projection operator

$$\hat{\mathbf{P}} = |I\rangle\rangle\langle\langle\rho| \tag{4.70}$$

and to divide $\langle\langle\alpha|$ into two components:

$$\langle\langle\alpha| = \langle\langle\alpha_1| + \langle\langle\alpha_2| = \langle\langle\alpha|\hat{\mathbf{P}} + \langle\langle\alpha|(1 - \hat{\mathbf{P}}). \tag{4.71}$$

The properties (4.65) and (4.66) may now be presented as

$$\hat{\mathbf{P}}\hat{\Gamma}^{(0)} = \hat{\Gamma}^{(0)}\hat{\mathbf{P}} = 0. \tag{4.72}$$

Without any restrictions we may shift the origin of the frequency scale to the centre of gravity of the spectrum:

$$\langle\langle\rho|\hat{\mathbf{\Omega}}|I\rangle\rangle = \sum_j \rho_j(2j+1)\Omega_j = \overline{\Omega_j} = 0, \tag{4.73}$$

where the line is centred after collapse. Since

$$\langle\langle\rho|I\rangle\rangle = \sum_j \rho_j(2j+1) = 1 \tag{4.74}$$

we mention that

$$\langle\langle\alpha_2(t)|I\rangle\rangle = \langle\langle\alpha(t)|I\rangle\rangle - \langle\langle\alpha(t)|I\rangle\rangle\langle\langle\rho|I\rangle\rangle = 0.$$

Using this result in Eq. (4.71) we obtain from Eq. (4.69)

$$I(\omega) = \mathrm{Re}\frac{1}{\pi}\int_0^\infty \exp(-\mathrm{i}\omega t)\langle\langle\alpha_1(t)|I\rangle\rangle\langle\langle I|\alpha\rangle\rangle\, \mathrm{d}t. \tag{4.75}$$

The properties (4.72) and (4.73) help to split the equation of motion (4.68) into two parts. Their Fourier-transforms are

$$\mathrm{i}\omega\langle\langle\alpha_1(\omega)| = \langle\langle\rho\alpha|\hat{\mathbf{P}} + \mathrm{i}\langle\langle\alpha_2(\omega)|\hat{\mathbf{\Omega}}\hat{\mathbf{P}} \tag{4.76a}$$

$$\mathrm{i}\omega\langle\langle\alpha_2(\omega)| = \mathrm{i}\langle\langle\alpha_1(\omega)|\hat{\mathbf{\Omega}} + \mathrm{i}\langle\langle\alpha_2(\omega)|\hat{\mathbf{\Omega}}(1-\hat{\mathbf{P}}) - \langle\langle\alpha_2(\omega)|\hat{\Gamma}^{(0)}. \tag{4.76b}$$

Their initial conditions are $\langle\langle\alpha_1(0)| = \langle\langle\rho\alpha|$ and $\langle\langle\alpha_2(0)| = 0$. It follows from (4.76), (4.70) and (4.74) that

$$\mathrm{i}\omega\langle\langle\alpha_1(\omega)|I\rangle\rangle = \langle\langle\rho\alpha|I\rangle\rangle + \langle\langle\alpha_2(\omega)|i\Omega|I\rangle\rangle\,, \quad (4.77a)$$

$$\langle\langle\alpha_2(\omega)|\left\{-\omega^2 + \mathrm{i}\omega\hat{\mathbf{\Gamma}}^{(0)} + \omega\hat{\mathbf{\Omega}}(1-\hat{\mathbf{P}}) + \hat{\mathbf{\Omega}}\hat{\mathbf{P}}\hat{\mathbf{\Omega}}\right\} = \mathrm{i}\langle\langle\rho\alpha|\hat{\mathbf{P}}\hat{\mathbf{\Omega}}\,. \quad (4.77b)$$

These results, first obtained by Kubo [188], permit us to show the general nature of spectral narrowing phenomena as arising from modulation which satisfies the demands of Eq. (4.72).

Following Kubo let us denote the orders of magnitudes of $\hat{\mathbf{\Gamma}}^{(0)}$ and $\hat{\mathbf{\Omega}}$ by τ_E^{-1} and ω_Q. Then the four terms on the left-hand side of Eq. (4.77b) are roughly ω^2, ω/τ_E, $\omega\omega_Q$ and ω_Q^2 respectively. We concentrate on the central part of the spectrum where

$$\omega \ll 1/\tau_E. \quad (4.78)$$

In the fast modulation limit

$$\omega_Q\tau_E \ll 1 \quad (4.79)$$

perturbation theory with respect to $\omega_Q\tau_E$ is valid. To lowest order in this parameter the first and third terms on the left-hand side of Eq. (4.77b) may be neglected. The equation reduces to

$$\langle\langle\alpha_2(\omega)|\left\{\mathrm{i}\omega\hat{\mathbf{\Gamma}}^{(0)} + \hat{\mathbf{\Omega}}\hat{\mathbf{P}}\hat{\mathbf{\Omega}}\right\} = \langle\langle\rho\alpha|\mathrm{i}\hat{\mathbf{P}}\hat{\mathbf{\Omega}}$$

or

$$\langle\langle\alpha_2(\omega)| = \frac{1}{\omega}\left[\langle\langle\rho\alpha|\hat{\mathbf{P}}\hat{\mathbf{\Omega}}\frac{1}{\hat{\mathbf{\Gamma}}^{(0)}} + \langle\langle\alpha_2(\omega)|i\hat{\mathbf{\Omega}}\hat{\mathbf{P}}\hat{\mathbf{\Omega}}\frac{1}{\hat{\mathbf{\Gamma}}^{(0)}}\right]. \quad (4.80)$$

Multiplying this equation by $\mathrm{i}\hat{\mathbf{\Omega}}|I\rangle\rangle$ we get

$$\begin{aligned}\langle\langle\alpha_2(\omega)|\mathrm{i}\hat{\mathbf{\Omega}}|I\rangle\rangle &= \frac{1}{i\omega}[\langle\langle\rho\alpha|I\rangle\rangle\langle\langle\rho|\mathrm{i}\hat{\mathbf{\Omega}}\frac{1}{\hat{\mathbf{\Gamma}}^{(0)}}\mathrm{i}\hat{\mathbf{\Omega}}|I\rangle\rangle \\ &\quad +\langle\langle\alpha_2(\omega)|\mathrm{i}\hat{\mathbf{\Omega}}|I\rangle\rangle\langle\langle\rho|\mathrm{i}\hat{\mathbf{\Omega}}\frac{1}{\hat{\mathbf{\Gamma}}^{(0)}}\mathrm{i}\hat{\mathbf{\Omega}}|I\rangle\rangle]\end{aligned} \quad (4.81)$$

or

$$\langle\langle\alpha_2(\omega)|\mathrm{i}\hat{\mathbf{\Omega}}|I\rangle\rangle = \frac{\langle\langle\rho\alpha|I\rangle\rangle\langle\langle\rho|\mathrm{i}\hat{\mathbf{\Omega}}\frac{1}{\hat{\mathbf{\Gamma}}^{(0)}}\mathrm{i}\hat{\mathbf{\Omega}}|I\rangle\rangle}{i\omega - \langle\langle\rho|\mathrm{i}\hat{\mathbf{\Omega}}\frac{1}{\hat{\mathbf{\Gamma}}^{(0)}}\mathrm{i}\hat{\mathbf{\Omega}}|I\rangle\rangle}. \quad (4.82)$$

Using this result in Eq. (4.77a) we obtain

$$\langle\langle\alpha_1(\omega)|I\rangle\rangle = \frac{\langle\langle\rho\alpha|I\rangle\rangle}{\mathrm{i}\omega - \langle\langle\rho|\mathrm{i}\hat{\mathbf{\Omega}}\frac{1}{\hat{\mathbf{\Gamma}}^{(0)}}\mathrm{i}\hat{\mathbf{\Omega}}|I\rangle\rangle}. \quad (4.83)$$

It follows from Eq. (4.75) and Eq. (4.83) that the centre of the collapsed spectrum has the Lorentzian shape

$$I(\omega) = \frac{1}{\pi}\mathrm{Re}\left[\frac{\langle\langle\rho\alpha|\alpha\rangle\rangle}{\mathrm{i}\omega + \langle\langle\rho|\hat{\mathbf{\Omega}}\frac{1}{\hat{\mathbf{\Gamma}}^{(0)}}\hat{\mathbf{\Omega}}|I\rangle\rangle}\right] = \frac{\overline{\alpha_j^2}}{\pi}\frac{\Gamma}{\omega^2+\Gamma^2}, \tag{4.84}$$

and its width

$$\Gamma = \langle\langle\rho|\hat{\mathbf{\Omega}}\frac{1}{\hat{\mathbf{\Gamma}}^{(0)}}\hat{\mathbf{\Omega}}|I\rangle\rangle = \int_0^\infty \langle\langle\rho|\hat{\mathbf{\Omega}}\mathrm{e}^{-\hat{\mathbf{\Gamma}}^{(0)}t}\hat{\mathbf{\Omega}}|I\rangle\rangle\mathrm{d}t = \omega_Q^2\tau_E, \tag{4.85}$$

where $\omega_Q^2 = \langle\langle\rho|\Omega^2|I\rangle\rangle = \overline{\Omega_j^2}$. This is a well-known expression for the width of a collapsed spectrum, which decreases as the rate of the energy relaxation increases.

The whole shape of the spectrum (before and after collapse) is described by a more general formula of quantum theory which follows from Eq. (4.55) and Eq. (4.62) [185, 186]. For $\alpha_j = inv$ the normalized spectral shape is

$$I(\omega) = \mathrm{Re}\frac{1}{\pi}\langle\langle\rho|(\hat{\mathbf{\Gamma}}^{(0)} + \mathrm{i}\mathbf{\Omega} - \mathrm{i}\omega\mathbf{I})^{-1}|I\rangle\rangle. \tag{4.86}$$

It is of no less interest to explore the shape of the spectral wings, especially beyond the low-frequency edge of the Q-branch where there are no rotational components. If the properties (4.72) are accepted, then the impact operator yields Eq. (4.65) and Eq. (4.66) and the far spectral wings have universal $1/\omega^4$ asymptotics like in Eq. (2.62). However, we have already seen that vibrational dephasing is not negligible, though small. It not only corrects the width of the collapsed line as in Eq. (3.49) but affects also the shape of the far wings, which is the same before and after collapse. To account for the non-zero dephasing rate γ_{dh} we have to keep in mind that Eq. (4.65) and Eq. (4.66) are valid if the intermolecular interaction commutes with variables that determine the spectrum. It is so in the basis of rotational states, which is complete. However, the extended basis includes combining vibrational states, which are only two from the full vibrational spectrum. Since the extended basis is not complete, the same time-dependent variables do not commute even with themselves. Therefore Eq. (4.65) and Eq. (4.66) were generalized in [61]:

$$\hat{\mathbf{\Gamma}}^{(0)}|I\rangle\rangle = \gamma_{\mathrm{dh}}|I\rangle\rangle, \quad \langle\langle\rho|\hat{\mathbf{\Gamma}}^{(0)} = \gamma_{\mathrm{dh}}\langle\langle\rho|. \tag{4.87}$$

Having included line shifts in the component frequencies, it can be regarded that $\mathrm{Im}\,\hat{\mathbf{\Gamma}}^{(0)} = 0$ and the asymptotic expansion of the denominator in Eq. (4.86) has the form

$$\text{Re } [\hat{\mathbf{\Gamma}}^{(0)} + \text{i}\mathbf{\Omega} - \text{i}\omega\mathbf{I}]^{-1} = \frac{\hat{\mathbf{\Gamma}}^{(0)}}{\omega^2} + \frac{\hat{\mathbf{\Gamma}}^{(0)}\mathbf{\Omega} + \mathbf{\Omega}\hat{\mathbf{\Gamma}}^{(0)}}{\omega^3} + \frac{\mathbf{\Omega}^2\hat{\mathbf{\Gamma}}^{(0)} + \mathbf{\Omega}\hat{\mathbf{\Gamma}}^{(0)}\mathbf{\Omega} + \hat{\mathbf{\Gamma}}^{(0)}\mathbf{\Omega}^2 - (\hat{\mathbf{\Gamma}}^{(0)})^3}{\omega^4} + \ldots \tag{4.88}$$

Substituting Eq. (4.88) into Eq. (4.86) and rearranging the expression, allowing for Eq. (4.87) and Eq. (4.73), we find that the second term vanishes and the rest transform to the following expression:

$$\pi I(\omega) = \frac{\gamma_{dh}}{\omega^2} + \frac{1}{\omega^4}\left(2\gamma_{dh}\omega_Q^2 + \sum_{jj'} \rho_j(2j+1)\Omega_j\Gamma^{(0)}_{jj'}\Omega_{j'} - \gamma_{dh}^3\right) + \ldots, \tag{4.89}$$

where

$$\omega_Q^2 = \alpha^2\overline{[J^2 - \overline{J^2}]^2} = \alpha^2[\overline{J^4} - (\overline{J^2})^2] = (\alpha_e kT/B)^2$$

with α_e introduced in Eq. (3.1). If the dephasing contribution to the spectral broadening is predominant, the spectral wings are Lorentzian. However, for diatomic molecules as a rule it is the opposite, which means that in a sufficiently large frequency range one may set $\gamma_{\text{dh}} = 0$ and consider $\mathbf{\Gamma}^{(0)}$ as a two-index operator. In this range the spectral wings are the following:

$$\pi I(\omega) = \frac{\alpha_e^2\langle\langle \rho J^2|\mathbf{\Gamma}^{(0)}|J^2\rangle\rangle}{\omega^4} = \frac{\omega_Q^2}{\omega^4\tau_E'}, \tag{4.90}$$

where [61]

$$\frac{1}{\tau_E'} = \frac{\langle\langle \rho J^2|\mathbf{\Gamma}^{(0)}|J^2\rangle\rangle}{\langle\langle \rho J^2|J^2\rangle\rangle - \langle\langle \rho J|J\rangle\rangle^2}. \tag{4.91}$$

Experimental verification of the universal wing shape (4.90) is not only an important way of checking the dominant role of spectral exchange but also an additional spectroscopic way to measure energy relaxation time even before collapse (in rare gases). Unfortunately it has not been done yet due to lack of accuracy far beyond the spectral edge.

5

Rotational relaxation: kinetic and spectral manifestations

Specification of the impact operator $\tilde{\mathbf{\Gamma}}^{(\kappa)}$ is a central problem of quantum theory. It may be approached *ab initio* or phenomenologically. The simplest solution of the problem from first principles is given by semiclassical approximations, which consider translational motion of colliding particles as independent of rotational transitions induced by the anisotropic part of the potential (Section 5.1). However, two important problems immediately arise in this approximation as to how to allow for (*a*) detailed balance and (*b*) adiabaticity of collisions. The latter may be simply ignored in the quantum J-diffusion model, which is purely non-adiabaic, but the former must be solved *a posteriori* (Section 5.2) to avoid the principal difficulties in calculation of angular momentum and rotational energy relaxation considered in Section 5.3. Using realistic atom–molecule interparticle potentials the impact operator is calculated semiclassically and corrected for detailed balance as in Section 5.4. It is done in a purely non-adiabatic infinite-order sudden (IOS) approximation as well as in centrifugal sudden (CS) approximation allowing for adiabatic corrections of the impact operator. The calculated magnitude and temperature-dependence of the rotational energy cross-section in N_2–Ar mixture is verified by experimental data in Section 5.5. Finally, rotational structure broadening and its collapse at higher pressures are described using the impact operator either calculated *ab initio* (Section 5.6) or introduced by a phenomenological 'fitting law' (Section 5.7).

5.1 The impact operator in semiclassical theory

Let us consider first quantum J-diffusion. It is carried out by purely non-adiabatic collisions realized for $\bar{\omega}\tau_c \ll 1$ where $\bar{\omega}$ is the average rotational frequency. A semiclassical analogue of the infinite-order sudden approximation (IOS) [189] was developed by Strekalov and Burshtein [185, 186] and used for calculating S-matrices and the impact operator (4.60). For $j_i = j_f$, $j_{i'} = j_{f'}$ it may be reduced to the following [61]:

$$\tilde{\mathbf{\Gamma}}^{(\kappa)}_{jj'} = (2j' + 1)^{3/2}(2j + 1)^{1/2}(-1)^{\kappa}$$
$$\sum_{J=0}^{\infty}(2J + 1)\left\{\begin{matrix} j & j & \kappa \\ j' & j' & J \end{matrix}\right\}\begin{pmatrix} j & J & j' \\ 0 & 0 & 0 \end{pmatrix}^2 W_J. \tag{5.1}$$

If dephasing of molecular vibration is negligible one may put $\mathbf{S}_i = \mathbf{S}_f$ as in Eq. (4.17). Then the relaxation parameters

$$W_j = \langle \delta_{j0} - |S_j|^2 \rangle \tag{5.2}$$

have the clear physical sense of rates, for non-adiabatic transitions from the Jth rotational level to $J = 0$. Factorization of this kind yields the entire matrix of rates in terms of a basic set of W_j which comprise only one row (or column). They are expressed via coefficients of $\mathbf{S}$-matrices expanded in Legendre polynomials

$$S_j = \frac{1}{2}\int_{-1}^{1} P_J(x) \exp\left(-\mathrm{i}\sum_k \delta_k P_k(x)\right) \mathrm{d}x. \tag{5.3}$$

The scattering phases

$$\delta_k = \frac{1}{\hbar}\int_{-\infty}^{\infty} V_k(r(t))\,\mathrm{d}t \tag{5.4}$$

are determined by the anisotropic part of the interparticle interaction

$$V(r, \gamma) = \sum_k V_k(r) P_k(\cos\gamma),$$

where γ is the angle between the axis $\boldsymbol{u}$ and the vector $\boldsymbol{r}$, joining the mass centres of colliding particles. In EFA approximation the classical trajectory $\boldsymbol{r}(t)$ is affected solely by the isotropic part of the same interaction $V_0(r)$.

The important relation resulting from $\mathbf{S}$-matrix unitarity

$$\sum_{J=0}^{\infty}(2J + 1)W_J = 0 \tag{5.5}$$

has an evident explanation. Since $\tilde{\mathbf{\Gamma}}^{(\kappa)}_{jj'}$ is calculated in EFA ('external field approximation') it corresponds to a high-temperature limit when the Boltzmann factor is the same for all rotational states. The product

$(2J+1)|W_J|$ becomes the rate of the Jth level relaxation taking account of degeneracy of the level. The conservation law established by Eq. (5.5) means that what goes up from the level $J=0$ with rate W_0 returns with the same total rate: $W_0 = \sum_{J=1}^{\infty}(2J+1)|W_J|$.

Interference in the Q-branch and population relaxation are ruled by the operator

$$\tilde{\mathbf{\Gamma}}^{(0)}_{jj'} = (2j'+1)\sum_J (2J+1)\begin{pmatrix} j & j' & J \\ 0 & 0 & 0 \end{pmatrix}^2 W_J . \tag{5.6}$$

Using the well-known properties of the $3j$ symbols

$$\sum_{j'} \begin{pmatrix} j & j' & J \\ 0 & 0 & 0 \end{pmatrix}^2 (2j'+1) = 1,$$

$$\sum_{j'} \begin{pmatrix} j & j' & J \\ 0 & 0 & 0 \end{pmatrix}^2 (2j'+1)j'(j'+1) = j(j+1)+J(J+1) \tag{5.7}$$

and relation (5.5) one can verify that in EFA the J-diffusion operator has the following important properties:

$$\tilde{\mathbf{\Gamma}}^{(0)}|I\rangle\rangle = 0, \tag{5.8}$$

$$\tilde{\mathbf{\Gamma}}^{(0)}|J^2\rangle\rangle = -\gamma|I\rangle\rangle, \tag{5.9}$$

where ${}_j|J^2\rangle\rangle = j(j+1)$ is an eigenvector of rotational energy and

$$\gamma = \sum_{J=1}^{\infty}(2J+1)J(J+1)|W_J| . \tag{5.10}$$

Eq. (5.8) formulates the same particle conservation law that was expected to hold for any $\hat{\mathbf{\Gamma}}^{(0)}$ in Eq. (4.65). The meaning of Eq. (5.9) becomes clear if one looks for rotational energy relaxation, which obeys the equation

$$\frac{d}{dt}|J^2(t)\rangle\rangle = -\tilde{\mathbf{\Gamma}}^{(0)}|J^2(t)\rangle\rangle . \tag{5.11}$$

The average rotational energy is

$$\bar{E}(t) = B\langle\langle\rho(0)|J^2(t)\rangle\rangle = B\langle\langle\rho(0)|\exp(-\tilde{\mathbf{\Gamma}}^{(0)}t)|J^2\rangle\rangle . \tag{5.12}$$

The energy is expected to relax from any initial state $\langle\langle\rho(0)|$ to equilibrium, but an entirely different result emerges if we employ in Eq. (5.12) an impact operator with properties (5.8) and (5.9). Considering the initial matrix as a normalized one ($\langle\langle\rho(0)|I\rangle\rangle = 1$), we obtain

$$\bar{E}(t) = E_o + B\gamma t, \tag{5.13}$$

where $E_o = B\langle\langle\rho(0)|J^2\rangle\rangle$ is the rotator initial energy. A divergence of the energy with time points to monotonic heating of the system due to collisions, which are functioning as a thermostat with $T = \infty$. This result is an innate defect of the relaxation operator calculated in external field approximation (EFA): it obeys the equation

$$\sum_j (2j+1)\tilde{\mathbf{\Gamma}}^{(0)}_{jj'} = \langle\langle\rho(\infty)|\tilde{\mathbf{\Gamma}}^{(0)} = 0\,, \tag{5.14}$$

which coincides with the necessary demand of Eq. (4.66) only at $T = \infty$. At any finite temperature an inconsistency exists between $\tilde{\mathbf{\Gamma}}^{(0)}$ properties and the recipe (4.55) where it must be used together with an equilibrium $\rho(T)$.

A similar defect is also inherent to the operator $\tilde{\mathbf{\Gamma}}^{(1)}$, which rules the angular momentum relaxation according to

$$\frac{\mathrm{d}}{\mathrm{d}t}|J(t)\rangle\rangle = -\tilde{\mathbf{\Gamma}}^{(1)}|J(t)\rangle\rangle\,. \tag{5.15}$$

It may be presented in the form

$$\begin{aligned}\tilde{\mathbf{\Gamma}}^{(1)}_{jj'} &= (2j'+1)\sum_J (2J+1)\begin{pmatrix} j & j' & J \\ 0 & 0 & 0 \end{pmatrix}^2 \\ &\quad \frac{j(j+1)+j'(j'+1)-J(J+1)}{2[j(j+1)j'(j'+1)]^{1/2}}\,W_J\,.\end{aligned}$$

Employing Eq. (5.7) we get

$$\tilde{\mathbf{\Gamma}}^{(1)}|J\rangle\rangle = \sum_{j'} \tilde{\mathbf{\Gamma}}^{(1)}_{jj'}[j'(j'+1)]^{1/2} = 0\,, \tag{5.16}$$

where $[j(j+1)]^{1/2} =_j |J\rangle\rangle$. Using this property in Eq.(5.15) we find

$$\frac{\mathrm{d}}{\mathrm{d}t}|J(t)\rangle\rangle = 0\,,$$

which is the conservation law for angular momentum averaged over collisions.

The nature of this artificial law is easily understood by considering relaxation of any of the momentum projections, e.g. J_z. Its equilibrium distribution is Gaussian with a width $(kT/B)^{1/2}$. The average J_z value relaxes to 0 at any finite width. However, at $T = \infty$ the width of the equilibrium distribution extends to infinity and it becomes homogeneous in J_z space with $\rho \to \mathbf{I}/Z = 0$. In this limit there is no preference to turn J_z by collisions to smaller or greater values. Random shifts of opposite sign but equal size are equally probable. Thus the distribution

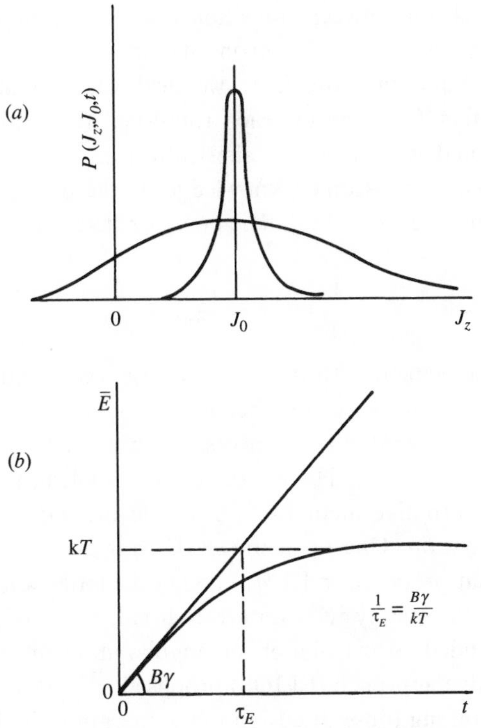

Fig. 5.1. The semiclassical description of angular momentum distribution relaxation (*a*) and rotational energy relaxation at $E_o = 0$ (*b*).

in J_z broadens to infinity but its maximum is not shifted in time and $\bar{J}_z = J_z(0) = \text{const}$ (Fig. 5.1). Moreover, at sufficiently long times $\overline{J_z^2} \propto t$ as it is for any one-dimensional diffusion in homogeneous space. Consequently, the rotational energy Eq. (5.13) linearly increases in time instead of reaching its equilibrium value kT (Fig. 5.1b). The only way to avoid artificial **J** conservation and the divergence of rotational energy with time is to take into account the finite temperature of the translational reservoir, giving up EFA.

The most simple way to accomplish this objective is to correct the external field operator *post factum*, as was repeatedly done in magnetic resonance theory, e.g. in [39]. Unfortunately this method is inapplicable to systems with an unrestricted energy spectrum. Neither can one use the method utilizing the Landau–Teller formula for an equidistant energy spectrum of the harmonic oscillator. In this simplest case one need correct for detailed balance only lowest transition W_{01} to bring in order all the rest $W_{nn'}$. The relation between transition rates in rotational spectra is much more complicated. Instead of one parameter determining energy relaxation of the quantum oscillator we deal with an infinite set of collision parameters W_J: one for each rotational transition. The only method to be noted is that which takes into account detailed balance in block $\mathbf{\Gamma}^{(0)}$, based on assumed knowledge of the eigenfunctions and eigenvalues of this operator [185]. The energy relaxation time found by this method was

$$\frac{1}{\tau_E} = \beta\gamma = \frac{B\gamma}{kT}\,. \tag{5.17}$$

This result is in agreement with the rate of energy rise in Eq. (5.13). If at the beginning $E(0) = 0$, the energy becomes as large as kT at time τ_E. A correction of the operator $\tilde{\mathbf{\Gamma}}^{(0)}$ is necessary in order to inhibit further rise of the energy, at $t > \tau_E$. Having faced this problem, the authors of [190] used two alternative methods to calculate the impact operator in the IOS approximation. One of them (DGIOS) is not consistent to the same extent as that proposed in [185]. Though the latter was successfully used to calculate the energy relaxation rate it still remains intuitive and may not be extended to calculation of angular momentum relaxation rate. The alternative approach (ELIOS) proposed in [190] is much better or even the best among those discussed in the literature. Its EFA version developed in [61] was successfully used for high-temperature estimation of any relaxation times. Subsequently the semiclassical CS method was developed to extend the theory to lower temperature and weakly adiabatic collisions [191]. The principles of the methods are described below.

5.2 Correction for detailed balance

To recognize what has to be done we must start from the exact definition of the impact operator [192]. Its off-diagonal part responsible for Q-branch interference and rotational transitions is given by the following expression [61]

$$\mathbf{\Gamma}^{(\kappa)}_{jj'} = -\frac{4\pi n\bar{v}\hbar^2}{2\mu(kT)^2}\int \exp\left(-E_k/kT\right)\,\mathrm{d}E_k \sum_{ll'J_iJ_f} \begin{Bmatrix} J_f & J_i & \kappa \\ j & j & l \end{Bmatrix} \begin{Bmatrix} J_f & J_i & \kappa \\ j' & j' & l' \end{Bmatrix}$$

$$\times(2J_i+1)(2J_f+1)(-1)^{j+j'+l+l'}\left(\frac{2j'+1}{2j+1}\right)^{1/2} T^{J_i*}_{j'l',jl}(E_k)\,T^{J_f}_{j'l',jl}(E_k)\,. \tag{5.18}$$

Here $\boldsymbol{J} = \boldsymbol{j} + \boldsymbol{l}$ is the total angular momentum of the system which is conserved during collisions.

T-matrix elements are taken from the relation

$$S^J_{j'l',jl} = \delta_{jj'}\delta_{ll'} - 2\mathrm{i}T^J_{j'l',jl}\,. \tag{5.19}$$

They have two very important properties

$$T^J_{j'l',jl}(E_k) \quad = T^J_{jl,j'l'}(E_k - \epsilon_{j'} + \epsilon_j) \qquad \text{at } E_k > \epsilon_{j'} - \epsilon_j \tag{5.20a}$$

$$T^J_{j'l',jl}(E_k) \quad = 0 \qquad \text{at } E_k < \epsilon_{j'} - \epsilon_j\,. \tag{5.20b}$$

The latter relation results from energy conservation and forbids rotational transitions when translational energy is deficient. The back processes with transfer of the rotational energy to translational energy are unrestricted. As a consequence, the lower limit of integration in Eq. (5.18) equals $\epsilon_{j'} - \epsilon_j$ at $j < j'$ and otherwise it is equal to 0. It is this very difference that leads to an exact relation between off-diagonal elements of the impact operator

$$\frac{\boldsymbol{\Gamma}^{(\kappa)}_{jj'}}{\boldsymbol{\Gamma}^{(\kappa)}_{j'j}} = \frac{2j'+1}{2j+1}\exp\left(-\frac{\epsilon_{j'} - \epsilon_j}{kT}\right)\,. \tag{5.21}$$

Being applied for the relaxation of populations ($\kappa = 0$), this equality expresses the demands of the detailed balance principle. This is simply a generalization of Eq. (4.25), which establishes the well-known relation between rates of excitation and deactivation for the rotational spectrum. It is much more important that equality (5.21) holds not only for $\kappa = 0$ but also for $\kappa = 1$ when it deals with relaxation of angular momentum $\mathbf{J}$ and the $\boldsymbol{\Gamma}^{(1)}_{jj'}$ elements should not be attributed any obvious physical sense. The non-triviality of this generalization is emphasized by the fact that it is impossible to extend it to the elements of the four-index operator $\boldsymbol{\Gamma}^{(\kappa)}_{j_ij_f,j'_ij'_f}$. It is unclear what the energy conservation law must consist of when two rotational transitions on different vibrational levels take place simultaneously. Apparently in this case a photon should be taken into account in the energy balance. This possibility may be treated in non-Markovian theory, but not in impact theory.

It follows from the definition of the impact operator and the S-matrices unitarity that $\hat{\boldsymbol{\Gamma}}^{(0)}$ obeys not only relation (4.65) but also Eq. (4.66), instead of Eq. (5.14) of EFA. Consequently we obtain an equilibrium (not equiprobable) distribution of populations. The property (5.9) as well as (5.16) are not confirmed. They are peculiar only to EFA and cannot be verified by a precise calculation. Instead it was shown in [61] that

$$\mathbf{\Gamma}^{(1)}_{jj} = \mathbf{\Gamma}^{(0)}_{jj} + \mathbf{\Gamma}_j , \tag{5.22}$$

where

$$\mathbf{\Gamma}_j = \left\langle \frac{4\pi}{k^2} \left(\sum_{Jll'} (-1)^{l+l'} \frac{2J+1}{2j+1} |T^J_{jl,jl'}|^2 - \sum_{ll' J_i J_f} (-1)^{l+l'} (2J_i+1)(2J_f+1) \right.\right.$$
$$\left.\left. \begin{Bmatrix} J_f & J_i & 1 \\ j & j & l \end{Bmatrix} \begin{Bmatrix} J_f & J_i & 1 \\ j & j & l' \end{Bmatrix} T^{J_i *}_{jl,jl'} T^{J_f}_{jl,jl'} \right) \right\rangle \tag{5.23}$$

is the rate of angular momentum reorientation due to '*m*-diffusion'. For impact theory this process is the only relaxation remaining in the adiabatic limit when $\hat{\mathbf{\Gamma}}^{(0)} = 0$ and

$$\mathbf{\Gamma}^{(1)}_{jj'} = \mathbf{\Gamma}_j \delta_{jj'} . \tag{5.24}$$

This relaxation proceeds without energy exchange between rotational and translational degrees of freedom and is supposed to be the same in EFA as in exact theory: $\tilde{\mathbf{\Gamma}}_j = \mathbf{\Gamma}_j$. With this assumption we obtain a result identical to the ELIOS approximation [190]:

$$\mathbf{\Gamma}^{(1)}_{jj} = \mathbf{\Gamma}^{(0)}_{jj} + (\tilde{\mathbf{\Gamma}}^{(1)}_{jj} - \tilde{\mathbf{\Gamma}}^{(0)}_{jj}) . \tag{5.25}$$

Thus the only thing we still have to find is the relation between off-diagonal elements $\mathbf{\Gamma}^{(\kappa)}_{jj'}$ and $\tilde{\mathbf{\Gamma}}^{(\kappa)}_{jj'}$. Then the way for necessary correction of the impact operator will be opened.

We make use of the assumption which is conventional in kinetic theory of the harmonic oscillator [193] as well as in energy-corrected IOS [194]. All the transition rates 'from top to bottom' in the rotational spectrum are supposed to remain the same as in EFA. Only transition rates 'from bottom upwards' must be corrected to meet the demands of detailed balance. In the same way the more general requirements expressed in Eq. (5.21) may be met:

$$\mathbf{\Gamma}^{(\kappa)}_{jj'} = \tilde{\mathbf{\Gamma}}^{(\kappa)}_{jj'} \text{ at } j > j' , \quad \mathbf{\Gamma}^{(\kappa)}_{jj'} = \tilde{\mathbf{\Gamma}}^{(\kappa)}_{jj'} \exp\left(-\frac{\epsilon_{j'} - \epsilon_j}{kT} \right) \text{ at } j < j' . \tag{5.26}$$

The impact operator corrected in such a way still remains semiclassical though the requirements of detailed balance are satisfied. It is reasonable provided that the change of rotational energy is small on average, relative to translational energy: $\overline{|\epsilon_{j'} - \epsilon_j|} \ll kT$, where the overbar means averaging performed over the 'distribution of products' after collision.

It is noteworthy that, having assumed Eq. (5.26), we no longer can keep the diagonal elements $\tilde{\mathbf{\Gamma}}^{(0)}_{jj}$ unchanged. They must be corrected to prevent breakdown of the particle conservation law. The requirement expressed in Eq. (4.65) has to be used to define diagonal elements via other already corrected elements:

$$\mathbf{\Gamma}^{(0)}_{jj} = -\sum_{j' \neq j} \mathbf{\Gamma}^{(0)}_{jj'}. \tag{5.27}$$

The right-hand side of this equation includes components with and without exponential Boltzmann factor but their sum equals the total flow of particles from the jth rotational level to the rest of the levels. After this correction both necessary demands, Eq. (4.65) and Eq. (4.66), are satisfied. This result is of great advantage since calculation of the impact operator with the rather simple semiclassical formula, Eq. (5.1), does not lead after correction to any principal difficulties. The set of equations (5.26) and (5.27) determine the operator $\hat{\mathbf{\Gamma}}^{(0)}$ consistently but not uniquely. Other recipes may be used as well (see Chapter 7 and [195]).

At high temperatures the corrected impact operator has the following important property [61]:

$$\hat{\mathbf{\Gamma}}^{(0)}|J^2\rangle\rangle = -\gamma|I\rangle\rangle \;+\; \beta\gamma|J^2\rangle\rangle \;\text{ at }\; \sqrt{\beta} = (B/kT)^{\frac{1}{2}} \ll 1, \tag{5.28}$$

which is essentially different from Eq. (5.9). It is this difference that eliminates the paradox connected with unrestricted increase of rotational energy. At high but finite temperature we obtain instead of Eq. (5.13)

$$\bar{E}(t) = (E_o - kT)\, \exp\,(-t/\tau_E) + kT, \tag{5.29}$$

where τ_E is that of Eq. (5.17). As to the corrected operator $\hat{\mathbf{\Gamma}}^{(1)}$ controlling the angular momentum relaxation, it should be thought of as obeying relation (5.25) with the proper $\hat{\mathbf{\Gamma}}^{(0)}$ found after correction of EFA. It was proved in [61] that it takes the following high-temperature limit:

$$\hat{\mathbf{\Gamma}}^{(1)}|J\rangle\rangle = \frac{1}{2}\beta\gamma|J\rangle\rangle. \tag{5.30}$$

This contrasts with relation (5.16), which led to a non-physical conservation law for $\boldsymbol{J}$. Eqs. (5.28) and Eq. (5.30) make it possible to calculate in the high-temperature limit the relaxation of both rotational energy and momentum, avoiding any difficulties peculiar to EFA. In the next section we will find their equilibrium correlation functions and determine corresponding correlation times.

5.3 Angular momentum and energy relaxation

The normalized correlation functions are defined as follows:

$$K_J(t) = \frac{\sum(2j+1)\rho_{jj}\langle j|J|j\rangle\langle j|J(t)|j\rangle}{\sum(2j+1)\rho_{jj}\langle j|J|j\rangle^2} = \frac{\langle\langle\rho J|J(t)\rangle\rangle}{\langle\langle\rho J|J\rangle\rangle}, \tag{5.31}$$

$$K_E(t) = \frac{\mathrm{Sp}[\rho J^2 J^2(t)]}{\mathrm{Sp}[\rho J^4]} = \frac{\langle\langle\rho J^2|J^2(t)\rangle\rangle}{\langle\langle\rho J^2|J^2\rangle\rangle}, \tag{5.32}$$

where $|J\rangle\rangle$ and $|J^2\rangle\rangle$ are the above-mentioned vectors in Liouville space and

$$\langle\langle\rho J|_j = (2j+1)\rho_{jj}[j(j+1)]^{1/2}, \ \langle\langle\rho J^2|_j = (2j+1)\rho_{jj}j(j+1)$$

are their covectors. As before the vectors $|J(t)\rangle\rangle$ and $|J^2(t)\rangle\rangle$ differ from $|J\rangle\rangle$ and $|J^2\rangle\rangle$ only in the time-dependence of the reduced elements. According to this definition,

$$\hbar^2\langle\langle\rho J|J\rangle\rangle = \hbar^2 Z^{-1}\sum(2j+1)\mathrm{e}^{-\beta j(j+1)}j(j+1) = \overline{J^2}, \tag{5.33}$$

$$\hbar^4\langle\langle\rho J^2|J^2\rangle\rangle = \hbar^4 Z^{-1}\sum(2j+1)\mathrm{e}^{-\beta j(j+1)}j^2(j+1)^2 = \overline{J^4}$$

are the mean squares of J^2 and J^4.

Though $K_J(t)$ decays from 1 to 0 it is in general non-exponential relaxation. Its conventually defined correlation time

$$\tau_J = \int_0^\infty K_J(t)\,\mathrm{d}t \tag{5.34}$$

may differ from that found from the equation

$$\frac{1}{\tau_J'} = \left.\frac{\mathrm{d}K_J}{\mathrm{d}t}\right|_{t=0}, \tag{5.35}$$

which implies that relaxation proceeds exponentially. The same is true for the correlation time of energy which is defined as

$$\tau_E = \int_0^\infty \frac{K_E(t) - K_E(\infty)}{1 - K_E(\infty)}\,\mathrm{d}t, \tag{5.36}$$

where $K_E(\infty) = \langle\epsilon\rangle^2/\langle\epsilon^2\rangle = \langle\langle\rho J|J\rangle\rangle^2/\langle\langle\rho J^2|J^2\rangle\rangle$. The experimentally available rate is usually defined as in Eq. (4.91):

$$\frac{1}{\tau_E'} = \frac{\langle\langle\rho J^2|\hat{\mathbf{\Gamma}}^{(0)}|J^2\rangle\rangle}{\langle\langle\rho J^2|J^2\rangle\rangle - \langle\langle\rho J|J\rangle\rangle^2} = -\frac{K_E'(0)}{1 - K_E(\infty)}. \tag{5.37}$$

If the rotational energy reaches its equilibrium value $\langle \epsilon \rangle = kT$ exponentially as in Eq. (5.29) then the times are the same:

$$\frac{1}{\tau_E'} = -\frac{1}{E_o - \langle \epsilon \rangle} \frac{\mathrm{d}\bar{E}}{\mathrm{d}t}\bigg|_{t=0} = \frac{1}{\tau_E}. \tag{5.38}$$

As we shall see, it is approximately true only in the high-temperature limit.

Using corrected operators in Eq. (5.11) and Eq. (5.15) and substituting their solutions into Eq. (5.31) and Eq. (5.32), we bring them into the form

$$K_J(t) = \langle\langle \rho J | \exp\left(-\hat{\Gamma}^{(1)}(t)\right) | J \rangle\rangle / \langle\langle \rho J | J \rangle\rangle, \tag{5.39a}$$

$$K_E(t) = \langle\langle \rho J^2 | \exp\left(-\hat{\Gamma}^{(0)}(t)\right) | J^2 \rangle\rangle / \langle\langle \rho J^2 | J^2 \rangle\rangle. \tag{5.39b}$$

Performing power series expansion of the exponents and using formulae (5.30) and (5.28) respectively we obtain

$$K_J(t) = \exp\left(-t/\tau_J\right) \tag{5.40}$$

$$K_E(t) = \frac{1}{2} + \frac{1}{2}\exp\left(-t/\tau_E\right). \tag{5.41}$$

It is implied here that $\overline{J^2} = \hbar^2/\beta$ and $\overline{J^4} = 2\hbar^4/\beta^2$, since the temperature is expected to be fairly high when Eq. (5.30) and Eq. (5.28) are used. It may be proved in the same way that Eq. (5.38) is also valid in this limit. As to the correlation times, they are determined from Eq. (5.34) and Eq. (5.36) and correspond as follows:

$$\frac{1}{\tau_E} = \frac{2}{\tau_J} = \frac{B\gamma}{kT}. \tag{5.42}$$

Remarkably, this result holds true irrespective of the kind of interparticle interaction that induces rotational relaxation.

This universality is peculiar for the high-temperature approximation, which is valid for $\sqrt{\beta} \ll 1$ only. For sufficiently high temperature the quantum theory confirms the classical Langevin theory result of J-diffusion, also giving $\tau_J = 2\tau_E$ (see Chapter 1). This relation results from the assumed non-adiabaticity of collisions and small change of rotational energy in each of them:

$$\frac{\overline{|\epsilon_{j'} - \epsilon_j|}}{kT} \approx \left(\frac{\overline{\Delta j}}{\bar{j}}\right)^{\frac{1}{2}} \ll 1. \tag{5.43}$$

When interparticle interaction is weak, only transitions to nearest levels are important and obviously $\overline{|\Delta j|} = 1 \ll \bar{j}$. However, 'multiroton' transitions were shown to be less probable for any strength of interaction. Requirement (5.44) is consistent with the weak collision limit of the classical J-diffusion model. The great changes in $\overline{\Delta j}$, which cannot be described semiclassically, take place in collisions of light molecules with light atoms (strong-collision limit). Semiclassical theory is addressed to the opposite situation and was successfully used to calculate relaxation times of nitrogen in argon [191].

In the high-temperature limit microscopic calculation [186] led to a formula quadratic in scattering phases:

$$\frac{1}{\tau_E} = n\beta \sum_l \frac{l(l+1)}{2l+1} \langle \delta_l^2 \rangle . \tag{5.44}$$

Such a construction is not a result of perturbation theory in δ_l, rather it appears from accounting for all relaxation channels in rotational spectra. Even at large δ_l the factor $\beta = B/kT \ll 1$ makes $1/\tau_E$ substantially lower than a collision frequency in gas. This factor is of the same origin as the factor $\hbar\omega/kT \ll 1$ in the energy relaxation rate of a harmonic oscillator, and contributes to the trend for increasing τ_E and τ_J with increasing temperature, which has been observed experimentally [81, 196].

5.4 Calculation methods

Factorization of the impact operator (5.1) greatly reduces the computational effort required in any variant of IOS [197]. When the inequality (5.43) holds, this factorization is acceptable but only for purely non-adiabatic relaxation as is J-diffusion. Though N_2–Ar collisions are mostly non-adiabatic, it would still be better to account for adiabaticity, which becomes more significant the higher the rotational quantum number j.

The original semiclassical version of the centrifugal sudden approximation (SCS) developed by Strekalov [198, 199] consistently takes into account adiabatic corrections to IOS. Since the orbital angular momentum transfer is supposed to be small, scattering occurs in the collision plane. The body-fixed correspondence principle method (BFCP) [200] was used to write the S-matrix for $|j_i - j_f| \ll j_i, j_f$ and arbitrary Massey parameter $\omega\tau_c$. At low quantum numbers, when $\omega\tau_c \to 0$, it reduces to the usual non-adiabatic expression, which is valid for any j_i, j_f. Though more complicated, this method is the necessary extension of the previous one adapted to account for adiabatic corrections at higher excitation level. To the extent that adiabatic corrections are taken into account, the J-diffusion relation (5.42) no longer holds and the two relaxation times τ_E and τ_J must be calculated separately. Below we concentrate on the energy relaxation cross-section although the cross-section of angular momentum relaxation was also obtained in a semiclassical approximation [201].

Calculating the rotational transition rates by the formula

$$| \tilde{\Gamma}^{(0)}_{jj'} | = n \frac{\bar{v}}{(kT)^2} \int_0^\infty \sigma_{jj'}(E) e^{-E/kT} E \, dE \,, \qquad (5.45)$$

we have to find the corresponding cross-sections $\sigma_{jj'}(E)$ for each kinetic energy E of the relative motion before collision. In SCS it was defined in [191] and [202] as

$$\sigma_{j_i j_f}(E) = \frac{v_f}{v_i} [j_f^2] \sum_L [L^2] \begin{pmatrix} j_i & L & j_f \\ 0 & 0 & 0 \end{pmatrix}^2 \int_0^\infty 2\pi b \, db \, |S_L(v, \omega_{if})|^2 \,, \quad (5.46)$$

where v_i and v_f are the initial and final translational velocities, $[j^2] = 2j+1$ and

$$S_L(v, \omega_{if}) = \frac{1}{2} \int_{-1}^{1} P_L(x) \exp \left(- \sum_{k=0}^{2} \delta_k(k\omega_{if}) P_k(x) \right) dx \,. \qquad (5.47)$$

The scattering phase

$$\delta_k(k\omega_{if}) = \frac{1}{\hbar} \int_{-\infty}^{\infty} V_k[r(t)] e^{-ik\omega_{if} t} \, dt = \frac{2}{\hbar v} \int_{r_t}^{\infty} \frac{V_k(r) \cos \, [k\omega_{if} t(r)]}{[1 - V_o(r)/E - b^2/r^2]^{1/2}} \, dr \qquad (5.48)$$

is calculated along the classical trajectory, which is deduced from the equation

$$dt = \frac{dr}{v[1 - 2V_o(r)/\mu v^2 - b^2/r^2]^{1/2}} \,, \qquad (5.49)$$

where $v = (v_i + v_f)/2$ and r_t is a turning point of the radial motion which is found as a positive root of the denominator in Eq. (5.49). As before $V_o(r)$ is the isotropic part of the interaction potential and $V_k(r)$ are the coefficients of its anisotropic part expended in Legendre polynomials.

As $V_o(r)$ decreases very rapidly with distance the major contribution to the scattering phases comes from a small part of the trajectory near the turning point. Expanding the isotropic potential in $z^2 = (r/r_t)^2 - 1 \ll 1$ one finds [202, 203]:

$$\delta_k(k\omega) = \frac{2r_t}{\hbar v^*} \int_0^\infty V_k[r_t(1+z^2)^{1/2}] \, \cos \, (k\omega r_t z / v^*) dz \,, \qquad (5.50)$$

where $v^* = [r_t \ddot{r}(0)]^{1/2}$. From the formal point of view the main difference between SCS and conventional IOS approximation [197, 204] is that $|S_L(v, \omega_{if})|^2$ can no longer be considered as the transition probability except for $j_f = 0$. The scattering phases must be calculated for each transition $i \to f$ separately using the frequency $\omega_{if} = B(j_i + j_f + 1)/\hbar$. The property of factorization of the impact operator is lost in the more adequate theory.

The 6–12 potential of Lennard-Jones is especially convenient for calculations of this kind. The simplest one is

$$V(r, \gamma) = 4\epsilon \left[\left(\frac{d}{r}\right)^{12} [1 + b_2 P_2(\cos\gamma)] - \left(\frac{d}{r}\right)^{6} [1 + a_2 P_2(\cos\gamma)] \right]. \tag{5.51}$$

Here ϵ is the depth of the well and d is the distance to zero crossing point of the isotropic potential whereas a_2 and b_2 are anisotropy parameters. Substituting this potential into Eq. (5.50) we get for each anisotropic term [202]

$$\delta_2(2\omega) = \frac{2^{1-p}\pi^{1/2}}{\Gamma(p + 1/2)} \left(\frac{C_n^2}{\hbar v^* r_t^{n-1}} \right) x^p K_p(x) , \tag{5.52}$$

where $C_6^2 = 4\epsilon d^6 a_2, C_{12}^2 = 4\epsilon d^{12} b_2$, $K_p(x)$ is a modified Bessel function $(p = (n-1)/2)$ of the argument $x = 2\omega r_t / v^*$ and

$$v^* = v \left[1 + 20\frac{\epsilon}{E} \left(\frac{d}{r_t}\right)^{12} - 8\frac{\epsilon}{E} \left(\frac{d}{r_t}\right)^{6} \right]^{1/2} .$$

In the purely non-adiabatic limit the phase (5.52) coincides with that calculated in [203] and for very long flights ($r_t \approx b, v^* \approx v$) or high energies ($E \gg \epsilon$) it reduces to what can be obtained from the approximation of rectilinear trajectories. However, there is no need for these simplifications. The SCS method enables us to account for the adiabaticity of collisions and consider the curvature of the particle trajectories. The only demerit is that this curvature is not subjected to anisotropic interaction and is not affected by transitions in the rotational spectrum of the molecule.

It is generally accepted that the centrifugal sudden (CS) approximation is the most reliable approximate method. Its results are usually very close to those obtained by *ab initio* ‘close coupling’ (CC) calculations. The integral and differential cross-sections of Ar inelastic scattering on nitrogen were performed for a few low-frequency rotational transitions and four different interaction potentials [205]. Much better agreement of CC with CS results was found than with IOS calculations performed in

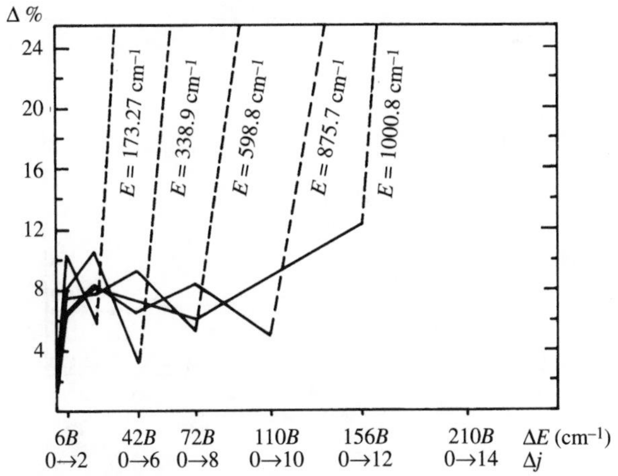

Fig. 5.2. Computed percentage error (absolute value) for the He–N_2 ($j_i = 0$) system using potential function HFD1. The state to state inelastic cross-sections are compared at several collision energies as a function of Δj transitions. The B value for N_2 is taken to be about 2 [207].

[206]. The authors claim that 'apparent exceptions, where IOS is more accurate than CS are presumably due to a fortuitous cancellation of errors'. The relative mistake in cross-section calculations was defined as

$$\Delta\% = \frac{\sigma^{\rm CS}_{j_i j_f} - \sigma^{\rm IOS}_{j_i j_f}}{\sigma^{\rm CS}_{j_i j_f}} \times 100$$

and estimated for some rotational transitions $j_i \to j_f$ and different collisional energies E in [207, 208]. The results presented in Fig. 5.2 show that the error increases with amount of energy transferred in collision. The average relative energy

$$\frac{\langle \Delta E_{\rm rot} \rangle_{j_i}}{E} = \frac{\sum_{j_f} |\epsilon_{j_i} - \epsilon_{j_f}| \sigma_{j_i j_f}}{E \sum_{j_f} \sigma_{j_i j_f}} \tag{5.53}$$

was calculated in both approximations and the results are shown in Fig. 5.3. The IOS approximation considerably overestimates the efficiency of collisions in changing rotational energy. In contrast to the conservation law, the change may be even greater than the collisional energy. Such an error originates from disregarding the change in relative velocity resulting from inelastic collision. As can be seen from Fig. 5.3 the factor (5.53)

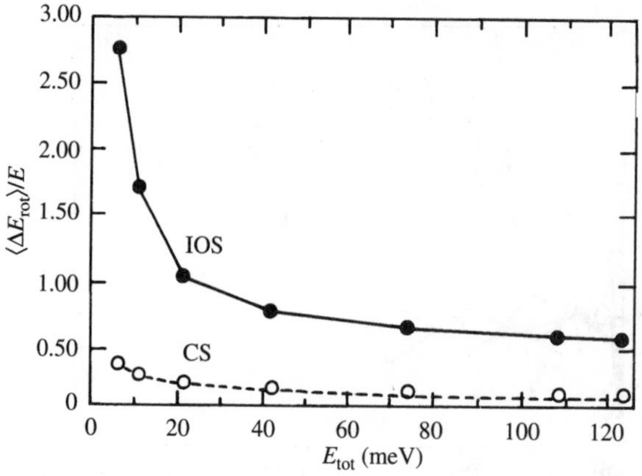

Fig. 5.3. Computed average relative energy transfer as a function of collision energy (meV) for the Ar–N_2 ($j_i = 0$) and BTT potential in IOS (solid line) and CS (broken line) approximations [208].

analogous to the classical conventional average $\int (J^2 - J'^2) f(\boldsymbol{J}, \boldsymbol{J}')\, d\boldsymbol{J} / J'^2$ is small enough at any J' before collision. This result confirms that $\boldsymbol{J}(t)$ is a correlated random process, which is carried out in the weak collision limit of J-diffusion ($\gamma \approx 1$). In this limit rotational energy relaxation may be described in a diffusional approximation [199].

In the semiclassical centrifugal sudden (SCS) approximation some additional simplifications were made, which permit us to estimate the scattering phase by Eq. (5.50). Therefore the accuracy of SCS has to be checked separately. Fortunately, for the Ar–N_2 system some cross-sections were calculated by the BFCP method [200] as well as by the CC method [206], which is considered to be the best. Using the same potential as in [209] the SCS cross-sections were found in [191] for fixed total energy of collisions E. The results are compared in Table 5.1.

As is seen, the results of BFCP and SCS are practically identical when $j_i, j_f \gg |j_i - j_f|$ but for the low-frequency transitions SCS yields a halved relative error. In the high-frequency region, the approximate theory provides only the order of magnitude since the basic criterion of SCS for semiclassical relative motion is violated.

The cross-sections for rotational de-excitation have a characteristic peculiarity: when $E \to 0$ they diverge (Fig. 5.4). The general reason for such behaviour is that the trajectory is greatly affected by attractive

Table 5.1. *Cross sections in Å^2 for rotational transitions at $E = 618$ K.*

Transition	CC	BFCP	SCS	Transition	CC	BFCP	SCS
0–2	16.0	18.8	14.4	0–6	11.4	11.7	10.3
2–4	15.5[a]	11.7	14.1	2–8	5.8	5.2	5.4
4–6	13.4	10.3	11.7	4–10	2.37	3.0	2.98
6–8	11.7	9.5	10.3	6–12	0.36	1.05	1.26
8–10	10.6	9.4	9.8	0–8	4.5	5.9	4.4
10–12	5.1	8.3	9.0	2–10	1.08	1.71	1.39
0–4	13.4	15.1	12.1	4–12	0.117	0.40	0.41
2–6	11.4	7.9	10.0	0–10	0.58	1.30	0.79
4–8	7.9	6.3	7.2	2–12	0.037	0.153	0.122
6–10	4.9	5.0	5.5	0–12	0.01	-	0.049
8–12	1.14	2.7	3.36				

[a] Data taken from [210].

forces when $E \ll \epsilon$. This divergence occurs in quantum theory as well but the oscillatory behaviour of cross-sections as $E \to 0$ is qualitatively different from the smoothed one in semiclassical centrifugal sudden approximation. If only small j terms contribute to the cross-section, the semiclassical criterion is the first to be violated. With increasing energy the cross-sections of low-frequency rotational transitions decrease while those of high frequency grow. This is not surprising. Since low-frequency transitions are surely non-adiabatic ($\omega\tau_c \ll 1$) the distant flights contribute significantly to the cross-section. The corresponding scattering phase decreases with increasing energy and cross-section as well. For high-frequency transitions only close flights are effective because their Massey parameter ($\omega\tau_c$) is minimal. For such trajectories, which are close to head-on, the higher the energy, the greater is the phase, i.e. cross-section grows with increasing energy. In the intermediate case, which is stretched out over a rather wide energy range (Fig. 5.4), the deactivation cross-sections are practically energy-independent. For comparison Fig. 5.4 presents also the energy-dependence of the cross-section $2 \to 4$ with a threshold typical for excitation transitions.

The principal advantage of SCS in comparison with IOS is that the adiabaticity of collisions may be taken into account. The difference between actual cross-sections and their purely non-adiabatic estimation is not large but increases with rotational frequency. As shown in Fig. 5.5 the adiabatic correction improves even qualitatively the high-frequency alteration of cross-sections by minimizing the discrepancy between SCS

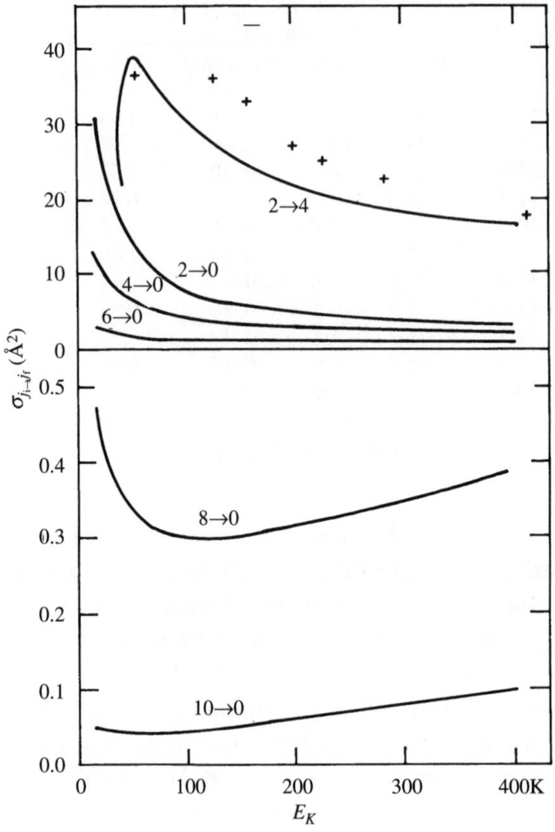

Fig. 5.4. Rotational transition cross-section for Ar–N_2 versus kinetic energy of collision (Kelvins) from [191]. (+) CC calculations of $\sigma_{2\to 4}$ from [211].

and CC results. Though the SCS is still an approximation, it is highly satisfactory and much simpler than any concurrent methods.

5.5 Experimental verification of SCS calculated rates

The simplicity of SCS is a great advantage when the problem becomes calculation of rates of transitions $\mathbf{\Gamma}^{(0)}_{jj'}$ or their rate constants

$$k_{jj'} = \mid \mathbf{\Gamma}^{(0)}_{jj'} \mid /n = \bar{v}\langle\sigma_{jj'}\rangle \,. \tag{5.54}$$

The full E-dependence of $\sigma_{jj'}(E)$ has to be found to calculate each of them from Eq. (5.45) and the huge matrix of such constants must be computed

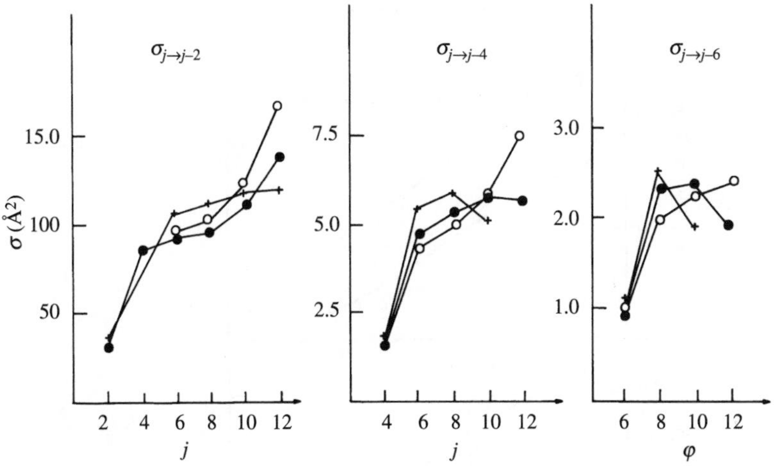

Fig. 5.5. Comparison of CC results (•) with SCS without adiabatic correction (∘) and with it (+) [191].

before looking for relaxation times, or contour transformation. Actually, the matrix of 20 × 20 dimensions ($j_i, j_f = 0, 2, ..., 38$) was calculated by the SCS method in [191] for some temperatures between 40 and 300 K. The diagonal cross-sections are much larger than off-diagonal ones. According to Eq. (5.26) they are the sum of all the rest in a full line. These cross-sections determine the j-component widths of the well-resolved Q-branch:

$$\mathbf{\Gamma}_{jj} = nk_{jj} = n\bar{v}\langle\sigma_{jj}\rangle\,. \tag{5.55}$$

Since good resolution of the Q-branch is hardly achievable by means of the usual Raman spectroscopy the first verification of this formula was carried out on side branches of anisotropic spectra which are easier to resolve (see Fig. 0.2 and Fig. 3.1). Generally speaking the right formula for component widths of these branches must be separately derived [212] but approximate estimation for the S-branch may be done as proposed in [213]:

$$\mathbf{\Gamma}_{jj} = n\kappa_j, \quad \kappa_j = \frac{k_{jj} + k_{j+2,j+2}}{2}\,. \tag{5.56}$$

It is reasonable when both vibrational and rotational dephasings are negligible. Using this approximation SCS estimations of the rate coefficients of line broadening were compared in [191] with the experimental j-dependence of κ_j in the S-branch of N_2–Ar mixture obtained in [214].

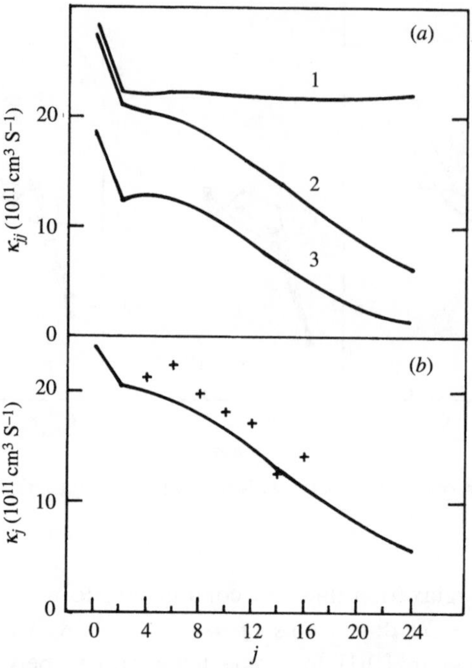

Fig. 5.6. Collisional broadening of N_2 rotational components. (*a*) In Q-branch, calculated by purely non-adiabatic theory at 300 K (1) and with adiabatic corrections at 300 K (2) and at 100 K (3) [215]. (b) In S-branch, calculated in [191] with adiabatic corrections using the recipe of Eq. (5.56). The experimental data (+) are from [214].

When calculating the rate constants, two potentials were used: the anisotropic 6–12 Lennard-Jones from [209] and the anisotropic Morse [216] for comparison. The results appeared to be very similar, thus indicating low sensitivity of the line widths to the potential surface details. The agreement with experimental data shown in Fig. 5.6(*b*) is fairly good. Moreover, the SCS approximation gives a qualitatively better approach to the problem than the purely non-adiabatic IOS approximation. As is seen from Fig. 5.6 the significant decrease of the experimental line widths with j is reproduced as soon as adiabatic corrections are made [215].

It is commonly accepted [217, 218] that rotational energy relaxation to its equilibrium value $\langle\epsilon\rangle = kT$ proceeds roughly exponentially as in Eq. (5.29). A current value of rotational energy $\bar{E} = \sum_j \epsilon_j N_j$ is obtained if the populations of any rotational levels N_j are measured. Such a measurement may be performed optically in different cross-sections of a supersonic beam moving with velocity u_x [191]. Then $\bar{E}$ yields the equation

$$\frac{d\bar{E}}{dt} = -\frac{\bar{E} - kT}{\tau'_E} = u_x \frac{d\bar{E}}{dx} . \tag{5.57}$$

The $\bar{E}(x)$ dependence on distance x from the nozzle source along the jet axis reflects the time evolution of rotational energy caused by cooling the gas mixture with expansion. Expansion into vacuum being supersonic, gas density as well as collision rate decrease so quickly that the rotational reservoir of N_2 has no time to reach equilibrium with the translational one. The further the rotating molecules are from the source, the greater $\bar{E}$ deviates from its equilibrium value kT corresponding to the translational temperature of Ar atoms. Therefore, measuring $N_j(x)$ along the jet axis allows one to determine the rates of rotational transitions, solving the inverse problem with Eq. (4.64).

Taking into account the gas density and assuming certain 'fitting laws' for the set of rate constants $k_{jj'}$ not were only these constants obtained in [191] but their temperature-dependence as well. In this manner τ'_E was found experimentally from the recipe which is identical to Eq. (5.37),

$$\frac{1}{\tau'_E} = n \sum_j^{\infty} \sum_{j<j'} \rho_j k_{jj'} \frac{(\epsilon_{j'} - \epsilon_j)^2}{\langle \epsilon^2 \rangle - \langle \epsilon \rangle^2} = n\bar{v}\sigma_{RT} . \tag{5.58}$$

An effective cross-section σ_{RT} is usually related to the translational velocity cross-sections σ_v and

$$Z_R = \frac{\sigma_{\mathrm{RT}}}{\sigma_v} \tag{5.59}$$

is the characteristic number of inelastic collisions necessary to establish equilibrium between rotational and translational degrees of freedom. For the Ar–N_2 system it has long been known [219] that $\sigma_v = \sigma_o(1 + C/T)$ where $\sigma_o = 30.2$ Å^2 and $C = 122$ K.

The recipe (5.58) is even more sensitive to the high-frequency dependence of $k_{jj'}$ than similar criterion (5.53), which was used before averaging over kinetic energy of collisions E. It is a much better test for validity of microscopic rate constant calculation than the line width's j-dependence, which was checked in Fig. 5.6. Comparison of experimental and theoretical data on Z_{R} for the Ar–N_2 system presented in [191] is shown in Fig. 5.7. The maximum value $Z_R = 22$ corresponding to point 3 at 300 K is determined from the rate constants obtained in [220].

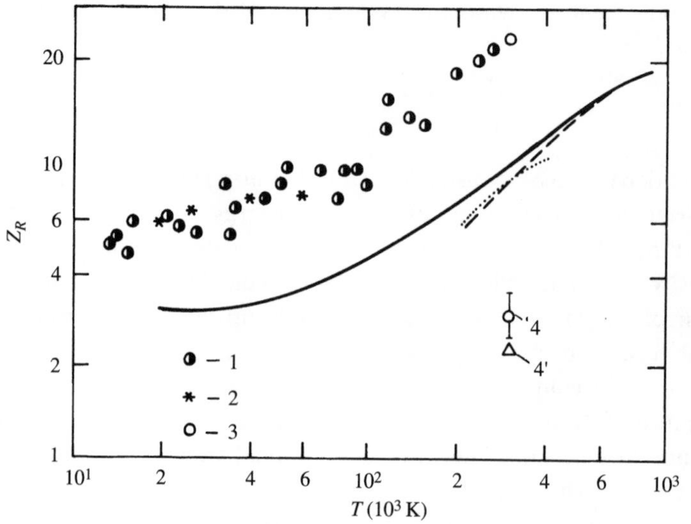

Fig. 5.7. Comparison of experimental data from [191] (1), [221] (2), [220] (3) and ultrasonic point [216] (4) with SCS calculation [191] (solid line) and classical trajectories calculations from [222] (dotted line), [223] (broken line) and [216] (4′).

This value agrees well with the rest found in [191] but not with point (4), which was shown to be $Z_R = 3.1$ in ultrasonic experiment [216]. No such discrepancy was found for pure nitrogen. Therefore it may be attributed to the low sensitivity of ultrasonic absorption measurements when nitrogen is present at small concentration in a gas mixture.

The important fact is that the number of collisions Z_R increases with temperature. It may be attributed to the effect of attraction forces. They accelerate the molecule motion along the classical trajectories favouring more effective R–T relaxation. This effect becomes relatively weaker with increase of temperature. As a result the effective cross-section σ_{RT} decreases monotonically [199], as was predicted for the quantum J-diffusion model in [186] ($\sigma_{RT} \propto 1/T$). As can be seen from Fig. 5.7 this conclusion is well verified not only by SCS results (solid line) but by classical trajectory calculations (dotted and broken lines) as well. At temperatures above 300 K both theoretical approaches are in satisfactory mutual agreement whereas some other approaches used in [224, 225] as well as SCS with attraction forces neglected [191] were shown to have the opposite temperature dependence for Z_R [191]. Thus SCS results with a reasonable potential provide qualitative or semiquantitative description of experimental data over a large temperature range. However, a systematic difference in numerical factor of 2–2.5 still remains. Of course the energy cross-section calculation is less accurate than that reached in SCS estimation of such transport properties as diffusion ($< 8\%$) and viscosity ($< 22\%$) [226]. This difference may arise because rotational relaxation is induced solely by the anisotropic part of the potential surface whereas transport coefficients are primarily sensitive to the isotropic part of the potential. However, the difference in accuracy may also result from some additional simplifications made in SCS as compared with original centrifugal sudden approximations. Another possibility is that relaxation experiments themselves are not as accurate as diffusion and viscosity measurements. In fact, agreement between different experimental data is no better than between experiment and theory. This disagreement may be objective, because velocity anisotropy of collisions in supersonic jets was neglected although it may affect the rate of rotational transitions.

Another principal demerit is that energy relaxation was considered to be exponential *ad hoc*, but it is not. In order to quantify the difference between τ_E' from Eq. (5.58) or Eq. (5.37) and τ_E from Eq. (5.36) the true correlation function of Eq. (5.32) was calculated in [215]:

$$K_E(t) = \frac{\sum_{j,j'} \rho_j j(j+1)[\exp(-\hat{\mathbf{\Gamma}}^{(0)}t)]_{jj'} j'(j'+1)}{\sum_{j,j'} \rho_j j^2(j+1)^2}. \tag{5.60}$$

Time evolution of this correlation function is ruled by the eigenvalues of operator $\hat{\mathbf{\Gamma}}^{(0)}$, which are determined by the equation

$$\sum_{j'} \hat{\Gamma}^{(0)}_{jj'} U^{(k)}_{j'} = \gamma_k U^{(k)}_j. \tag{5.61}$$

Possessing operator $\hat{\mathbf{\Gamma}}^{(0)}$, one is able to find its eigenvalues γ_k and corresponding eigenvectors $U^{(k)}_j$, from which $K_E(t)$ may be reduced to the form

$$K_E(t) = \sum_k P_k \exp(-\gamma_k t) \tag{5.62}$$

where P_k are expressed via the eigenvalues by the relation

$$P_k = \frac{[\sum_j \rho_j j(j+1) U^{(k)}_j]^2}{[\sum_j \rho_j (U^{(k)}_j)^2] \sum_j \rho_j j^2(j+1)^2}$$

and $\sum_k P_k = 1$. In Fig. 5.8(*a*) the eigenvalue spectrum of the $\hat{\mathbf{\Gamma}}^{(0)}$ is shown with corresponding weightings P_k. Actually, instead of a $\hat{\mathbf{\Gamma}}^{(0)}$ matrix, the

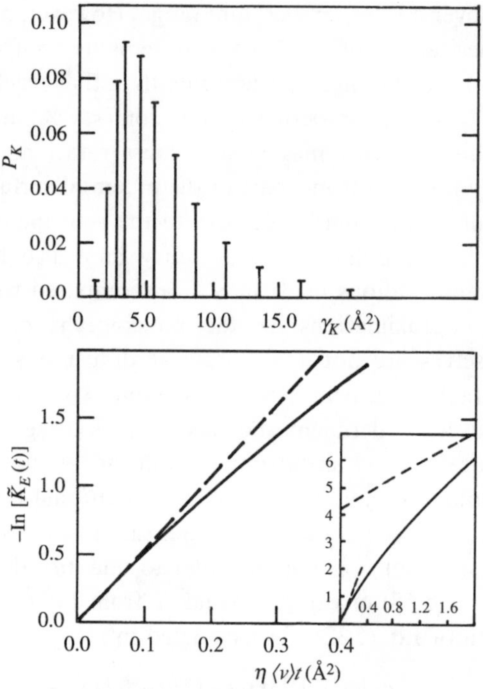

Fig. 5.8. (*a*) The weights of the eigenvalues of energy relaxation operator and (*b*) energy correlation function behaviour at short and long times (in inset) from [215].

corresponding cross-section matrix partly presented in Table 5.2 was used so that the γ_k are measured in ångström units. Since the spectrum is rather narrow it is clear that energy relaxation is not far from exponential.

To be more precise, the integrand of Eq. (5.36) may be defined as

$$\tilde{K}_E = \frac{K_E(t) - K_E(\infty)}{1 - K_E(\infty)} = \frac{\sum_{k\neq 0} P_k e^{-\gamma_k t}}{1 - P_0}, \tag{5.63}$$

where

$$P_0 = K_E(\infty) = \frac{[\sum_j \rho_j j(j+1)]^2}{\sum_j \rho_j j^2(j+1)^2} = \frac{\langle J^2\rangle^2}{\langle J^4\rangle} \approx \frac{1}{2} \quad \text{at } B/kT \ll 1$$

is the weighting of a minimum eigenvalue γ_0 which is 0 as a consequence of the property (5.27). The correlation function (5.63) decreases with time from 1 to 0 but is not necessarily mono-exponential and therefore $\tau_E = \int \tilde{K}_E(t)\,dt$ may differ from τ_E'. For the particular case considered in

Table 5.2. *Cross-sections* $\langle\sigma_{jj'}\rangle$ *given in* $Å^2$ *at* $T = 295$ K.

	j'					
j	0	2	4	6	8	10
0	80	24	27	18	8.1	2.6
2	5.2	63	26	19	9.2	3.3
4	3.6	17	60	20	12	5.1
6	2.1	10	18	58	16	8.3
8	0.96	5.1	10	16	55	13
10	0.36	2.2	5.3	10	16	51

[215] the relaxation rate decreases with time as shown in Fig. 5.8b and consequently the corresponding cross-sections are slightly different:

$$\sigma_E = 4.28\ Å^2\,, \quad \sigma_E' = 5.54\ Å^2\,. \tag{5.64}$$

This difference is primarily an effect of partial adiabaticity of collision. If it is completely ignored as in the J-diffusion limit then decay is practically mono-exponential so that $\sigma_E = 10.04\ Å^2$ and $\sigma_E' = 10.07\ Å^2$ are almost the same. However, these cross-sections are nearly twice those represented in Eq. (5.64), which proves that adiabatic correction of the J-diffusion model (IOS approximation) is significant, at least at $T = 300$ K.

5.6 Broadening and collapse of the isotropic Q-branch

The best resolution of Q-branch rotational structure in a N_2–Ar mixture was achieved by means of coherent anti-Stokes/Stokes Raman spectroscopy (CARS/CSRS) at very low pressures and temperatures (Fig. 0.4). A few components of such spectra obtained in [227] are shown in Fig. 5.9. A composition of well-resolved Lorentzian lines was compared in [227] with theoretical description of the spectrum based on the secular simplification. The line widths (5.55) are presented as

$$\Gamma_{jj} = \gamma_j^{(0)} p_0 + \gamma_j^{(1)} p_1\,, \tag{5.65}$$

where p_0 is the argon pressure, p_1 is the pressure of N_2 and the pressure coefficients $\gamma_j^{(0)}$, $\gamma_j^{(1)}$ are fitting parameters. Using a least-squares analysis these values were determined for even j (from 2 to 18) at pressures 50–760 Torr and two temperatures of 295 and 140 K. Data on broadening coefficients for $j = 0$ and 1 should be regarded as estimates as these

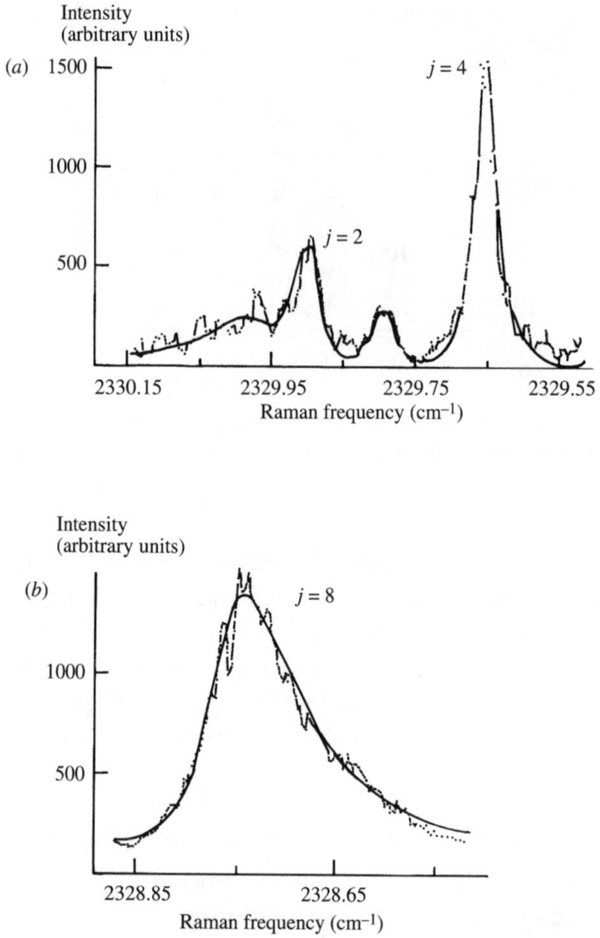

Fig. 5.9. Examples of experimental and fitted CSRS spectra of nitrogen in argon from [227] at 140 K under total pressure 200 Torr (a) and 600 Torr (b), N_2 : Ar= 1 : 10: (···) observed; (——) calculated.

lines are badly resolved even at the lowest pressures. Fig. 5.10 represents experimental results found in [227] together with those published in [228]. Both sets of data available at 295 K agree within the indicated error limit. Theoretical calculations of broadening coefficients give almost the same results whatever potential is used: either PLBC from [209] or KDV from [216].

Though the functional j-dependence is reproduced theoretically rather

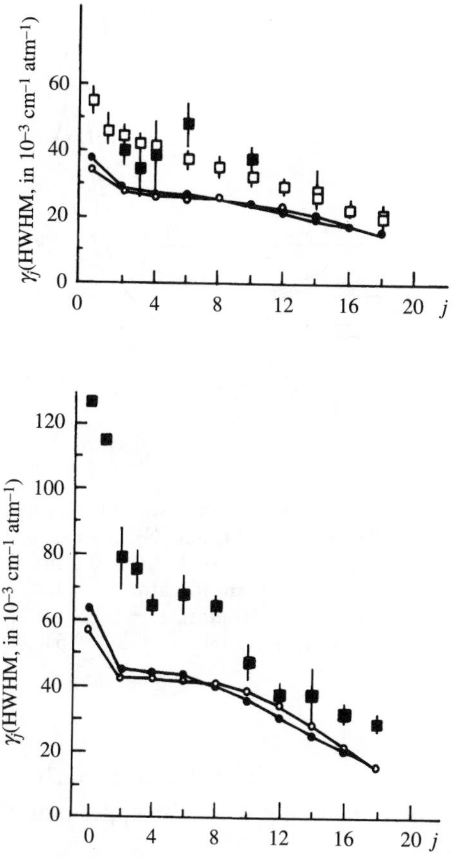

Fig. 5.10. Argon collisional broadening coefficients for nitrogen Q-branch components measured in [227] (■), [228] (□) and calculated with PLBC potential (•) and KDV potential (○).

well, numerical agreement with experimental values of $\gamma_j^{(1)}$ is less good than it was for the S-branch rate constants shown in Fig. 5.6. The discrepancy is objective since the CSRS experiment is more accurate than normal Raman spectroscopy and the theory of the Q-branch is free from the uncontrolled assumption used to estimate rate constants in Eq. (5.56). At room temperature the theoretical results are 20% less than experimental values and the difference becomes twice as large at 140 K. Calculations with the more anisotropic KDV potential provided practically the same results. Even taking into account the higher order anisotropy term in the potential, namely $P_4(\cos\theta)$ in the BTT model [229], changes the results within 10% limits [227] and the agreement

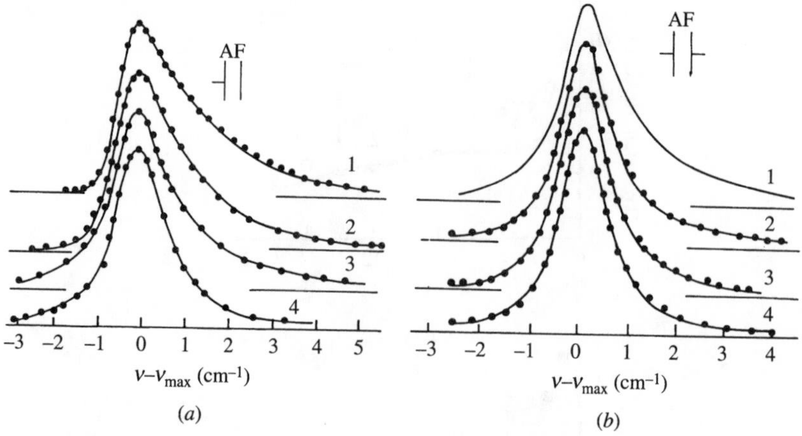

Fig. 5.11. Classical description (dots) of the isotropic Raman spectra (solid curves) of CO in strong collision model (*a*) and N_2 in weak collision model (*b*) at 295 K and different $\Gamma = 1/\omega_Q\tau_J$ [230]. (*a*): 1) CO (10 atm), $\Gamma = 0.23$; (2) CO (15 atm), $\Gamma = 0.32$; (3) CO (15 atm) + He (60 atm), $\Gamma = 0.8$; (4) CO (15 atm) + He (90 atm), $\Gamma = 1.44$. (*b*): (1) N_2 (10 atm), $\Gamma = 0.12$; (2) N_2 (20 atm) + Ar (40 atm), $\Gamma = 0.44$; (3) N_2 (20 atm) + Ar (80 atm), $\Gamma = 0.64$; (4) N_2 (20 atm) + Ar (130 atm), $\Gamma = 0.94$.

with experiment becomes even worse. Hence the conclusion is that SCS calculations with PLBC potential provide the upper limit for broadening coefficients.

At higher pressures only Raman spectroscopy data are available. Because the rotational structure is smoothed, either quantum theory or classical theory may be used. At a mixture pressure above 10 atm the spectra of CO and N_2 obtained in [230] were well described classically (Fig. 5.11). For the lowest densities (10–15 amagat) the band contours have a characteristic asymmetric shape. The asymmetry disappears at higher pressures when the contour is sufficiently narrowed. The decrease of width with $1/\tau_J$ measured in [230] by NMR is closer to the strong collision model in the case of CO and to the weak collision model in the case of N_2. This conclusion was confirmed in [215] by presenting the results in universal coordinates of Fig. 5.12. It is also seen that both systems are still far away from the fast modulation (perturbation theory) limit where the upper and lower borders established by alternative models merge into a universal curve independent of collision strength.

Using quantum theory instead of classical we have to describe the

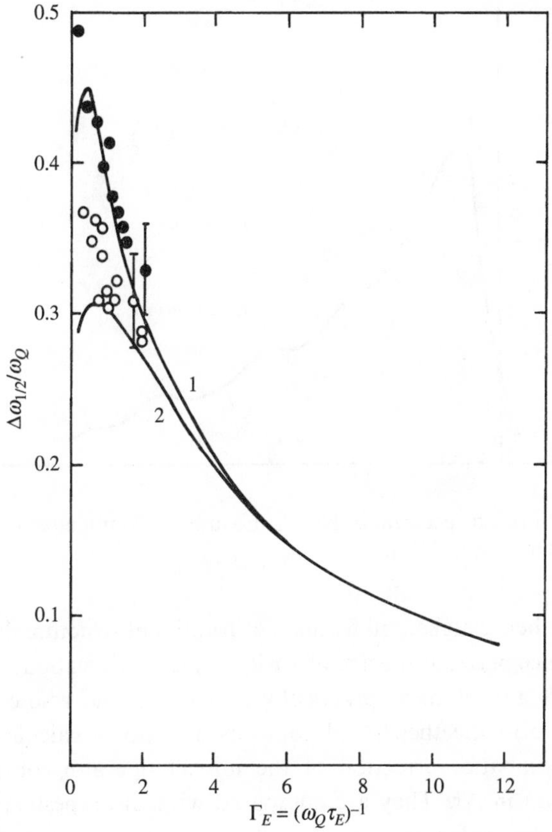

Fig. 5.12. Q-branch narrowing in classical J-diffusion theory in strong collision (1) and weak collision (2) models [215]. The widths are taken from experimental spectra shown in Fig. 5.11 for systems CO–He (•) and N_2–Ar (○).

spectral shape at moderate pressures by general formula (4.86). With known rate constants, calculation of the spectral shape may be performed numerically at any pressure. The spectrum shown in Fig. 5.13 was obtained in [215] by summation of *ortho* and *para* modifications of N_2 taken with proper weights. As can be seen, at argon pressure $p_0 = 10$ atm a high-frequency wing is still resolved in contrast to the pure nitrogen spectrum at the same pressure presented by curve 1 in Fig. 5.11(*b*). It may be a direct indication of higher efficiency of spectral exchange for N_2–N_2 collisions due to supplementary R–R energy transfer, which does not occur for N_2–Ar collisions.

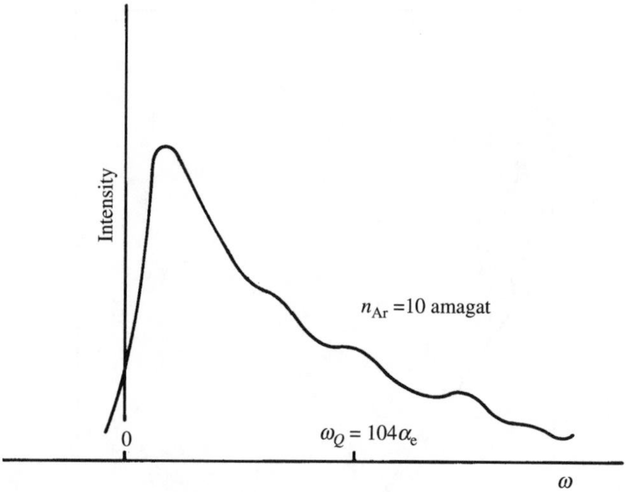

Fig. 5.13. The Q-branch spectrum in N_2–Ar mixture at room temperature calculated with quantum theory [215].

The pressure being higher, all features of rotational structure disappear and the difference between spectra of various spin modifications becomes so smooth that any of them practically reproduces the whole contour shape. In Fig. 5.14 the theoretical contours are shown calculated with and without adiabatic correction of the impact operator for an ideal nitrogen solution in Ar. They are compared with the experimental one related to the same value of

$$\frac{1}{\tau_E} = \sigma_E^{(0)}\langle v\rangle n_0 + \sigma_E^{(1)}\langle v\rangle n_1 = 2\sigma_J^{(0)}\langle v\rangle n^*,$$

where indices 0 and 1 indicate Ar and N_2 components as before. It was in fact assumed in [230] that 'equivalent' buffer gas concentration in mixtures may be determined as $n^* = n_0 + (\sigma_J^{(1)}/\sigma_J^{(0)})n_1$. Consequently, not only is the purely non-adiabatic J-diffusion model supposed to be valid but also collisions with nitrogen and argon are assumed to be equally weak:

$$\sigma_E^{(0)}/\sigma_J^{(0)} = \sigma_E^{(1)}/\sigma_J^{(1)} = 1 + \gamma \approx 2\,.$$

Such a rough comparison of real mixtures with ideal solutions is definitely not perfect but it allows the authors of [230] to proceed using conventional theory. The general conclusion following this comparison is that the quantum J-diffusion model just slightly differs from its

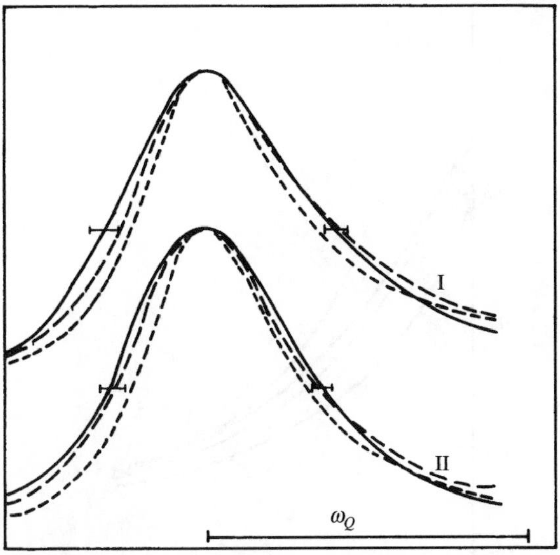

Fig. 5.14. Quantum calculations of the spectral shape with (- - -) and without (— — —) adiabatic corrections for argon density $n_0 = 40$ amagat (I) and $n_0 = 130$ amagat (II). Corresponding experimental contours (solid lines) are taken from [230].

classical analogue, which practically coincides with experiment. Being adiabatically corrected, it leads to a noticeably narrower contour.

The same is clearly seen from Fig. 5.15, which shows different theoretical approaches to Q-branch narrowing. Quantum theory allows one to calculate *ab initio* both the spectral width $\Delta\omega_{1/2}$ and the energy relaxation rate $1/\tau_E$ linear in gas density. The latter is excluded in Fig. 5.15, which plots dimensionless width $\Delta\omega_{1/2}/\omega_Q$ against $\Gamma = 1/\omega_Q\tau_E$. It is this dependence, universal in the classical J-diffusion model, that is a basis of its experimental verification (see Chapter 3). There is nothing surprising in the coincidence of purely non-adiabatic quantum theory with the weak collision limit of classical J-diffusion because an average change of rotational energy in collisions (5.53) was shown to be small (Fig. 5.3). The pronounced difference is obtained only with adiabatically corrected quantum theory. There is no question that such a correction is necessary even though it makes worse the agreement with experiment.

It is remarkable that, after correcting the theory, the difference between its result and the fast modulation asymptotic is preserved for very high densities. This difference is an indication that the ultimate pertur-

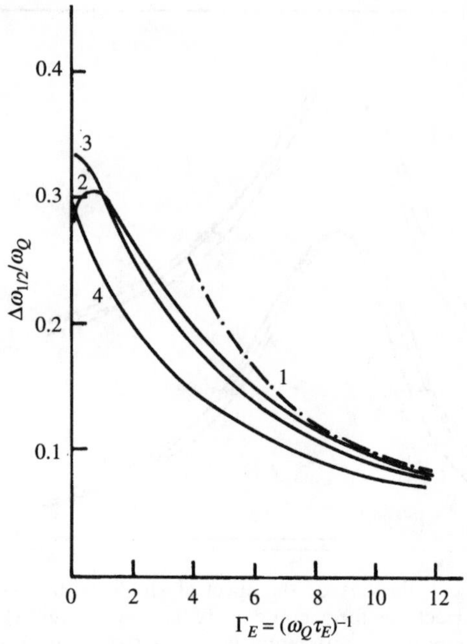

Fig. 5.15. Theoretical dependences of HWHM on the rate of rotational energy relaxation: perturbation theory asymptotics (1), classical weak-collision *J*-diffusion model (2), quantum theory without (3) and with (4) adiabatic correction.

bation theory which leads to Eq. (3.16) or Eq. (4.85) becomes valid at significantly larger Γ. At moderate densities $1/\tau_E$ may not be used as a universal parameter to characterize the rate of exchange for any part of the spectrum. Whereas the front spectral components have already collapsed, those of the high-frequency wing, engaged in comparably weak exchange, collapse later. Therefore effective frequency dispersion of the spectral centrum involved in collapse at this stage is less than the real one. As a result the collapsed width is also smaller than in the true perturbation theory limit, when all the lines have already participated in collapse.

5.7 The fitting laws

Direct calculation of the whole matrix of rate constants is a rather difficult problem, even if the intermolecular potential is well known. Actually, it was done only once for a N_2–Ar mixture in the semiclassical centrifugal sudden approximation [191, 215]. To deal with other systems a lot of the 'fitting laws' were proposed instead. The aim was to describe the resolved Q-branch phenomenologically, eluding *ab initio* calculations of rate constant matrix $k_{jj'}$. If all the constants are considered as unknown, the number of parameters to be determined greatly exceeds that which experiment is really able to present. This leads to the necessity of introducing a 'fitting law', which must approximate the dependence of $k_{jj'}$ on quantum numbers disposing of only few parameters. The unknown parameters are usually adjusted to obtain the best fit to experimental halfwidths Γ_j of Q-branch rotational components equalizing them to nk_{jj} [231–237].

Alternatively one may start from the known set of $k_{jj'}$ and try to approximate it by a suitable fitting law with minimum free parameters. This was done with the *ab initio* calculated matrix $k_{jj'}$ for a N_2–Ar mixture for a broad range of temperatures [215]. The following fitting law for $j > j'$ was proposed:

$$k_{jj'} = a(2j' + 1)\mathrm{e}^{-c(j^2-j'^2)^{\frac{1}{2}}(j+j')} \sum_J (2J+1) \begin{pmatrix} j & j' & J \\ 0 & 0 & 0 \end{pmatrix}^2 \mathrm{e}^{-bJ-b_1J^2} . \tag{5.66}$$

Here $a(T)$ sets the characteristic value of the rate constants, $b(T)$ and $b_1(T)$ determine the strength of collisions and $c(T)$ takes adiabaticity into consideration. For $j \to 0$ transitions a rather simple expression appears:

$$k_{j0} = a(T)\ \exp\{-b(T)j - [b_1(T) + c(T)]j^2\} . \tag{5.67}$$

Possessing a set of k_{j0} (theoretical or experimental) one can find a, b and $b_1 + c$ approximating the dependence $\ln k_{j\to 0}$ by a parabola. On the other hand, the high-j asymptotic behaviour of $\ln k_{j\to j-\Delta}$ with Δ fixed, considered as a function of $(j^2 - j'^2)^{\frac{1}{2}}(j + j')$, is a straight line, according to Eq. (5.66). In the non-adiabatic limit ($c = 0$) the line is horizontal but generally its slope allows one to estimate the adiabatic parameter $c(T)$. The exponential adiabatic correction introduced in Eq. (5.66) is physically more appropriate to real collisions than the power law assumed in [232]. The latter is acceptable for small Massey parameter $\Delta\omega_{jj'}\,\tau_c$ but not to its higher values. This extension is rather voluntary as it implies abrupt change of $V_k(r(t))$ during collision (like $\exp(-|t|/\tau_c)$. Such perturbation is non-adiabatic in principle and does not actually exist.

From SCS calculated $k_{jj'}$, the following parameters were extracted for N_2–Ar at 300 K: $a = 6.10$ Å^2 $\bar{v}$, $b = 0.19$, $b_1 = 0.018$, $c = 0.0016$,

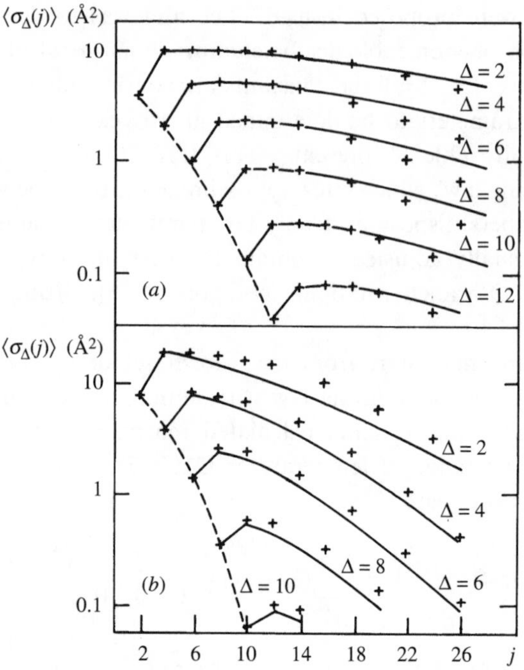

Fig. 5.16. Effective cross-sections $\langle\sigma_\Delta(j)\rangle = k_{j\to j-\Delta}/\bar{v}$ for nitrogen in argon at (*a*) T=300 K, (*b*) T=100 K: SCS calculated (+) and approximated by fitting law (solid line).

and at 100 K: $a = 11.6$ Å^2 $\bar{v}$, $b = 0.11$, $b_1 = 0.036$, $c = 0.0053$. It is noteworthy that adiabatic parameter $c(T)$ is extracted with the largest error. It is shown in Fig. 5.16 how the fitting law (5.66) with abovementioned parameters approximates the original SCS data as functions of j and Δ. Since agreement is rather good and the physical background is fairly reasonable, application of this fitting law to other gases seems to be promising.

The simple fitting procedure is especially useful in the case of sophisticated nonlinear spectroscopy such as time domain CARS [238]. The very rough though popular 'strong collision' model is often used in an attempt to reproduce the shape of pulse response in CARS [239]. Even if it is successful, information obtained in this way is not useful. When the fitting law is used instead, both the finite strength of collisions and their adiabaticity are properly taken into account. A comparison of

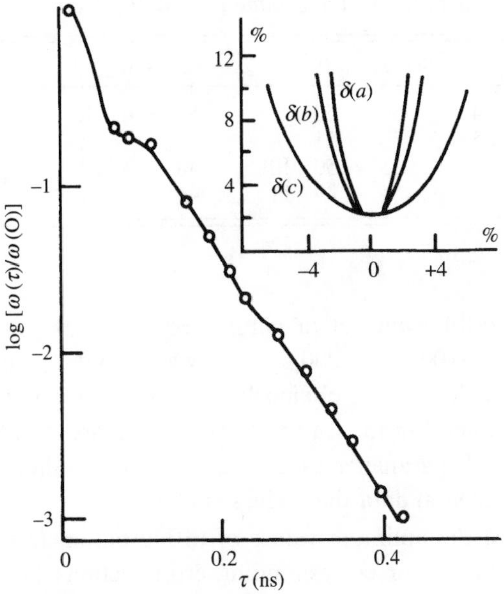

Fig. 5.17. Time domain CARS of nitrogen under normal conditions. Points designate experimental data, solid line calculation with $a = 6.0$ Å, $b = 0.024$, $c = 0.0015$. The insert depicts the dependences of the relative mean-square deviation on each of the parameters a, b and c, the other two being fixed at their optimum values. The deviations are expressed as percentage of optimum parameters.

theoretical prediction with experimental data yields fitting parameters, which, in fact, restore the whole matrix of rate constants.

This programme realized in [237] was based on a simplified fitting law obtained by asymptotic summation of that proposed in [234] provided $\langle\sigma_{j0}\rangle \propto \exp\,[-bj(j+1)]$. The matrix of the effective cross-sections $\langle\sigma_{jj'}\rangle = k_{jj'}/\bar{v}$ acquires the form (for $j > j'$)

$$\langle\sigma_{jj'}\rangle = a(2j'+1)\exp\,[-c(\epsilon-\epsilon')^{\frac{1}{2}}(\sqrt{\epsilon}+\sqrt{\epsilon'})-b(\epsilon+\epsilon')]\; I_o(2b(\epsilon\epsilon')^{\frac{1}{2}}), \quad (5.68)$$

where $\epsilon = j(j+1)$, $\epsilon' = j'(j'+1)$ and I_o is the modified Bessel function. Here the j-dependence is determined only by two parameters. One of them, $b(T)$, sets the strength of collisions, and the other, $c(T)$, their adiabaticity. The 'strong collision' model is a limit where $b = \infty$, which is rather unrealistic. The result of simulation of the time domain CARS signal at T=295 K and $n = 2.5 \times 10^{19}$ cm^{-3} is shown in Fig. 5.17. The minimum value of the mean square relative deviation δ is 1.8%. It

Table 5.3. *Rate constants of line broadening in 10^{-3} cm^{-1} amagat^{-1}.*

j	Theory	Exp.	j	Theory	Exp.	j	Theory	Exp.	j	Theory	Exp.
0	76	71	4	56	52	8	52	49	12	44	44
1	66	59	5	56	51	9	50	48	13	42	42
2	60	57	6	55	50	10	48	47	14	40	40
3	57	55	7	54	49	11	46	45	15	38	38

corresponds to the following set of parameters: $a = 6.0 \pm 0.3$ Å^2, $b = 0.024 \pm 0.003$, $c = 0.0015 \pm 0.0002$. As b was found to be so small, nitrogen–nitrogen collisions are obviously weak, in accordance with what we inferred in Chapter 3 in the frame of classical theory. Fig. 5.17 shows the accuracy of each parameter estimated with this method, when the accuracy of experimental data themselves is about 10%.

Having obtained the optimum values of fitting parameters the whole matrix of rate constants or corresponding cross-sections $\langle\sigma_{jj'}\rangle = k_{jj'}/\bar{v}$ for N_2–N_2 collisions may be restored [237]. Part of it is shown in Table 5.2. One should be reminded that the diagonal elements $\langle\sigma_{jj}\rangle$ determine the components' widths of the resolved isotropic Raman spectrum. In Table 5.3 their broadening coefficients measured in [240] are compared with the corresponding rate constants obtained from data of Table 5.2. Agreement is quite satisfactory considering that experimental results of different authors agree between themselves only to 5% as in [241] and [234]. Though different authors proposed different fitting laws [194, 231–237], all of them concentrated on the j-dependence of spectral widths (5.55) which is

$$\mathbf{\Gamma}_{jj} = n\bar{v} \sum_{s=-j}^{\infty} \langle\sigma_{j\to j+s}\rangle(1 - \delta_{s0}). \tag{5.69}$$

For homonuclear molecules $s = j' - j$ takes only even values whereas j is even for *para* modification and odd for *ortho* modification of the molecules. With a proper choice of fitting parameters any fitting law reproduces experimental line width rather well. Hence the good fit to their j-dependence may not be considered as a criterion of quality of a fitting law. To discriminate between models it is necessary to gain agreement with experimental data on τ_E or τ'_E, which are much more sensitive to peculiarities of the $k_{jj'}$ matrix. According to Eq. (5.58)

$$\sigma_E' = \frac{1}{2}\beta^2 \sum_{j=0}^{\infty}(2j+1)e^{-\beta j(j+1)} \sum_{s=-j}^{\infty}\langle\sigma_{j\to j+s}\rangle[(2j+1)s+s^2]^2, \tag{5.70}$$

where $\beta = B/kT \ll 1$. This cross-section is expressed via higher moments $\overline{s^2}, \overline{s^3}$ and $\overline{s^4}$ of a distribution $\langle\sigma_{j\to j+s}\rangle/\langle\sigma_{jj}\rangle$ whereas the width (5.68) is just the zero moment of the same distribution. Therefore σ_E' is much more sensitive to multiquantum transitions than the line widths.

Using cross-sections from Table 5.2 the following result was obtained in [237] for $T = 295$ K:

$$\sigma_E' = 12 \pm 5 \text{ Å}^2.$$

Despite the fact that relaxation of rotational energy in nitrogen has already been experimentally studied for nearly 30 years, a reliable value of the cross-section is still not well established. Experiments on absorption of ultrasonic sound give different values in the interval 7.7–12.2 Å^2 [242]. As we have seen already, data obtained in supersonic jets are smaller by a factor two but should be rather carefully compared with bulk data as the velocity distribution in a jet differs from the Maxwellian one. In the contrast, the NMR estimation of $\sigma_J = 30$ Å^2 in [81] brought the authors to the conclusion that $\sigma_E' = 40$ Å^2 in the frame of classical J-diffusion. As the latter is purely nonadiabatic it is natural that the authors of [237] obtained a somewhat lower value by taking into account adiabaticity of collisions by non-zero parameter b in the fitting law.

Slightly different are results obtained with other fitting laws. Most of the laws may be arranged in four groups:

(i) polynomial energy gap law (PEG) [232],
(ii) exponential gap law (EGL) and modified exponential gap law (MEG) [231, 243],
(iii) energy corrected sudden laws: (ECS-P) [235] and (ECS-E) [237],
(iv) statistical power energy gap law (SPEG) [236].

Each fitting law is utterly defined by the excitation rate constants $k_{jj'} = \bar{v}\sigma_{j\to j+s}$ for $s > 0$. Their cross-sections

$$\sigma_{j\to j+s} = \sigma A(j,s)\, \exp\,(-b\Delta E/kT) \tag{5.71}$$

are exponential functions of the transition energy

$$\Delta E = E_{j'} - E_j = B[(2j+1)s + s^2]\,.$$

Here σ and b are considered as fitting parameters depending on temperature. De-excitation rate constants ($s < 0$) are obtained from the detailed balance principle. All fitting laws differ in the pre-exponential factor in Eq. (5.70). In the PEG model

$$A^{\mathrm{PEG}} = (2j + 2s + 1)\rho_j \sum_k C_k(\Delta E)^{-k} \tag{5.72}$$

and $b = 1$. In the SPEG approximation

$$A^{\mathrm{SPEG}} = \left(\frac{B}{\Delta E}\right)^{\delta}, \tag{5.73}$$

where δ and b are variable parameters. For different exponential gap laws

$$A^{\mathrm{EGL}} = 1, \quad A^{\mathrm{MEG}} = \left(\frac{1 + \alpha j(j+1)}{1 + \beta j(j+1)}\right)^2 \tag{5.74}$$

not only b but also α and β are fitting parameters (proportional to $1/T$). Dynamically based fitting laws following from the ECS approximation use

$$A^{\mathrm{ECS}} = (2j + 2s + 1)\sum_L (2L + 1)\begin{pmatrix} j & L & j+s \\ 0 & 0 & 0 \end{pmatrix}^2 \left(\frac{\Omega_{j+s}}{\Omega_L}\right)^2 A_{L\to 0}\,. \tag{5.75}$$

where Ω-factors are adiabatic corrections introduced in [235] and basis vector $A_{L\to 0}$ is assumed to yield either a power law (ECS-P)

$$A^{P}_{L\to 0} = [L(L+1)]^{-\gamma} \tag{5.76}$$

or an exponential law (ECS-E)

$$A^{E}_{L\to 0} = \exp\left[-\Theta L(L+1)\right]. \tag{5.77}$$

In both cases $b = 1$. The most sophisticated fitting law of this kind, known as ECS-EP [244], uses the advantages of both exponential and polynomial modelling and has five fitting parameters. As was shown in [245] it is no better than three-parameter SPEG within experimental accuracy.

The results of energy transfer cross-section calculations using the above fitting laws [246] are given in Table 5.4 together with experimental data and classical trajectory (CT) calculations with non-empirical potential energy surface. The fitting parameters are taken as being the same as in original articles referred to in Table 5.4. According to the degree of consistency with experimental data the PEG model is the worst, whereas

Table 5.4. *Rotational cross-sections for N_2 near 295 K in* $\AA^2$. $X = (\sigma'_E - \sigma_E)/\sigma_E$.

Source	Reference	σ_E	σ'_E	X
PEG	[232]	34.5	34.9	0.01
EGL	[231]	16.7	17.9	0.07
MEG	[243]	12.9	15.2	0.18
ECS-P	[235]	12.3	12.4	0.01
ECS-E	[237]	10.8	12.6	0.17
SPEG	[245]	10.6	12.7	0.20
CT calculation	[248]	7.5	8.9	0.18
Acoustic	[242]	7.6 $\pm$0.8		
Free jet	[248]	7		
Free jet	[249]	10		

the SPEG model is the best. It is not surprising that PEG and EGL, which are just statistical models, are worse than the others, which are both dynamically and statistically based laws. The success of dynamically based fitting laws is rather natural since the applicability conditions for ECS approximation are well satisfied for nitrogen [191]. MEG is not much worse because it was actually derived from ECS approximation [231, 243]. However, SPEG is the best even between energy corrected laws and it is the only one whose parameters have been adjusted in [245] not only to the set of line widths $\Gamma_{jj} = n\sum_{j'\neq j} k_{jj'}$ but also to a few rate constants $k_{jj'}$ directly measured in [247]. Combination of two fitting procedures has permitted minimization of the uncertainty to $\delta = 0.11 \pm 0.03$. When only line widths were used, the uncertainty was even greater than the value of δ itself.

As can be seen from the above, the shape of the resolved rotational structure is well described when the parameters of the fitting law were chosen from the best fit to experiment. The values of σ_E presented in Table 5.4 were estimated from the rotational width of the collapsed Q-branch $\omega_Q^2\tau_E$. Therefore the models giving the same σ_E are indistinguishable in both low-density and high-density limits. One may hope to discriminate between them only in the intermediate range of densities where the spectrum is unresolved but has not yet collapsed. The spectral shape in this range may be calculated only numerically from Eq. (4.86) with impact operator $\Gamma^0_{jj'}$ linear in n. Of course, it implies that binary theory is still valid and that vibrational dephasing is not yet

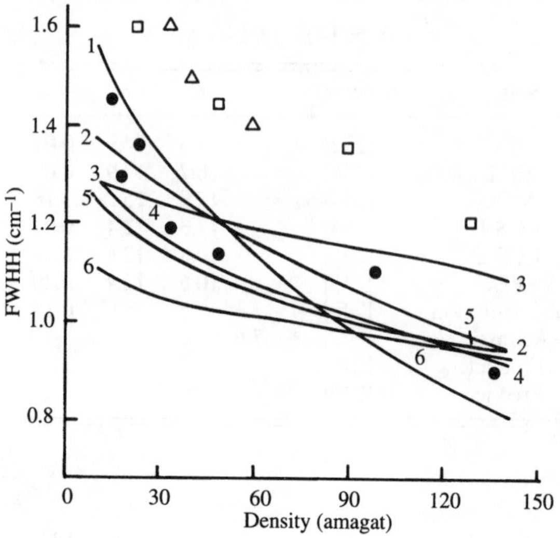

Fig. 5.18. Line width calculated with PEG (1), EGL (2), ECS-P (3), MEG (4), SPEG (5), ECS-E (6) [246] in comparison with old experimental data [89] (□), [250] (△) and rather new data [232], [235] (•).

comparable with rotational broadening. This is exactly the interval where experimental data are grouped in Fig. 5.12 ($\alpha_e/\omega_Q < \Gamma_E < 2$).

For nitrogen this interval is the pressure range between 1 and 150 amagat shown in Fig. 5.18. The major problem is uncertainty of experimental data at these densities, which is greater than the discrepancy between all fitting law predictions. The old data are 1.5 times greater than relatively new ones whereas, at high pressures, they are almost the same. As a result the old data exhibit steeper narrowing than the new ones. In this respect the PEG model fits better to the old data over the whole density region. The most modern data (between 400 and 800 amagat) presented in Fig. 5.19 were also found to be in better agreement with the PEG fitting law than with other models fitted to them (ECS-EP, MEG and ECS-P being the worst) [251]. The PEG model is purely non-adiabatic and has the largest energy relaxation cross-section (34.5 Å), i.e. the strongest spectral exchange, which efficiently narrows the spectrum. It is remarkable that this cross-section is almost the same as that found in [135] (39 Å) when the classical J-diffusion model (also non-adiabatic) was used to fit to the same data (see Chapter 3).

However, we are now inclined to trust more the new low-pressure

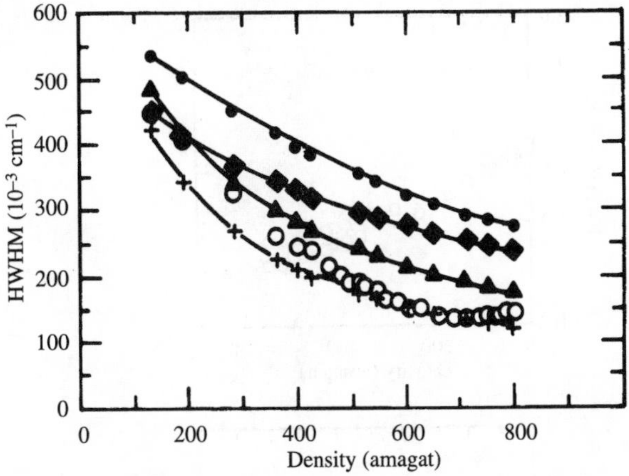

Fig. 5.19. Experimental line width and calculated line widths predicted by the fitting laws in binary collision approximation [251]: (○) experimental; (+) PEG; (▲) ECS-EP; (◆) MEG; (●) ECS-P.

data. They agree better with other moderately adiabatic models and give more reasonable energy relaxation cross-sections, presented in Table 5.4. One must not attach importance to their failure to describe high-density data in Fig. 5.19. At these densities the impact operator linear in n, used in fitting laws, significantly underestimates the rate of collisions, producing spectral exchange. Therefore the rotational width of collapsed spectra in the binary approximation may be greater than in reality. Some new CARS data on nitrogen obtained in [14] confirm this conclusion. Although appropriate at low pressures, the MEG model significantly overestimates the spectral width above 300 bar (Fig. 5.20). As a result CARS temperature measurements at such pressures are in error by −150 K at 295 K when the binary MEG model is used with temperature as a variable parameter. Let us note that numerous CARS investigations of the Q-branch of nitrogen over a broad range of temperatures have aimed at employing N_2 as a probe for combustion diagnostics. Below 200 amagat agreement between measured and actual temperature within 20 K was found in [14].

It was underlined twice in [251] that 'the use of fitting laws ties in with the validity of the impact theory, and the present results have to be interpreted cautiously'. At high pressure nonlinear increase of rates of energy relaxation and vibrational dephasing with gas density must be

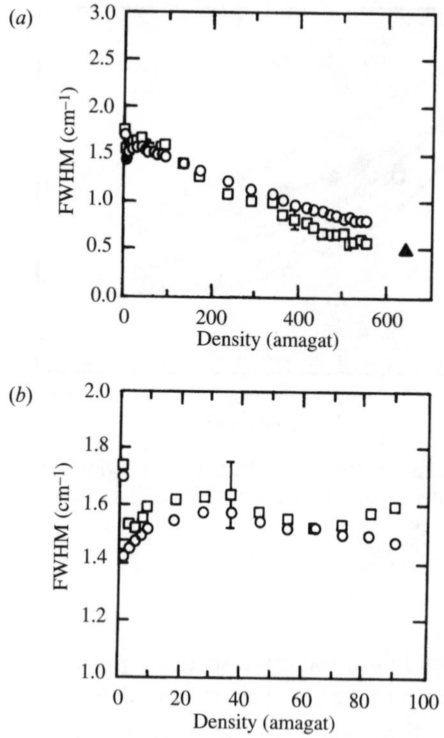

Fig. 5.20. Full width of nitrogen Q-branch CARS spectra measured at 295 K versus densities (squares) and calculated width using the MEG law (circles) [14]. Shown also are the error bar and the width measured in liquid nitrogen (triangle). (*a*) Density range up to 700 amagat. (*b*) Density range up to 100 amagat showing part of Fig. 5.20(*a*) in more detail

taken into account in some way. For instance, by substituting $g(\eta)n$ for n, as was done in Eq. (1.119), one may hope that this estimate at room temperature works better than it did for liquid nitrogen (see Fig. 1.23). Even so, it is inconsistent to account for attraction in a fitting law and then completely ignore it when calculating collision frequency in the hard sphere approximation. The authors of [251] found a better way using a modified Schweizer–Chandler model, which takes into account both attractive forces and the soft-core, hard sphere approximation. Although this model is very sensitive to the choice of hard sphere diameter (the quality of the fit depends critically upon varying this parameter ± 0.01 nm [252]) it describes semiquantitatively the interplay between rotational

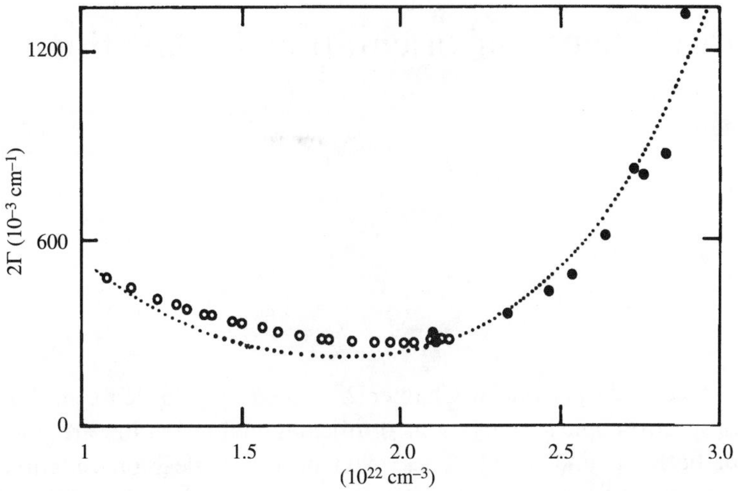

Fig. 5.21. Isotropic Q-branch width as a function of density: (∘) experimental data from [251], (•) experimental data from [253], (· · · ·) theoretical curve [251].

and vibrational contributions in the width of the collapsed spectrum (Fig. 5.21):

$$\Gamma = \omega_Q^2 \tau_E(n) + \Gamma_v(n) . \tag{5.78}$$

Residual rotational broadening dominates in the density range 400–800 amagat but at higher densities studied in [252, 253] (800–1000 amagat) vibrational dephasing prevails. Since the cross-section of dephasing increases with temperature, when it is high enough [203] the contribution of vibrational broadening to the width (5.77) also increases at any densities. That is why at 700 K the actual width may be significantly greater not only than the rotational contribution to Γ but even than the MEG overestimation of this contribution [14]. As was shown in Chapter 3, for some other molecules dephasing becomes important even at lower pressures (see Fig. 3.9 and Fig. 3.13). Ignorance of the dephasing contribution in Q-branch broadening is a common demerit of all the fitting laws discussed above.

6

Impact theory of orientational relaxation

Debye's theory, considered in Chapter 2, applies only to dense media, whereas spectroscopic investigations of orientational relaxation are possible for both gas and liquid. These data provide a clear presentation of the transformation of spectra during condensation of the medium (see Fig. 0.1 and Fig. 0.2). In order to describe this phenomenon, at least qualitatively, one should employ impact theory. The first reason for this is that it is able to describe correctly the shape of static spectra, corresponding to free rotation, and their impact broadening at low pressures. The second (and main) reason is that impact theory can reproduce spectral collapse and subsequent pressure narrowing while proceeding to the Debye limit.

The above capabilities of impact theory are illustrated in preceding chapters by consideration of the isotropic scattering spectrum, which consists of one Q-branch. The peculiarity of the present problem is that in the spectrum of orientational relaxation there are always several branches, and, generally speaking, one cannot consider their transformation independently. The very first attempt [2] to build a quasi-classical impact theory of rotational structure drew one's attention to this fact as being of principal importance. It made the theory similar to the quantum theory of unresolved atomic spectra [254, 181, 255], whose Stark or Zeeman components interfere with each other during collisions. Interference of the same nature takes place between rotational branches of vibrational spectra, described classically. Increase of collisional frequency causes spectral collapse, but very rarely does an atomic spectrum narrow afterwards [176, 256]. In contrast, for molecular spectra the case of narrowing is the most frequent one. The necessary condition for non-adiabaticity of collisions $\bar{\omega}\tau_c \ll 1$ is at the same time a sufficient condition for narrowing of a rotational spectrum after collapse.

The rotational phase shift δ, which cannot exceed a mean angle of a molecular rotation during collisional time ($\bar{\omega}\tau_c$), is certainly small in the case of non-adiabatic collisions. This condition is exactly that needed for anisotropic scattering (or IR absorption) spectrum narrowing, just as vibrational dephasing must be weak for an isotropic spectrum to narrow.

In Chapter 2 we did not allow the molecular axis to turn during collisions, assuming $\delta = 0$, as is usually done in various models of J-diffusion [31, 29]. We retain this simplifying assumption since the high-frequency spectral wings, affected by m-diffusion, do not contribute to collapse, which is taking place in the central part of the spectrum. Assuming that their role is negligible, we shall consider the whole spectrum as involved in j-diffusion to an extent determined by the Keilson–Storer kernel.

Firstly, we are going to demonstrate how branch interference may be taken into account within the quasi-classical impact theory. Then we shall analyse a quasi-static case, when the exchange frequency between branches is relatively small. An alternative case, when exchange is intensive and the spectrum collapses, has been already considered in Chapter 2. Now it will be shown how the quasi-static spectrum narrows with intensification of exchange. The models of weak and strong collisions will be compared with each other and with experimental data. Finally, the mutual agreement of various theoretical approaches to the problem will be considered.

6.1 Interference of rotational branches in molecular spectra

According to Eq. (2.13), the spectra we are interested in are given by a Fourier transform of the orientational correlation functions of the corresponding order ℓ. Similarly to what was done in Chapter 3, the correlation functions for linear and spherical molecules may be represented as a superposition of the partial (marginal) components

$$K_\ell(t) = \int_0^\infty K_\ell(t, J)\, \mathrm{d}J\,. \tag{6.1}$$

Physical substantiation of this relationship is discussed in Appendix 6. Here, it is enough to mention that

$$K_\ell(t, J) = \sum_{q=-\ell}^{\ell} K_q^\ell(t, J)\varphi_B(J) = \sum_{q=-\ell}^{\ell} (-1)^q d_q^\ell(t, J) d_{-q}^\ell(0) \tag{6.2}$$

consists of correlation functions, corresponding to the Q-branch of the spectrum. Any of these functions is a multiplication of d_q^ℓ. One of them depends on time and should satisfy the initial condition, identical to Eq. (3.27):

$$d_q^\ell(0,J) = d_q^\ell(0)\varphi_B(J) = D_{oq}^\ell\left(0,-\frac{\pi}{2},0\right)\varphi_B(J)\,. \tag{6.3}$$

Time evolution of this component, as is proven in Appendix 7, is described by the kinetic equation of impact theory, which is a generalization of Eq. (3.26):

$$\frac{\partial}{\partial t}d_q^\ell = \mathrm{i}\Omega_q d_q^\ell - \frac{1}{\tau_0}d_q^\ell + \frac{1}{\tau_0}\sum_{q'=-\ell}^{\ell}\int \mathrm{d}J'\;\Phi_{q'q}^\ell(J',J)d_{q'}^\ell(t,J')\,. \tag{6.4}$$

Its first term corresponds to free rotation with a given J. A frequency

$$\Omega_q = \alpha J^2 + qJ/I \tag{6.5}$$

determines the position of spectral components in all branches of a static spectrum (herein for linear molecules). The structure of the integral term is something other than that used in Chapter 3. The kernel of this term is a matrix of rank $2l+1$, which controls both frequency exchange and exchange of branches with different index q:

$$\Phi_{q'q}^\ell\left(J',J\right) = \int f\left(J',J,\alpha\right)D_{q'q}^\ell\left(-\frac{\pi}{2},\alpha,\frac{\pi}{2}\right)\,\mathrm{d}\alpha\,. \tag{6.6}$$

As is shown in Appendix 6, $f(J',J,\alpha)$ determines the probability of angular momentum becoming equal to $J = |\boldsymbol{J}|$ after collision and being declined by angle α from the preceding value $\boldsymbol{J}'$. $D_{q'q}^\ell$ is the Wigner matrix, which turns axis z, connected with $\boldsymbol{J}$, by angle α when the molecular axis, directed along x, does not shift during collision ($\delta = 0$). Since this matrix is off-diagonal, all d_q^ℓ vary jointly, and Eq. (6.4) is a system of coupled integro-differential equations.

In order to make this system more concrete, let us employ the Keilson–Storer model for definition of its kernel

$$f\left(\boldsymbol{J}-\gamma\boldsymbol{J}'\right) = \prod_{i=1}^{2} f\left(J_i-\gamma J_i'\right) = f\left(J',\varphi';J,\varphi\right)\,. \tag{6.7}$$

In this case, it is expressed via polar coordinates of the angular momentum vector in a plane, perpendicular to the diatomic molecular axis. The function, which is present in (6.6), can be determined by (6.7) as follows:

$$f\left(J',J,\alpha\right) = \int_0^{2\pi}\mathrm{d}\varphi\int_0^{2\pi}\mathrm{d}\varphi'\;\delta\left(\varphi-\varphi'-\alpha\right)f\left(J',\varphi';J,\varphi\right)\,.$$

Because of space isotropy $f(J',\varphi';J,\varphi) = f(J',J;\varphi-\varphi')$. Taking this into account, one can easily find the precise form:

$$f\left(J',J,\alpha\right) = \frac{J}{\pi d(1-\gamma^2)} \exp\left(-\frac{J^2+\gamma^2 J'^2 - 2\gamma J'J\cos\alpha}{2d\,(1-\gamma^2)}\right). \qquad (6.8)$$

Using it, kernel (6.6) can be calculated, provided turning of $\boldsymbol{J}$ is performed by the succession of rotations:

$$\Phi^{\ell}_{q'q}\left(J',J\right) = \sum_{q''} \int D^{\ell}_{q'q''}\left(0,\frac{\pi}{2},0\right) D^{\ell}_{q''q}\left(0,-\frac{\pi}{2},0\right) f\left(J',J,\alpha\right) \mathrm{e}^{-iq''\alpha}\,\mathrm{d}\alpha\,. \qquad (6.9)$$

Since $f(J',J,\alpha)$ is even over α, one must only find the following integral:

$$\int_0^{\pi} \exp\left(\frac{\gamma J'J\cos\alpha}{d\,(1-\gamma^2)}\right) \cos\left(q''\alpha\right)\,\mathrm{d}\alpha = \pi\, I_{q''}\left(\frac{\gamma J'J}{d\,(1-\gamma^2)}\right),$$

where $I_{q''}(...)$ is a modified Bessel function [37]. Finally we have

$$\begin{aligned}\Phi^{\ell}_{q'q}\left(J',J\right) &= \sum_{q''} D^{\ell}_{q'q''}\left(0,\frac{\pi}{2},0\right) D^{\ell}_{q''q}\left(0,-\frac{\pi}{2},0\right) \frac{J}{d(1-\gamma^2)} \\ &\qquad \exp\left(-\frac{J^2+\gamma^2 J'^2}{2d\,(1-\gamma^2)}\right) \times I_{q''}\left(\frac{\gamma J'J}{d\,(1-\gamma^2)}\right). \qquad (6.10)\end{aligned}$$

With $\ell = 0$ the present expression reduces to the result obtained in Eq. (3.28). If, e.g., $\ell = 2$, then spectral exchange takes place both within the branches of an isotropic scattering spectrum (Fig. 6.1) and between them. The latter type of exchange is conditioned by collisional reorientation of the rotational plane, whose position is determined by angle α. As a result, the intensity of adsorbed or scattered light is redistributed between branches. In other words, exchange between the branches causes amplitude modulation of the individual spectral component, which accompanies the frequency modulation due to change of rotational velocity.

When accelerated sufficiently, amplitude–frequency modulation in the absence of dephasing results in signal monochromatization, just like in the case of pure frequency modulation. Before the spectrum collapses, exchange between branches causes their broadening, but after collapse it provides their coalescence into a single line at frequency

$$\langle\Omega\rangle = (2\ell+1)^{-1} \sum_{q=-\ell}^{\ell} \langle\Omega_q\rangle = \omega_Q\,. \qquad (6.11)$$

Narrowing of this quasi-Lorenzian line with increase of density has been

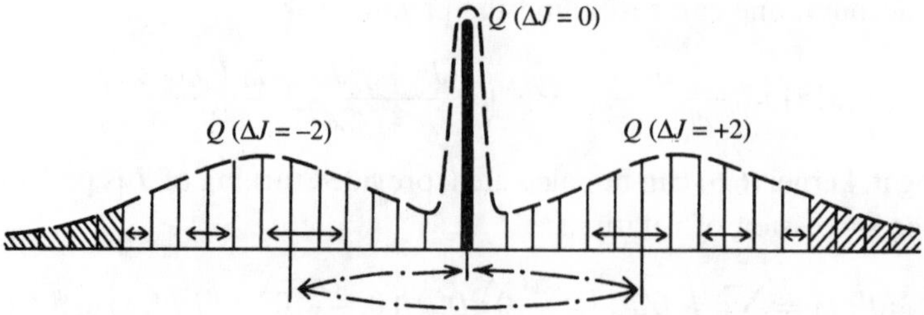

Fig. 6.1. A spectral exchange scheme between components of the rotational structure of an anisotropic Raman spectrum of linear molecules. The adiabatic part of the spectrum is shadowed. For the remaining part the various spectral exchange channels are shown: (← · →) between branches; (⟷) within branches.

considered in Chapter 2, and spectral broadening, which precedes it, is discussed in the next section.

6.2 The quasi-static approximation

Exchange between branches leads to frequency migration with a step approximately equal to their splitting. Its mean value is

$$\langle|\Omega_{q+\ell} - \Omega_q|\rangle = \langle J\rangle\, \ell/I \approx \ell(k\,T/I)^{\frac{1}{2}}\,. \qquad (6.12)$$

As long as the collisional frequency measured by $1/\tau_J$ is less than this value, i.e.

$$1/\tau_J \ll (k\,T/I)^{\frac{1}{2}} = \bar{\omega}\,, \qquad (6.13)$$

the shape of broad side branches varies just slightly, while the narrow central Q-branch broadens with increasing density (Fig. 6.2). Comparison of Eq. (6.13) and Eq. (2.30) shows that quasi-static transformation of the spectrum precedes its collapse. At this stage, broadening of its central part due to exchange between branches favours their coalescence.

In order to describe quantitatively the Q-branch transformation at this stage, one should use the 'secular simplification' of the problem [133, 257]. It neglects completely the off-diagonal elements of the kernel of integral operator (6.9), i.e. terms taking into account transfer from the other branches in equation (6.4):

$$\frac{\partial}{\partial t}\, d_0^\ell\,(t,J) = \mathrm{i}\,\alpha J^2 d_0^\ell(t,J) - \frac{1}{\tau_0} d_0^\ell(t,J) + \frac{1}{\tau_0}\int \mathrm{d}J'\; \Phi_{00}^\ell(J',J) d_0^\ell(t,J')\,. \qquad (6.14)$$

This approximation is valid when inequality (6.13) holds, and, of course,

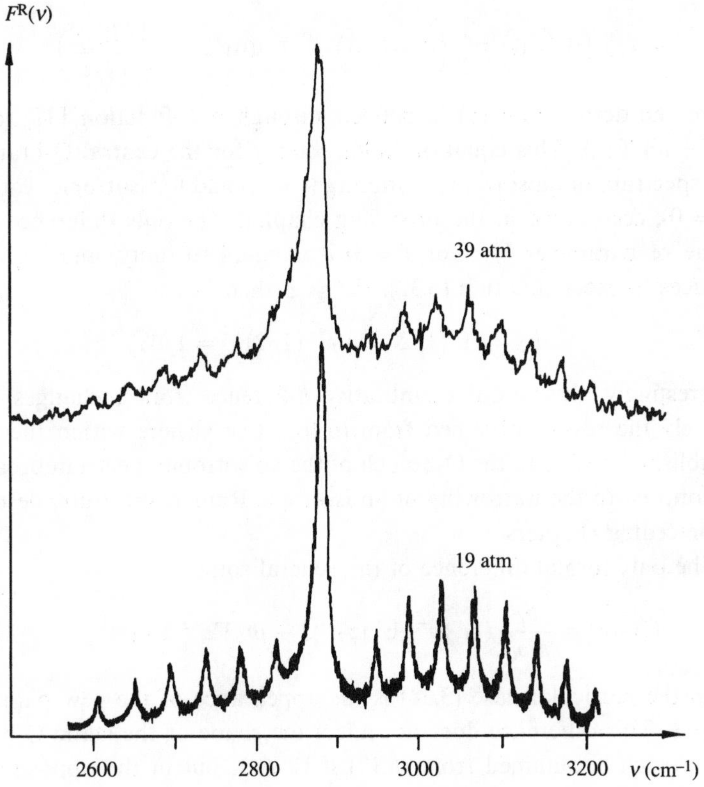

Fig. 6.2. The broadening of the depolarized Raman Q-branch of HCl with increase of density [258].

only when the Q-branch is really present in the orientational spectrum of the system. The analysis of its shape

$$G_\ell^Q(\omega) = \frac{1}{\pi}\,\mathrm{Re}\int_0^\infty \mathrm{e}^{-\mathrm{i}\omega t}\,\mathrm{d}t \int_0^\infty \mathrm{d}J\;\mathrm{d}_0^\ell(t,J)d_0^\ell(0) \qquad (6.15)$$

simplifies it essentially. The variation of shape with density increase is described in the models of both weak and strong collisions.

In the case of strong collisions, the integral part of Eq. (6.14) becomes even simpler and has the form

$$\frac{\partial}{\partial t}d_0^\ell(t,J) = \mathrm{i}\alpha J^2 \mathrm{d}_0^\ell(t,J) - \frac{1}{\tau_J}d_0^\ell(t,J) + \frac{b_\ell}{\tau_J}\varphi_B(J)\int_0^\infty \mathrm{d}J'\,d_0^\ell(t,J), \quad (6.16)$$

where

$$b_\ell = D^\ell_{00}\left(0,\frac{\pi}{2},0\right) D^\ell_{00}\left(0,-\frac{\pi}{2},0\right) = (P_\ell(0))^2 = \left(\frac{(1/2)_{\ell/2}}{(\ell/2)\,!}\right)^2 . \quad (6.17)$$

Here the designation (α) is defined through a Γ-function [37] $(\alpha)_m = \Gamma(\alpha+m)/\Gamma(\alpha)$. This equation holds equally for the central Q-branch in the spectrum of anisotropic scattering ($\ell = 2$) and for isotropic scattering ($\ell = 0$), considered in the preceding chapter. The only difference is the value of parameter b_ℓ. For $\ell = 0$ it is equal to unity, and Eq. (6.16) reduces to exact equation (3.32). If $\ell = 2$, then

$$b_2 = \left[\Gamma\left(1/2+1\right)/\Gamma\left(1/2\right)\right]^2 = 1/4. \quad (6.18)$$

Interestingly, this minor quantitative difference from 1 changes qualitatively the results obtained from (6.16). Everywhere within the limits established by (6.13), the Q-branch of the anisotropic spectrum broadens, in contrast to the narrowing of an isotropic Raman spectrum, described in preceding chapters.

The only formal difference of the general solution

$$G^Q_\ell(\omega) = \frac{b_\ell}{\pi\omega_Q} Im\left\{e^{-z}\,\mathrm{Ei}(z)/\left[1+ib_\ell\Gamma e^{-z}\,\mathrm{Ei}(z)\right]\right\} \quad (6.19)$$

from the particular case (3.34) is the appearance of the new parameter $b_\ell \neq 1$. This difference does not affect the shape of the static Q-branch (3.6) which is obtained from (6.19) if $\Gamma \ll 1$, but in the opposite limit $\Gamma \gg 1$, the spectrum transforms into

$$G^Q_\ell(\omega) = \frac{b_\ell}{\pi\omega_Q}\frac{(1-b_\ell)\Gamma+\Gamma/(x^2+\Gamma^2)}{\left[x-1-x/(x^2+\Gamma^2)\right]^2+\left[(1-b_\ell)\Gamma+\Gamma/(x^2+\Gamma^2)\right]^2}. \quad (6.20)$$

In any case, it is symmetrical, and the central part becomes Lorentzian with width

$$\Delta\omega^\ell_{1/2}/\omega_Q = 1/\Gamma + (1-b_\ell)\Gamma \approx (1-b_\ell)\Gamma \quad (6.21)$$

and integral intensity

$$\int_{-\infty}^{\infty} G^Q_\ell(\omega)\,\mathrm{d}\,\omega = \left(d^\ell_0\right)^2 = b_\ell . \quad (6.22)$$

The latter value is equal to 1 for the isotropic spectrum. For the anisotropic one it is only 1/4, because the remaining intensity of the normalized spectrum is distributed between the side branches.

Spectral exchange between branches reduces the residence time in each of them. In the quasi-static approximation, frequency migration may be considered as irreversible, since its return from other branches is neglected. As a result, in accordance with the uncertainty principle, an additional broadening appears (if compared with the isotropic case), which is presented by the second term in Eq. (6.21). One can easily see that narrowing of the isotropic spectrum results from the fact that with $\ell = 0$ ($b_0 = 1$), the second term is absent and Eq. (6.21) reduces to Eq. (3.37). The present case is different, however. At $\Gamma \gg 1$, i.e. when the isotropic line narrows, the anisotropic line broadens. This arises because the first term in Eq. (6.21) is negligible and

$$\Delta \omega_{1/2}^{an} = 3/4\tau_J . \tag{6.23}$$

One may say that this broadening acts in Eq. (6.21) like the dephasing in Eq. (3.49), but in this case it is not small and fully disguises the narrowing action of exchange inside the Q-branch.

Let us demonstrate that the tendency to narrowing never manifests itself before the whole spectrum collapses, i.e. that the broadening of its central part is monotonic until Eq. (6.13) becomes valid. Let us consider quantity $\tau_{\theta,\ell}$, denoting the orientational relaxation time at $\ell = 2$. If rovibrational interaction is taken into account when calculating $K_\ell(t)$ it is necessary to make the definition of $\tau_{\theta,\ell}$ given in Chapter 2 more precise. Collapse of the Q-branch rotational structure at $\Gamma = 1/\omega_Q\tau_J \gg 1$ shifts the centre of the whole spectrum to frequency ω_Q. It must be eliminated by the definition

$$\tau_{\theta,\ell} = \int_0^\infty K_\ell(t) \mathrm{e}^{-\mathrm{i}\omega_Q t}\, \mathrm{d}t = \pi G_\ell\left(\omega_Q\right) \approx \pi G_\ell^Q\left(\omega_Q\right) . \tag{6.24}$$

Variation of the relaxation times subject to $\tau_J^* = 1/\Gamma$ is shown in Fig. 6.3. It is really monotonic, and after the Q-branch collapse it can be described by a very simple formula

$$\tau_{\theta,\ell} = \begin{cases} \left(\omega_Q^2\tau_J\right)^{-1} & \ell = 0 \\ \tau_J/3 & \ell = 2 \end{cases} \quad \text{at } \omega_Q \ll \frac{1}{\tau_J} \ll \bar{\omega} . \tag{6.25}$$

The last result was obtained independently in [27, 269]. In the logarithmic scale of Fig. 6.3 the dependence (6.25) is linear in both cases, but its slope in the isotropic case is opposite to that in the anisotropic case. This difference makes it possible to perform self-consistent verification of the theories. Unfortunately, independent information on τ_J is rather rare. It can be obtained from NMR investigations, or from analysis of the wings of the spectrum (6.20). Since both tasks are rather complex,

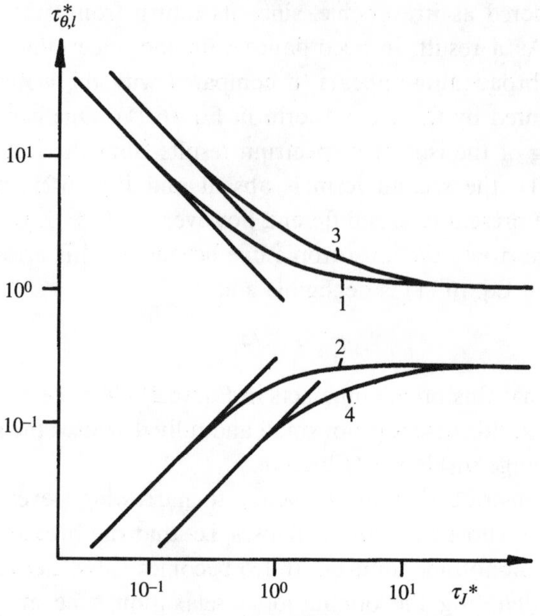

Fig. 6.3. Quasi-static behaviour of relaxation times $\tau_{\theta,2}$ (upper curves) and $\tau_{\theta,1}$ in the case of strong (1,2) and weak (3,4) collisions. The straight lines are the asymptotics of the curves after Q-branch collapse.

we can eliminate τ_J by direct comparison of times $\tau_{\theta,2}$ and $\tau_{\theta,0}$. After symmetrization of the Q-branch in the region of linear behaviour of the curves (Fig. 6.3), the well-known relation follows from Eq. (6.25):

$$\tau_{\theta,2} \cdot \tau_{\theta,0} = (1/3)\,\omega_Q^{-2}. \tag{6.26}$$

Since $\tau_{\theta,2} \propto \tau_J$, it is analogous to the Hubbard relation; however, it is not universal. The numerical coefficient provides information on the strength of collisions: the value 1/3 is peculiar to strong collisions.

Weak collisions are quite different. According to Appendix 4, the kinetic equation corresponding to such collisions has the form:

$$\frac{\partial}{\partial t} d_0^\ell = \mathrm{i}\alpha J^2 d_0^\ell + \frac{1}{\tau_J}\left(1 + \frac{a_\ell d}{J^2}\right) d_0^\ell + \frac{J}{\tau_J}\left(1 - \frac{d}{J^2}\right)\frac{\partial}{\partial J} d_0^\ell + \frac{d}{\tau_J}\frac{\partial^2}{\partial J^2} d_0^\ell. \tag{6.27}$$

It differs from its analogue (3.39) only in numerical parameter

$$a_\ell = 1 - \ell(\ell+1)/2 = \begin{cases} 1 & \ell = 0 \\ -2 & \ell = 2 \end{cases}. \tag{6.28}$$

Since its solution is rather complex, let us restrict ourselves to consideration of a collapsed spectrum at $\Gamma \gg 1$, when it is already symmetrical with a centre shifted to frequency ω_Q=0. As we are interested only in its broadening, we may neglect the rotational structure of the Q-branch in Eq. (6.27) assuming

$$\alpha = 0 . \tag{6.29}$$

Naturally, this simplification is valid only for the anisotropic Q-branch (ℓ=2) while it broadens. In this limit the general formula (A.5.12) of Appendix 5 is simplified by assuming B=1. Using the relation [37]

$${}_2F_1\left(-k, s_2+\frac{5}{4}, 2s_2+\frac{3}{2}, 1\right) = \frac{\Gamma(2s_2+3/2)\,\Gamma(k+s_2+1/4)}{\Gamma(2s_2+3/2+k)\,\Gamma(s_2+1/4)} \tag{6.30}$$

one can obtain the solution in the form of the series

$$\begin{aligned} G_2^{\mathrm{Q}}(\omega) &= \frac{b_2}{\pi\tau_J}\frac{\Gamma^2(s_2+5/4)}{\Gamma(2s_2+3/2)} \\ &\quad \sum_{k=0}^{\infty}\frac{[(s_2+1/4)_k]^2}{k!(2s_2+3/2)_k}\frac{(2s_2+1/2+2k)}{[\omega^2+(2s_2+1/2+2k)^2/\tau_J^2]} . \end{aligned} \tag{6.31}$$

The solution is a sum of Lorentzian curves, centred at the same frequency, but having different widths. The total width of this spectrum, as well as the width of each of its components, increases linearly with $1/\tau_J$ [257]

$$\Delta\omega_{1/2}^{\mathrm{an}} \approx 2/\tau_J . \tag{6.32}$$

The orientational relaxation time, according to (6.24), (6.29) and (6.31), is determined by the formula derived in [85]:

$$\tau_{\theta,2} = \pi G_2^{\mathrm{Q}}(0) = \left(\frac{\tau_J\Gamma^2(s_2+5/4)}{4\Gamma(2s_2+3/2)}\right)\sum_{k=0}^{\infty}\frac{\left[(s_2+1/4)_k\right]^2}{k!(2s_2+3/2)_k\,(2s_2+1/2+2k)} . \tag{6.33}$$

Series (6.33) converges rather slowly, but numerical summation of a large enough number of terms allows one to calculate the coefficient in a direct proportion

$$\tau_{\theta,2} = \tau_J/10.06 . \tag{6.34}$$

This proportion describes the straight part of the corresponding curve in Fig. 6.3.

As can be seen, the difference in behaviour of orientational relaxation times $\tau_{\theta,2}$ in models of weak and strong collisions is manifested more strongly than in the case of isotropic scattering. Relation (6.26) is reproduced in the weak collision limit, but with a different numerical coefficient:

$$\tau_{\theta,2} \cdot \tau_{\theta,0} = \frac{1}{5.03\ \omega_{\mathrm{Q}}^2} . \tag{6.35}$$

By measuring the value of this coefficient, one can judge, at least qualitatively, which model is more realistic. Analogous information can be obtained from comparison of the isotropic spectrum half-width with that of the anisotropic one. To do this, relations similar to (6.26) and (6.35) are essential, because they determine the upper and lower limits for the product of the half-widths [257]:

$$\frac{3}{4} \le \Delta\omega_{1/2}^{\mathrm{is}} \cdot \Delta\omega_{1/2}^{\mathrm{an}} \le 1 . \tag{6.36}$$

The lower boundary corresponds to strong collisions, and the upper one to weak collisions. This conclusion can be confirmed by experiment. According to [259], nitrogen dissolved in SF_6 has a symmetrical spectrum of isotropic scattering, indicating that collapse of the spectrum has already occurred. At the same densities, the Q-branch of the anisotropic spectrum is still well separated from the side branches, and in [259] the lower bound for its half-width is estimated as 5 cm^{-1}. So,

$$\Delta\omega_{1/2}^{\mathrm{is}} \cdot \Delta\omega_{1/2}^{\mathrm{an}} \left(\omega_Q\right)^{-2} \ge 0.5 ,$$

which does not contradict Eq. (6.36).

The non-model relation (6.36) may also be used in a slightly different manner. In [260] (6.36) was checked out for the case of liquid oxygen. The average frequency of the Q-branch was estimated as a shift of the collapsing Q-branch centre relative to a known frequency of the basic vibrational transition. This yielded $v_Q(O_2) \approx 4.6\ \mathrm{cm}^{-1}$. Using the value of $\Delta v_{1/2}^{\mathrm{is}} = 0.23\ \mathrm{cm}^{-1}$ [261] and the measured $\Delta v_{1/2}^{\mathrm{an}} \approx 53\ \mathrm{cm}^{-1}$, the authors of [260] obtained for the ratio in (6.36) 0.6 ± 0.3. On this ground, they concluded that strong collisions of liquid oxygen molecules prevailed. After this they were able to estimate the rotational relaxation time for the liquid oxygen molecule at 77 K as $\tau_J \approx 0.5$ ps.

6.3 Approaches of weak and strong collisions

Pressure induced broadening and narrowing of a whole spectrum are described by the quasi-static approximation and the perturbation theory, correspondingly. Comparing inequalities (6.13) and (2.53) one can see that the border between the stages is determined by the criterion $\bar{\omega}\tau_J \sim 1$, common to both. In its vicinity both approximations become invalid. In order to describe the transition from broadening to narrowing, one has to find a method of solution, which does not contain any restrictions on $\bar{\omega}\tau_J$. Such a method is of special importance for linear dipole molecules, e.g., HCl or CO. The Q-branch is absent in the absorption spectra of such systems, and it is impossible to employ the quasi-static approximation for description of their central part at low pressures (at $\bar{\omega}\tau_J \gg 1$).

In fact, such a method was proposed by Sack in the classical work [99], which was far ahead of its time. This method provides the general solution of Eq. (6.4) in the form of a continuous fraction, which is, however, rather difficult to analyse. In the case of weak collisions, there is no good alternative to this method, but for strong collisions, the solution can be found analytically. Let us first consider this case.

We need only one simplification: neglect of the Q-branch rotational structure. It is conditioned by Eq. (6.29). This assumption restricts the buffer gas densities, which should not be too low:

$$1/\tau_J \gg \omega_Q \,. \tag{6.37}$$

In other words, the following description of the orientational relaxation spectrum is valid at such pressures when the Q-branch has already collapsed and starts to broaden.

Substituting $\alpha = 0$ into (6.5) and $\gamma = 0$ into (6.10) and taking into account that $I_{q''}(0) = \delta_{q''0}$ we get from (6.4) the following equation:

$$\begin{aligned}\frac{\partial}{\partial t} d_0^\ell(t, J) &= \mathrm{i}q \left(\frac{J}{I}\right) d_q^\ell(t, J) - \frac{1}{\tau_J} d_q^\ell(t, J) \\ &+ \frac{1}{\tau_J} \varphi_B(J) \sum_{q'} D_{q'0}^\ell\left(0, \frac{\pi}{2}, 0\right) D_{0q}^\ell\left(0, -\frac{\pi}{2}, 0\right) d_{q'}^\ell(t),\end{aligned} \tag{6.38}$$

where

$$d_q^\ell(t) = \int d_q^\ell(t, J)\, \mathrm{d}J \tag{6.39}$$

is the completely averaged response. The correlation function (6.1) is expressed as

$$K_\ell(t) = \sum_{q=-\ell}^{\ell} (-1)^q d_q^\ell(t) d_{-q}^\ell(0) \,. \tag{6.40}$$

Thus defined, the problem can be solved by the standard method, which was used already in Section 3.4 and in [27]. However, there is a more elegant method, allowing one to obtain the result in the form that is most popular in the literature.

Using properties

$$d_q^\ell(0) = D_{0q}^\ell\left(0,-\frac{\pi}{2},0\right)\;;\; D_{q'0}^\ell\left(0,\frac{\pi}{2},0\right) = D_{0q'}^\ell\left(0,-\frac{\pi}{2},0\right) = (-1)^{q'} d_{-q'}^\ell(0) \tag{6.41}$$

one can perform summation over q′ in (6.38) and express the result via the correlation function (6.40):

$$\frac{\partial}{\partial t} d_q^\ell(t,J) = \mathrm{i}q\frac{J}{I} d_q^\ell(t,J) - \frac{1}{\tau_J} d_q^\ell(t,J) + \frac{1}{\tau_J}\varphi_B(J) d_q^\ell(0) K_\ell(t). \tag{6.42}$$

Using the Fourier transforms

$$\tilde{d}_q^\ell(\omega,J) = \int_0^\infty \mathrm{e}^{\mathrm{i}\omega t} d_q^\ell(t,J)\,\mathrm{d}t;\quad \tilde{K}_\ell(\omega) = \int_0^\infty K_\ell(t)\mathrm{e}^{\mathrm{i}\omega t}\,\mathrm{d}t \tag{6.43}$$

and taking into account the initial condition (6.3), one can easily find from Eq. (6.42)

$$\tilde{d}_q^\ell(\omega,t) = \left(1+\frac{1}{\tau_J}\tilde{K}_\ell(\omega)\right)\frac{\varphi_B(J) d_q^\ell(0)}{\mathrm{i}\omega - \mathrm{i}qJ/I + 1/\tau_J}\,. \tag{6.44}$$

Multiplying this relation by $(-1)^q d_{-q}^\ell(0)$ and summing according to (6.40), we obtain a closed equation for $\tilde{K}_\ell(\omega)$:

$$\tilde{K}_\ell(\omega) = \left(1+\frac{1}{\tau_J}\tilde{K}_\ell(\omega)\right)\sum_q \left[d_q^\ell(0)\right]^2 \int \frac{\varphi_B(J)\,\mathrm{d}J}{\mathrm{i}\omega - \mathrm{i}qJ/I + 1/\tau_J}\,.$$

Solving it, we find, taking account of Eq. (2.13)

$$G_\ell(\omega) = \frac{\operatorname{Re}\tilde{K}_\ell(\omega)}{\pi} = \frac{1}{\pi}\operatorname{Re}\left\{\frac{\sum_q \left[d_q^\ell(0)\right]^2 A_q}{1-\frac{1}{\tau_J}\sum_q \left[d_q^\ell(0)\right]^2 A_q}\right\}, \tag{6.45}$$

where

$$\left(d_{\pm1}^1\right)^2 = \frac{1}{2};\ \left(d_{\pm2}^2\right)^2 = \frac{3}{8};\ \left(d_0^2\right)^2 = \frac{1}{4} \tag{6.46}$$

are statistical weights of the branches, allowed by the selection rules for free rotators. The disappearance of the side branches with increasing density, as well as the appearance of the forbidden Q-branch in the linear molecule's IR spectrum, depend on the alteration of integral

$$\begin{aligned} A_q &= \int \frac{\varphi_B(J)\,\mathrm{d}J}{\mathrm{i}\omega - \mathrm{i}qJ/I + 1/\tau_J} \\ &= \int_0^\infty \exp\left[-\left(\mathrm{i}\omega+\frac{1}{\tau_J}\right)t\right]\mathrm{d}t \int_0^\infty \exp\left(\frac{\mathrm{i}tqJ}{I}\right)\varphi_B(J)\,\mathrm{d}J \end{aligned} \tag{6.47}$$

with increase of collisional frequency $1/\tau_0 = 1/\tau_J$. Substituting equilibrium distribution (3.27) into this expression, we have

$$\begin{aligned} A_0 &= \left(\mathrm{i}\omega + \frac{1}{\tau_J}\right)^{-1} \\ A_{-q} &= \frac{\mathrm{i}I}{q}\left(\frac{\pi}{2d}\right)^{\frac{1}{2}} + \frac{zI}{2q}\left(\frac{\pi}{2d}\right)^{\frac{1}{2}} \mathrm{e}^{-z^2}\left[1-\Phi(\mathrm{i}z)\right] \\ &- \frac{\mathrm{i}zI}{2q}\left(\frac{1}{2d}\right)^{\frac{1}{2}} \mathrm{e}^{-z^2}\,\mathrm{Ei}\left(z^2\right), \quad q \neq 0, \end{aligned} \tag{6.48}$$

where

$$z = \frac{1}{q}\left(\frac{1}{2d}\right)^{\frac{1}{2}}\left(I\omega - \frac{\mathrm{i}I}{\tau_J}\right).$$

Let us note now, that the results obtained can be easily expressed via the static correlation function

$$k_\ell = \sum_q \left[d^\ell_q(0)\right]^2 \int_0^\infty \exp\left(\frac{itqJ}{I}\right)\, \varphi_B(J)\, \mathrm{d}J\,. \tag{6.49}$$

At $\tau_J = \infty$ its Fourier transform describes the spectral contour, which is not altered by collisions. The quantities A_q, according to Eq. (6.47), determine the Laplace transform of all the components of (6.49) with $p = \mathrm{i}\omega + 1/\tau_J$. Therefore general formula (6.45) can be written in the form

$$\tilde{K}_\ell(\omega) = \frac{k_\ell\left(\mathrm{i}\omega + 1/\tau_J\right)}{1 - \left(1/\tau_J\right) k_\ell\left(\mathrm{i}\omega + 1/\tau_J\right)}, \tag{6.50}$$

which may be easily identified with the results obtained in [32, 29, 262].

The same result can be also reproduced in Mori's formalism, according to which

$$\frac{\partial}{\partial t} K_\ell(t) = -\int_0^t R\left(t-t'\right) K_\ell\left(t'\right)\, \mathrm{d}t' \tag{6.51}$$

as found in (A.3.5). Unfortunately, the kernel of equation (6.51), as well as its Laplace transform

$$\tilde{R}\left(\mathrm{i}\omega\right) = \frac{1 - \mathrm{i}\omega\tilde{K}_\ell\left(\mathrm{i}\omega\right)}{\tilde{K}_\ell\left(\mathrm{i}\omega\right)} \tag{6.52}$$

are unknown. It can be easily found only for the static limit (at $\tau_J \to \infty$) when $K_\ell \to k_\ell$ and

$$\tilde{r}\left(\mathrm{i}\omega\right) = \lim_{\tau_J\to\infty} \tilde{R}\left(\mathrm{i}\omega\right) = \frac{1 - \mathrm{i}\omega\tilde{k}_\ell}{\tilde{k}_\ell}. \tag{6.53}$$

In [263] it was postulated that

$$\tilde{R}(p) = \tilde{r}\,(p + 1/\tau_J)\ . \tag{6.54}$$

Substituting (6.52) and (6.53) into it, one can easily reproduce (6.50). In other words, assuming that the kernel of Mori's equation

$$R(t) = r(t)\,\mathrm{e}^{-t/\tau_J}\,, \tag{6.55}$$

where $r(t)$ is its well-known static analogue, one can easily obtain from (6.51) a solution of (6.50), valid for any $1/\tau_J$. In the same manner K_ℓ can be calculated for symmetrical tops [264]. Moreover the quantum correlation function for free rotators being used in Eq. (6.53) instead of the quasi-classical one allows the theory to be extended to very low densities, where rotational structure is resolved [265]. These advantages result from the assumed form of the kernel (6.55). This success would hardly be possible, had the final result (6.50) not been known. Mori's theory does not provide a constructive method for derivation of the closed form of the kernel. For example, it is still unknown for the model of weak collisions. The existing solution in this case cannot be represented naturally in the form (6.50).

The impact theory defines uniquely the spectral transformation in the limit of weak collisions. Expanding in a series over $J' - J$ the integrand of Eq. (6.4) one can obtain at $\Gamma = 1/\omega_Q\tau_J \gg 1$

$$\begin{aligned}
\frac{\partial}{\partial t} d_q^\ell(t,J) &= \mathrm{i}\frac{J}{I} q\, d_q^\ell(t,J) + \frac{1}{\tau_J} \\
&\quad \sum_{q'} \left[\left(1 + \frac{d}{J^2}\right)\delta_{q'q} - \frac{d}{J^2}\left(\hat{L}_x^2\right)_{q'q}\right] d_{q'}^\ell(t,J) \\
&\quad + \frac{J}{\tau_J}\left(1 - \frac{d}{J^2}\right)\frac{\partial}{\partial J} d_q^\ell(t,J) + \frac{d}{\tau_J}\frac{\partial^2}{\partial J^2} d_q^\ell(t,J)\,,
\end{aligned} \tag{6.56}$$

where $\hat{L}_x$ is an operator of infinitesimal rotation over axis x [23]. The matrix $\left(\hat{L}_x^2\right)_{q'q}$ in Eq. (6.56) governs the exchange between branches, as did the bilinear operator in the Wigner matrix in Eq. (6.38). It links a set of $\ell + 1$ differential equations of second order.

The solution of system (6.56) is a very complicated mathematical problem; it definitely needs numerical calculations on some stages of processing. At least two successful attempts to overcome these difficulties are well known in the literature. The first method was put forward by Sack and expresses the solution through a continuous fraction. The second was proposed by Fixman and Rider [29], and deals with a kinetic equation, similar to Eq. (6.56). The equation written down in a molecular reference system was adopted for successive application of numerical methods of matrix diagonalization. The eigenvalue with the smallest real part determines the long-time behaviour of the correlation function. In [29] the eigenfunctions of the differential operator of Eq. (6.56) were found and shown to be proportional to the generalized Laguerre polynomials. It is noteworthy that the functions, differing only by a factor, are eigenfunctions of the whole right-hand side of kinetic equation (A.5.1) including a dynamical term, which is quadratic over J. In the case of an anisotropic spectrum, the dynamical term is linear over J, but has off-diagonal elements in the same basis set of eigenfunctions. Consequently, it is necessary to invert matrices of infinite rank. Powles and Rickayzen have shown that, at least for linear rotators, the inversion procedure is mathematically identical to the continuous fraction representation of the solution given by Sack.

6.4 Orientational relaxation times

The results obtained allow one to follow the collapse of a static contour and its further narrowing by collisions. The limiting cases are the simplest to describe. In particular, from (6.45) it can be easily obtained that as $1/\tau_J \to 0$

$$\begin{aligned} G_\ell(\omega) &= \sum_q \left[d^\ell_q(0)\right]^2 \lim_{\tau_J\to\infty} \frac{1}{\pi} \int \frac{1/\tau_J}{\left[\omega - (J/I)q\right]^2 + (1/\tau_J)^2}\, \varphi_B(J)\, \mathrm{d}J \\ &= \sum_q \left[d^\ell_q(0)\right]^2 \frac{I}{q} \varphi_B\left(\frac{I\omega}{q}\right), \qquad \frac{\omega}{q} > 0\,. \end{aligned} \tag{6.57}$$

Each term of this sum corresponds to a static contour of the corresponding branch, except the Q-branch, which by virtue of (6.29) is a δ-function. In the opposite limiting case (as $1/\tau_J \to \infty$)

$$A_q \approx \frac{1}{\mathrm{i}\omega + 1/\tau_J} \left[1 + \mathrm{i}\frac{\langle J\rangle}{I}\, q\, \frac{1}{\mathrm{i}\omega + 1/\tau_J} - \frac{\langle J^2\rangle}{I^2}\, q^2\, \frac{1}{(\mathrm{i}\omega + 1/\tau_J)^2}\right]. \tag{6.58}$$

Substituting this expansion into (6.45) and taking into account that

$$\sum_q \left[d^\ell_q(0)\right]^2 q^i = \sum_{q,q'} D^\ell_{0q}\left(0, -\frac{\pi}{2}, 0\right) \left(\hat{L}^i_z\right)_{qq'} D^\ell_{q'0}\left(0, \frac{\pi}{2}, 0\right) = \left(\hat{L}^i_x\right)_{00},$$

$$\left(\hat{L}^i_x\right)_{00} = \begin{cases} 0, & i = 1 \\ \ell(\ell+1)/2, & i = 2 \end{cases}$$

we find

$$G_\ell(\omega) = \frac{w_\ell\ \left(1+\omega^2\tau_J^2\right)}{\pi\ \left[\omega^2\ \left(1+\omega^2\,\tau_J^2\right)^2 + w_\ell^2\right]}\,, \tag{6.59}$$

$$w_\ell = \frac{\ell(\ell+1)kT}{I}\ \tau_J\,. \tag{6.60}$$

Eq. (6.59) describes a collapsed spectrum, narrowing with increase of $1/\tau_J$. This is the Rocard formula derived in Chapter 2 by means of perturbation theory.

During transition from one limiting case to the other the change of shape is most significant in the central part of the spectrum. The Q-branch is much narrower than other ones observed in the IR spectrum and it broadens initially, as described in Section 2 of the present chapter.

The IR spectra of linear molecules at low pressure do not contain a Q-branch at all. The intensity increases with $1/\tau_J$ in the central part of this spectrum exclusively due to the exchange between P- and R-branches (Fig. 6.4). The secular simplification is inapplicable in this case. In order to describe the rise of intensity in a gap of the IR spectrum with increase of density, one has to know the exact solution of the problem, e.g. (6.45)–(6.47). Using it, one can calculate

$$\tau_{\theta,1} = \pi G_1(0) = \frac{1}{\tau_J}\int \frac{\varphi_B(J)\,\mathrm{d}J}{\left(1/\tau_J\right)^2 + J^2/I^2}\,, \qquad \bar{\omega}\tau_J \gg 1 \tag{6.61}$$

as a function of τ_J. Substituting equilibrium distribution φ_B for linear rotators and proceeding to dimensionless variables

$$\tau^*_{\theta,\ell} = \tau_{\theta,\ell}\ (kT/I)^{\frac{1}{2}}\,,\ \ \tau^*_J = \tau_J\ (kT/I)^{\frac{1}{2}} \tag{6.62}$$

we obtain after integration

$$\tau^*_{\theta,1} = -\frac{1}{2\tau^*_J}\ \exp\left(\frac{1}{2\tau_J^{*2}}\right)\ \mathrm{Ei}\ \left(-\frac{1}{2\tau^*_J}\right) \approx \frac{1}{\tau^*_J}\ln\ \tau^*_J\,, \qquad \tau^*_J \to \infty\,. \tag{6.63}$$

Kluk was the first to obtain these results [269]. They differ drastically from (6.25). When the medium becomes more rarefied, intensity in the central part of the IR spectrum decreases to zero. Hence, $\tau^*_{\theta,1}$ shortens, unlike $\tau^*_{\theta,2}$, which lengthens. The time $\tau^*_{\theta,1}$ behaves in the same manner for the case of weak collisions, though a formula quantitatively analogous to (6.63) is not found for this case. One can refer only to numerical calculations based on the general formulae by Sack or by Fixman and Rider. These calculations provide identical results [85]. Fig. 6.5 shows that, in rarefied media, the difference between weak and strong collision

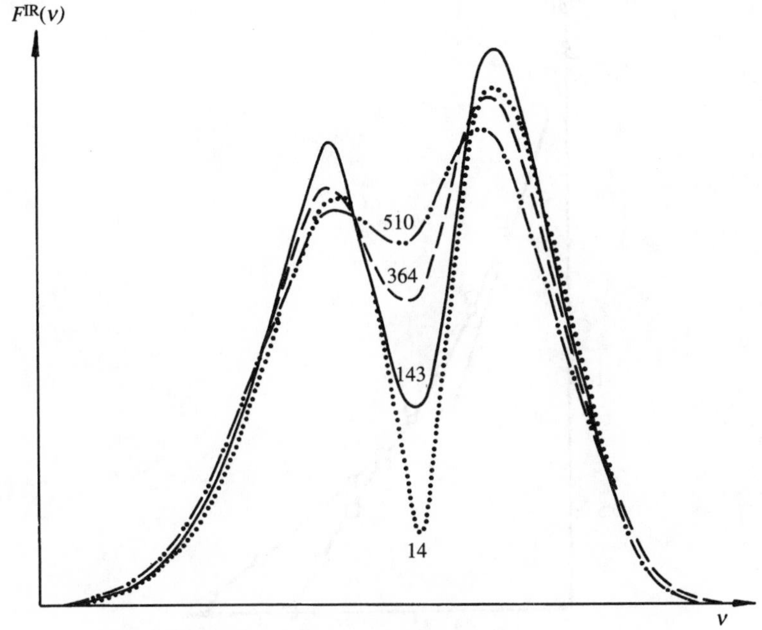

Fig. 6.4. The recovery of the gap in the IR spectrum of CO in argon [266]. The numbers denote the density of argon in amagat.

models is rather pronounced, but disappears after spectral collapse, which reflects the fact that at $\tau_J^* \ll 1$ the universal Hubbard relation (2.28) holds. This relation can be easily obtained from (6.59) and in dimensionless coordinates has the form

$$\tau_{\theta,\ell}^* \cdot \tau_J^* = \frac{1}{\ell\ (\ell+1)} \,. \tag{6.64}$$

So, $\tau_{\theta,\ell}$ increases monotonically with medium density, but in the beginning the increase corresponds to gap filling, and in the end to general narrowing of the spectrum.

The behaviour of $\tau_{\theta,2}\,(\tau_J)$ is qualitatively different. In the dense media this dependence also satisfies the Hubbard relation (6.64), and in logarithmic coordinates of Fig. 6.6 it is rectilinear. As τ_J^* increases, it passes through the minimum and becomes linear again when results (6.25) and (6.34) hold, correspondingly, for weak and strong collisions:

$$\tau_{\theta,2}^* = \begin{cases} \tau_J^*/3 & \text{at} \quad \gamma = 0 \\ \tau_J^*/10 & \text{at} \quad \gamma \to 1\,. \end{cases} \tag{6.65}$$

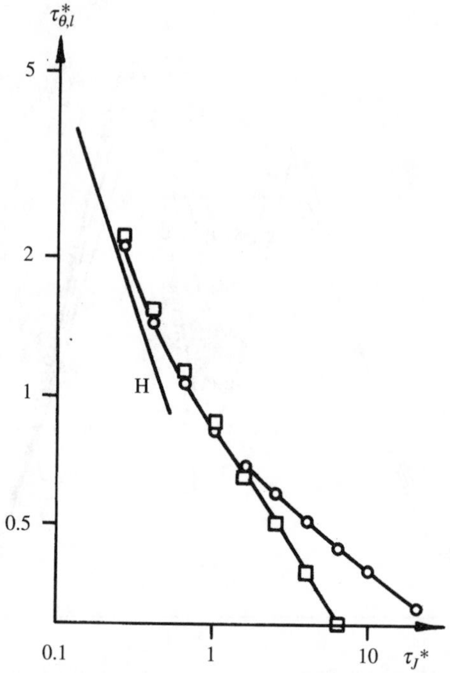

Fig. 6.5. The dependence of $\tau^*_{\theta,1}$ on τ^*_J in the case of strong (□) and weak (○) collisions [85]. The straight line (H) corresponds to the Hubbard relation.

The Hubbard straight line corresponds to rotational diffusion, and quasi-static straight lines (6.65) to quasi-free rotation. One type of motion substitutes for the other in the vicinity of the minimum point of curve $\tau^*_{\theta,2}(\tau^*_J)$ and is accompanied by collapse of the anisotropic scattering spectrum.

Impact theory is able to describe the transition from one limiting case to the other; however, this does not mean that this description is correct. Spectral collapse is within the impact theory limits, if inequalities (1.59) and (2.30) are compatible. This is only true when the region

$$\tau_c \ll \tau_J \ll 1/\bar{\omega} \tag{6.66}$$

can be found. In this region the results of impact theory overlap the results of perturbation theory, if inequality (1.58) is strong enough. In other words, when $\bar{\omega}\tau_c < 10^{-1}$, the model of J-diffusion is self-consistent, i.e. it takes into account not only the exchange between branches, but describes also the collapse caused by this exchange. If the model fails in

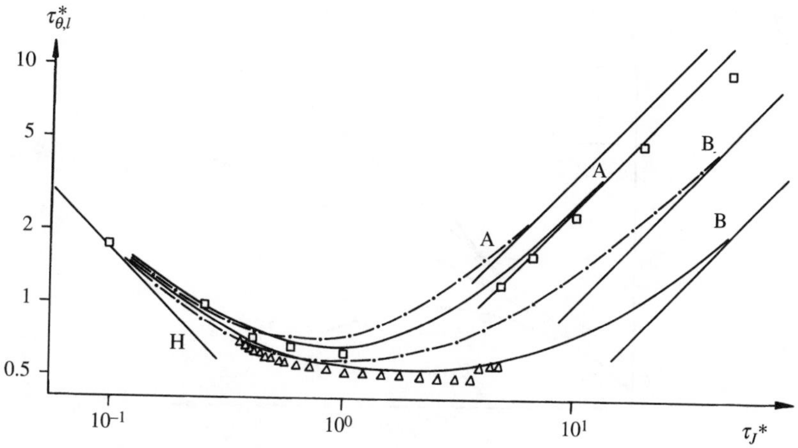

Fig. 6.6. The dependence of $\tau^*_{\theta,2}$ on τ^*_J in the case of strong (A) and weak (B) collisions for linear (—·—) and spherical (——) rotators. Experimental values of the relaxation times: (△) N_2 [71]; (□) CF_4 [124], [267]. The straight lines are the asymptotics of corresponding curves at high and low pressures.

the region where the Hubbard relation has already held, the results are not affected, because the Hubbard relation is indifferent to the models of molecular motion.

Diffusion description of orientational relaxation is only possible when

$$D_\theta \, \tau_J = \frac{\langle \omega^2 \rangle}{r} \tau_J^2 = \frac{kT}{I} \tau_J^2 \ll 1 \, . \tag{6.67}$$

Inequality (6.67) is the softest criterion of perturbation theory. Its physical meaning is straightforward: the reorientation angle (2.30) should be small. Otherwise, a complete circle may be accomplished during the correlation time of angular momentum and the rotation may be considered to be quasi-free. Diffusional theory should not be extended to this situation. When it was nevertheless done [268], the results turned out to be qualitatively incorrect: orientational relaxation time $\tau_{\theta,2}$ remained finite for $\tau_J \to \infty$. In reality $\tau_{\theta,2}$ tends to infinity in this limit [27, 269].

It is hardly possible to extrapolate the results of perturbation theory to the region of more rarefied media, taking into account terms of higher orders. Fig. 6.7 demonstrates that this series converges very slowly to an exact result, corresponding to strong collisions. An approximate solution including first-order correction of the Hubbard relation is qualitatively better than that including also second- and third-order ones. It is

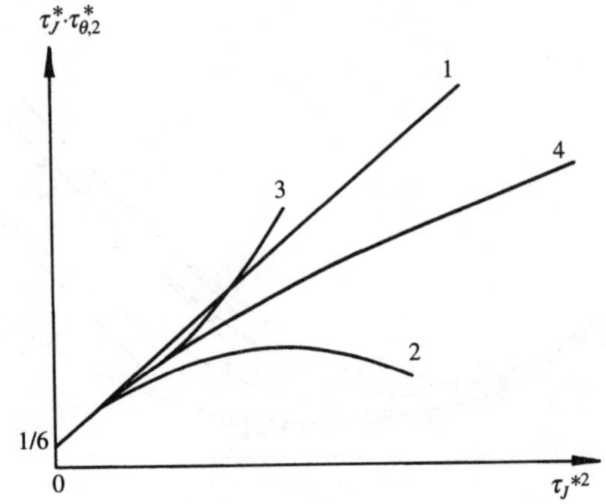

Fig. 6.7. The first-order (curve 1), second-order (curve 2) and third-order (curve 3) approximations to the exact dependence $\tau_{\theta,2}^*(\tau_J^*)$ in the strong collision model (curve 4).

impossible to reproduce quasi-static result (6.65) by summation of a finite number of terms of this series.

In the detailed review by Powles and Rickayzen [85], from which the data on the functional dependence $\tau_{\theta,\ell}^*(\tau_J^*)$ are borrowed, in line with the Langevin model (weak collisions) and an extended diffusion model (strong collisions), some other models are discussed. All of them are considered equally consistent with impact theory, except the *m*-diffusion model, which is counted as physically inconsistent. In fact, the other models are no better. We have proved already that a so-called friction model, based on the Fokker–Planck equation with a variable diffusion coefficient (2.40), is equivalent to the lowest order of perturbation theory. Hence, it cannot yield anything but the Hubbard relation. The 'strong collision' model of Sack [99] is self-contradictory, as it supposes that non-adiabatic collisions are able to break completely the phase δ, while there is no dephasing during J-diffusion at all. In this sense this model is no better than *m*-diffusion. It can predict only pressure broadening of the spectrum, and fails to describe the observed narrowing in dense media. Finally, Ivanov's model, improved by Kluk, reproduces qualitatively the transition from the Hubbard straight line to the quasi-static one for $\tau_{\theta,2}(\tau_J)$ dependence but it predicts the same behaviour for $\tau_{\theta,1}(\tau_J)$ which contradicts reality. The reason for this contradiction is the artificial improvement of the Ivanov–Valiev model of random walks in angular space. Let us consider this in more detail.

It is well known [11] that the reorientation rate in the jump model

$$1/\tau_{\theta,\ell} = \frac{1-\lambda_\ell}{\tau} \tag{6.68}$$

is not connected with τ_J directly. The quantity

$$\lambda_\ell = \int P_\ell(\cos\theta)\,\psi(\theta)\ \mathrm{d}\theta\,, \tag{6.69}$$

which determines the angular scale of a jump, may be expressed through the probability $\psi(\theta)$ for a molecule to turn by angle θ during rotation over barriers. If the mean duration of this molecular motion is τ' (time of a jump), then

$$\psi(\theta) = \int \mathbf{P}(\boldsymbol{u}, \boldsymbol{u}_0, t)\,\delta\left[\theta - \arccos\,(\boldsymbol{u}\cdot\boldsymbol{u}_0)\right]\ \mathrm{d}\boldsymbol{u}\ \mathrm{d}\boldsymbol{u}_0\ \mathrm{e}^{-t/\tau'}\ \mathrm{d}t/\tau'\,. \tag{6.70}$$

Here $\mathbf{P}(\boldsymbol{u}, \boldsymbol{u}_0, t)$ is determined by Eq. (2.36) and is the two-dimensional probability of molecular reorientation during time t. Substituting (6.70) into (6.69) and taking into account the definition of K_ℓ given in (2.10), we find

$$\lambda_\ell = \int_0^\infty K_\ell(t)\mathrm{e}^{-t/\tau}\ \mathrm{d}t/\tau'\,. \tag{6.71}$$

Even in this presentation λ_ℓ is not connected with τ_J uniquely, rather further physical assumptions are necessary. Those made by Kluk reduce to the following: motion over barriers is free rotation; and the following relation exists between the times introduced in the model:

$$\tau = \tau' = \tau_J\,. \tag{6.72}$$

This identifies the time between jumps τ and the time of a jump τ' and breaks the initial assumption of the model, which considers jumps as instantaneous: $\tau' \ll \tau$. This connection of τ and τ' with τ_J is necessary for the Hubbard relation to be reproduced. However, the means are not justified by the aim. To prove this, we shall reproduce the Hubbard relation without contradiction to the jump model.

Let us assume that motion over the barrier is free:

$$\tau' = \tau_J^0\,. \tag{6.73}$$

Here τ_J^0 is, as in Eq. (2.95), the life time of molecules in a state where their energy exceeds the height of the barrier. Equalizing K_ℓ to k_ℓ from (6.49) and taking into account (6.68), we find, as in [269]

$$\lambda_\ell = \frac{1}{\tau_J^0} \sum_{q=-\ell}^{\ell} \left[d_q^\ell(0)\right]^2 A_q(0)\,. \tag{6.74}$$

Substituting A_q from Eq. (6.48), we obtain

$$\lambda_\ell = \left[d_0^\ell(0)\right]^2 + \frac{1}{2\left(\tau_J^{0*}\right)^2} \sum_{q=1}^{\ell} \frac{\exp\left(\frac{1}{2\left(\tau_J^{0*}\right)^2 q^2}\right)}{q^2} \mathrm{Ei}\left(-\frac{1}{2\left(\tau_J^{0*}\right)^2 q^2}\right)\left[d_q^\ell(0)\right]^2\,. \tag{6.75}$$

In this result one can easily see two limiting cases, for small and large jumps, correspondingly:

$$\lambda_\ell = 1 - \ell\,(\ell+1)\,\left(\tau_J^{0*}\right)^2, \qquad \ell\,(\ell+1)\,\left(\tau_J^{0*}\right)^2 \ll 1, \tag{6.76}$$

$$\lambda_\ell = \left[d_0^\ell\,(0)\right]^2, \qquad \left(\tau_J^{0*}\right)^2 \gg 1\,. \tag{6.77}$$

In the first case the substitution of (6.76) into (6.68) yields formula (2.93) with the same $\langle\theta^2\rangle$ as in (2.94), and after substitution of (2.94) and (1.123) it reduces to the Hubbard relation. In the second case after substitution of (6.76) into (6.67), it appears that

$$1/\tau_{\theta,\ell} = \left\{1 - \left[d_0^\ell(0)\right]^2\right\}/\tau \tag{6.78}$$

does not depend on τ_J. In Fig. 6.5 and Fig. 6.6 the large jumps must be described by a horizontal line instead of a descending or ascending curve, observed for $\tau_{\theta,1}$ and $\tau_{\theta,2}$ correspondingly. In other words, the jump model, even improved correctly, is able to reproduce only the Hubbard relation, which appears in the limit of jumping diffusion. However, according to Chapter 2, the Hubbard relation reflects diffusion in angular phase space, and does not depend on the mechanism of an elementary angular jump, e.g., interrupted free rotation or instantaneous hopping. When the angle of elementary rotations becomes too large and their succession is not a diffusion (and this is the case for rarefied media), the model fails to describe the process in gases.

Fig. 6.8 plots experimental data on nitrogen in the gas phase. Intensive NMR measurements have recently turned nitrogen into the system for which the advantages of independent NMR probing of different relaxational processes for the same molecule are being taken in full. This highly informative separation can be achieved if a molecule with at least two different nuclei is available. The nucleus of spin 1/2 contributes to relaxation mainly through spin–rotational interaction, and the nucleus of spin 1 contributes through quadrupolar relaxation. Correspondingly, rotational and orientational relaxations are probed. The other possibility is to make an isotope substitution, which changes the spin, preserving almost the same molecular dynamics. The latter possibility was realized for the first time in the gas phase in [81]. These data have been discussed partly in Chapter 1 (Fig. 1.25).

Now we proceed to comparison between measured τ_J and $\tau_{\theta,2}$. The main basic formula extracting the relaxation time $\tau_{\theta,2}$ from spin relaxation times is well known [83, 39]

$$\frac{1}{T_1} = \frac{1}{T_2} = \frac{3}{8}(eqQ/\hbar)^2\tau_{\theta,2}\,, \tag{6.79}$$

where $eqQ/\hbar$ is the intramolecular quadrupolar coupling constant and nuclear spin $I = 1$. The data reported in [81] are plotted in Fig. 6.8 in rectangles. They are related to different experimental conditions: $(\times) - T = 300$ K, ($\rho = 4, 8, 16, 30$ Amagat); ($\square$) $T = 220$ K ($\rho = 16, 30, 50$ Amagat); ($\bullet$) $T = 150$ K ($\rho = 4, 8, 16$ Amagat) (see [81]). At $T = 220$ K no measurements were carried out for $^{15}N_2$, τ_J^* being calculated from $\sigma_J(T)$ from Fig. 1.25. It is seen from Fig. 6.8, that all the points corresponding to different temperatures belong, in fact, to the same straight line, giving an approximate ratio

$$\tau_{\theta,2} = \frac{1}{5}\tau_J\,. \tag{6.80}$$

As this kind of verification of classical J-diffusion theory is crucial, the remarkable agreement obtained sounds rather convincing. From this point of view any additional experimental treatment of nitrogen is very important. A vast bulk of data was recently obtained by Jameson *et al.* [270] for pure nitrogen and several buffer solutions. This study repeats the gas measurements of [81] with improved experimental accuracy. Although in [270] T_1 was measured, instead of T_2 in [81], at 150 amagat and 300 K and at high densities both times coincide within the limits of experimental accuracy.

The $\tau_{\theta,2}$ data from [270] were recalculated by us and the results are plotted in Fig. 6.8 together with original data. The reason for this correction is an additional factor of 1/4 introduced into the right-hand

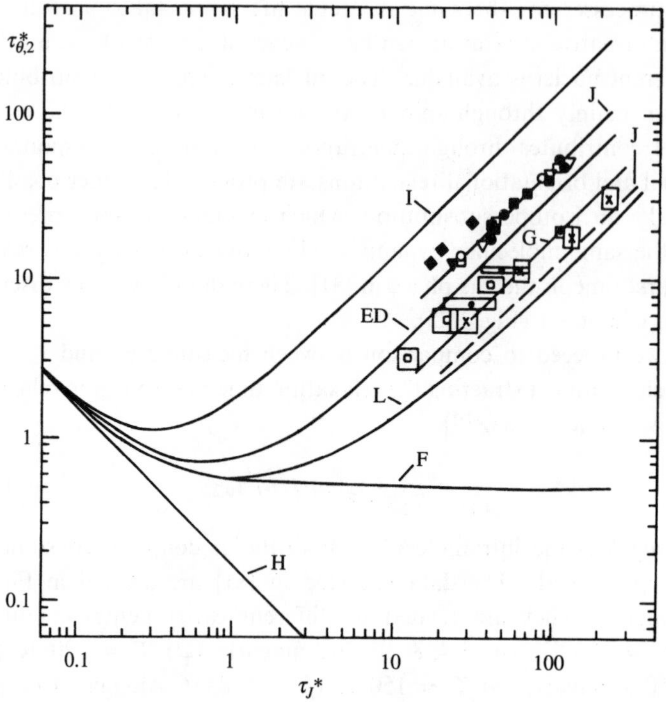

Fig. 6.8. The dependence of $\tau^*_{\theta,2}$ on τ^*_J by the 'Ivanov model' (I) and 'friction model' (F) in comparison with predictions of the extended J-diffusion (ED) and Langevin (L) models for linear molecules. The line (H) corresponds to the Hubbard inverse proportionality between $\tau_{\theta,2}$ and τ_J at very high densities. Experimental data from [81] are in rectangles around line G with the length of their vertical and horizontal sides being equal, correspondingly, to the experimental errors in $\tau^*_{\theta,2}$ and τ^*_J measurements. Experimental data from [270] (J) are shown both in original position and shifted down by a factor of four (broken line).

side of (6.79) in [270]. After this (6.79) defines a different relaxation time τ_2 [83], which is appropriate for interpretation of depolarized Rayleigh scattering but not for our goal. Insofar as all theoretical curves in Fig. 6.8 present the dependence of $\tau_{\theta,2}$ on τ_J, they should evidently be compared with measured values of the same physical quantities. After reduction of the ordinates all points from [270] are successfully caught within the theoretically prescribed region. It should be noted that false substitution of $\tau_{\theta,2}$ by τ_2 has a rather long history. For example, the authors of [272] were the predecessors of [270] in this. In [272] the ratio $\tau_J/\tau_{\theta,2} = 1.38$ for ClF diatomics was reported. After the analogous correction we have $\tau_{\theta,2}^{\mathrm{ClF}} = (1/5.52)\tau_J$ and with the announced [272] uncertainty 15% it again demonstrates excellent agreement with theory.

7
Rotation and libration in a fluctuating cell

It was demonstrated in Chapter 6 that impact theory is able to describe qualitatively the main features of the drastic transformations of gas-phase spectra into liquid ones for the case of a linear molecule. The corresponding NMR projection of spectral collapse is also reproduced qualitatively. Does this reflect any pronounced physical mechanism of molecular dynamics? In particular, can molecular rotation in dense media be thought of as free during short time intervals, interrupted by much shorter collisions?

It seems that an affirmative answer is hardly possible on the contemporary level of our general understanding of condensed matter physics. On the other hand, it is necessary to find a reason for numerous successful expansions of impact theory outside its applicability limits.

One possibility for this was demonstrated in Chapter 3. If impact theory is still valid in a moderately dense fluid where non-model stochastic perturbation theory has been already found applicable, then evidently the 'continuation' of the theory to liquid densities is justified. This simplest opportunity of unified description of nitrogen isotropic Q-branch from rarefied gas to liquid is validated due to the small enough frequency scale of rotation–vibration interaction. The frequency scales corresponding to IR and anisotropic Raman spectra are much larger. So the common applicability region for perturbation and impact theories hardly exists. The analysis of numerous experimental data proves that in simple (non-associated) systems there are three different scenarios of linear rotator spectral transformation. The IR spectrum in rarefied gas is a P–R doublet with either resolved or unresolved rotational structure. In the process of condensation the following may happen.

(i) The spectrum can transform into a relatively narrow structureless line, centred at the frequency of vibrational transition. This transformation is natural for weakly dipolar molecules with a large moment of inertia.

(ii) Molecules with a large rotational constant (e.g., hydrogen halides and their deuterium derivatives) yield spectra of a qualitatively different type. In inert solvents like He and SF_6 the doublet of P and R-branches is reproduced with resolved *j*-components, centred at the rotational frequencies of a free rotator. In fact, these spectra differ from gas-like ones only in the essentially larger width of their components and their temperature-dependence.

(iii) In other solvents a more or less intense Q-branch appears in the centrum. The triplet IR spectrum of linear molecules is the most interesting example. It is direct evidence of the inapplicability of the impact description of rotation. The Q-branch, which is centred at the vibrational frequency, is forbidden by the selection rules for a free linear rotator. In impact theory the Q-branch only may appear instead of a P–R doublet as a result of spectral collapse. The spectral exchange operator of rank 2 has only two roots, setting frequencies and widths of branches. The simultaneous registration of three lines is forbidden either before or after collapse. Hence the observation of a P–Q–R triplet in IR spectra of linear molecules is in qualitative disagreement with the impact theory.

The decisive step to overcome this problem was made by Robert and Galatry [17]. They put forward the idea of reducing the influence of the liquid neighbourhood to an 'orienting field'. This field requantizes the rotation of molecules with low energies, transforming it to libration and causes the appearance of a central Q-branch as part of the Stark structure. The motion of molecules with high rotational energies is perturbed just slightly, and therefore the P and R components are conserved at the periphery of the spectrum. Slight fluctuations of field amplitude lead to broadening of rotational components, and isotropy of the sample is provided by isotropic distribution of the orienting field directions. Though very successful for description of the IR spectrum of HCl in CCl_4 [17], the model corresponds, in fact, to a glass-like state, not to the liquid phase, where cage geometry as well as orienting field fluctuate with time. Calculations by the methods of molecular dynamics show that, at liquid-phase densities, the target molecule spends at least half of the time in the field induced by the environment. In the hopping model of orientational relaxation [11] it is assumed that transition time between different directions of stable librations τ_s is negligible in comparison with residence time of this direction τ_c,

$$\tau_s \ll \tau_c .$$

This hierarchy of times is characteristic for the solid state. Though ultimate hopping models utterly neglect the librational contribution to reorientation, they are widely used for analysis of IR and Raman spectra in liquids. When comparing experimental data with theory, they are considered as a natural alternative to gas-like models of reorientational relaxation. However, these models are hardly natural for liquids. The mechanism of rotation inside the solid and liquid cage is drastically different. In the first case, the orienting field of a crystal is constant in time, and the potential barriers created by it are overcome by rare fluctuations of kinetic rotational energy. In liquids, on the contrary, this field is on average equal to 0 and appears only as a result of breaking of the spherical symmetry of the environment. In other words, the orienting field in liquids constantly fluctuates in time, so that the molecule has to rotate in a randomly changing potential and is able to reorient even lacking sufficient kinetic energy. The jumps that the orienting field is subjected to are infinitely small [273] or finite [18].

The present chapter proposes a mathematical treatment adequate to the model of a fluctuating liquid cage [274, 275, 19]. In the frame of this formalism, reduction [18] of the rotational spectrum to a few-level model is no longer decisive. The consideration will be rather general and valid outside the limits of perturbation theory [273]. For this, let us introduce the time of free rotation of the target molecule τ_0. We restrict ourselves to the situation when the 'residence' time of a given cage configuration τ_c is large enough in comparison with both τ_s and τ_0:

$$\tau_s\,,\ \tau_0 \ll \tau_c\,. \tag{7.1}$$

These times are supposed to be small enough to prevent significant reorientation of the molecule during free rotation

$$\bar{\omega}\ \tau_0 \ll 1\,, \tag{7.2}$$

and during rearrangement of the neighbourhood

$$\bar{\omega}\ \tau_s \ll 1\,. \tag{7.3}$$

Notice the similarity of criteria (7.3) and (1.58): the latter gives the non-adiabaticity of angular momentum changes due to collisions, the former defines the effective non-adiabaticity of cage changes. In its turn, the meaning of inequality (7.2) is opposite to that of (6.13) and reflects the qualitative difference between impact approximation and fluctuating liquid cage model. In the latter orientational relaxation takes place mainly through librational motion, governed by the hopping potential. All correlation functions and spectra of this liquid reorientational process are derived within the present treatment as rigorously as was done earlier in the frame of impact theory.

Let us assume that there is a Poissonian distribution of the successive moments when the anisotropic potential abruptly changes (though this assumption is not obligatory). Then the time evolution of the potential $V(t)$ becomes a Markovian random process. This is the next logical step in deepening the microscopic level of modelling. In the Valiev–Ivanov theory Ω is considered as a random Markovian variable whereas Gordon's j-diffusion model treats angular momentum as Markovian random perturbation. In this chapter both orientational and angular momentum motion are a system's response to intermolecular interaction, which is a random Markovian process. An averaging of the linear system response to a Markovian noise is feasible in the frame of sudden modulation theory. It makes it possible to write down the corresponding set of kinetic equations and investigate their solutions in particular cases. At the end of this chapter an important problem similar to that met in Chapter 4 is considered. Since coupling between the target molecule and its environment is considered as a time-dependent random process, the kinetic operator, governing rotational relaxation, does not satisfy the detailed balance principle. It implies neglect of the rotator feed-back to the bath, which, in its turn, produces infinite heating of rotational degrees of freedom. Very recent original ideas for suppressing this difficulty are demonstrated in this chapter, both in quantum and classical versions of the theory. After corresponding corrections for the detailed balance to be valid, the possibility is demonstrated of liquid phase P–R doublet collapse of non-impact origin. Simultaneously, the Hubbard relation is reproduced to confirm its general non-model nature.

7.1 The fluctuating cell model

Let us imagine that the liquid cage is a spherical cavity in a continuous medium. When the molecule is in its centre, the orienting field is equal to zero. At this point the anisotropic part of the rotator–neighbourhood interaction appears only in the case of asymmetrical 'breathing' of the

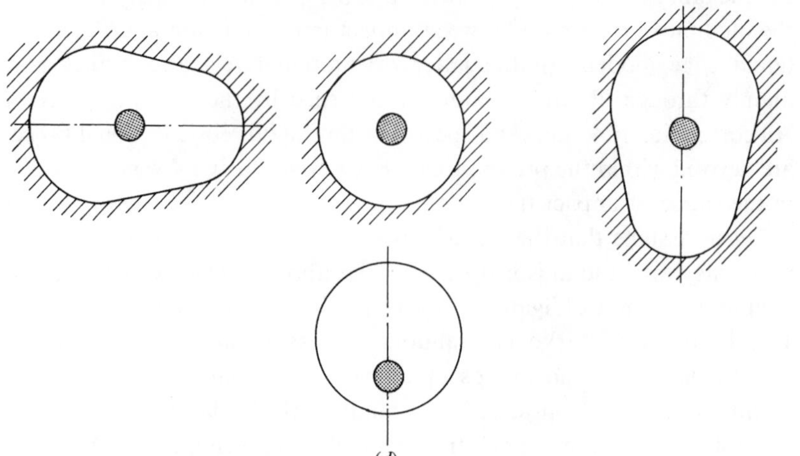

Fig. 7.1. The direction of the orienting field (dot–dashed line) in a fluctuating cage, when the molecule (shadowed circle) is in its centre (*a*)–(*c*), or shifted toward the cell edge (*d*).

cage (Fig. 7.1). In the other limiting case, when rotator translations inside the cage (from well to well) are faster than cage deformations, anisotropy arises even in a spherical cage. The latter possibility is considered in [17, 6]. The interaction with the orienting field $\boldsymbol{E}$, directed along the radius $\boldsymbol{r}$ (R is the radial size of the cage), is represented as [17, 273]:

$$V(\boldsymbol{r}) = -\boldsymbol{E}(\boldsymbol{r}) \cdot \boldsymbol{d} . \tag{7.4}$$

It is usually assumed that E increases rather sharply, reaching its maximum value $\lambda = |E(R)d|$ near the wall.

It is easy to describe qualitatively the orienting field fluctuations due to molecular translations inside the fixed cage. In the process of quasi-free movement the molecule crosses the cage along nearly straight lines. If the isotropic part of the potential $V(r)$ has a local maximum in the centre of a cell [219, 276, 277], which is typical for the liquid phase, then the trajectory of the motion slightly bends, rounding it. In any case the only influence of translational motion is that the orienting field, affecting the molecule, now weakens, now strengthens, changing at the same time its direction. As an average over time, the orienting field is equal to zero.

It is more difficult to estimate the variation of the orienting field, caused by fluctuations of the neighbourhood configuration. In [18] it was assumed that the field can have the values, differing only in sign:

$$V = \pm\lambda . \tag{7.5}$$

Each of them is conserved during time τ_c, then 'switches off' for time τ_0, and then acquires the opposite sign. This simplification of the random process $V(t)$ is also possible for a 'fixed cage, with rotator itinerant inside it'. For this, it is enough to separate the central part of the cage and its periphery (Fig. 7.2). After this, one has to neglect the orienting field in the central part of the cage and to assume that it has an equal magnitude near the wall

$$\boldsymbol{E}(\boldsymbol{r}) = \begin{cases} 0 & 0 < r < R_0 \\ E\boldsymbol{e} & R_0 < r < R, \end{cases} \tag{7.6}$$

where $\boldsymbol{e} = -\boldsymbol{r}/|\boldsymbol{r}|$. The main achievement of the authors of [18] is that their model takes into account potential fluctuations, comparable to the potential's magnitude, unlike in [17] where the spectrum was broadened by small fluctuations of the potential

$$|\Delta V(t)| \ll V . \tag{7.7}$$

The rectangular approximation (7.6) of dependence $\boldsymbol{E}(\boldsymbol{r})$ implies that $\tau_s = 0$. This simplification being valid only for non-adiabatic interaction, exact knowledge of the time-dependence $V(t)$ is not obligatory. Random walk approximation is quite acceptable. The value R_0/R is a free parameter of the model ($|R_0/R| \leq 1$) and makes it possible to vary the ratio of times $0 \leq \tau_c/\tau_0 \leq \infty$. This interval falls into two regions: one of them corresponds to impact theory ($0 \leq \tau_c/\tau_0 \leq 1$), and the other ($1 \leq \tau_c/\tau_0 \leq \infty$) to the fluctuating liquid cage. In the first case non-adiabaticity of the process is provided by the condition

$$\bar{\omega}_j \tau_c \ll 1 \tag{7.8}$$

and in the second case it is provided by (7.3). In impact theory, neglecting the collision duration τ_c, its effect $V\tau_c$ is conserved, which may be large, and is certainly crucial. In the present approach τ_0 and the variation of the system at this time are neglected (7.2). In line with (7.3) this is the so-called model of 'structurally limited' liquids [278], which is an alternative to the collisional model. In fact, one might create a theory for arbitrary values of ratio τ_c/τ_0 and parameter $\bar{\omega}_j\tau_s$ taking into account molecular rotation when the field is switching or absent. The only restriction is that switching must be sudden

$$\tau_s \ll \tau_0 , \ \tau_c . \tag{7.9}$$

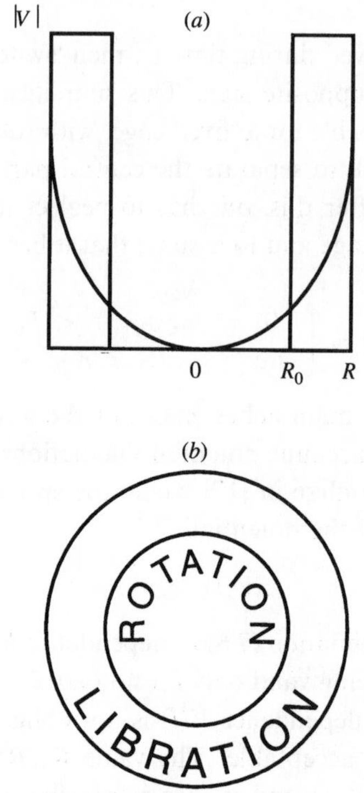

Fig. 7.2. The radial dependence of the anisotropic part of the intermolecular potential: (*a*) variation of height of the librational barrier in any diametrical cross-section of the cage and its rectangular approximation; (*b*) the corresponding rectangular approximation of $|V(r)|$ separation between the region of libration and that of free rotation inside the cage.

However, in the following we shall restrict ourselves solely to the structural limit.

7.2 Kinetic equation of the model

The IR spectrum of linear molecules

$$G^{\mathrm{IR}}(\omega) = \frac{1}{\pi}\mathrm{Re}\int K_{\theta,1}(t, \boldsymbol{e})\,\mathrm{e}^{\mathrm{i}\omega t}\,\mathrm{d}t\,\mathrm{d}\boldsymbol{e} \qquad (7.10)$$

is expressed via

$$K_{\theta,1}(t,e) = \mathrm{Tr}\left\{\sum_{q=-1}^{1}(-1)^q \rho d_q(e,t)\,\mathrm{d}_{-q}(0)\right\}. \tag{7.11}$$

An arbitrary rank of the orientational relaxation and the orienting field of more general form than in (7.5) will be considered below. Corresponding to (7.4) interaction

$$V(t) = -E\,e(t)\,\boldsymbol{d} \tag{7.12}$$

is a time-dependent purely discontinuous random process. Though the modulus $|\boldsymbol{E}(t)|$ is conserved, the value of $V(t)$ varies due to instantaneous reorientations of $\boldsymbol{e}(t)$, as the dipole moment does not change during such jumps. Sudden modulation theory allows one to write down a closed equation for the partial dipole moment operator

$$\frac{\partial}{\partial t}d_q(t,e) = \mathrm{i}[H(e)\,d_q(t,e)] - \frac{1}{\tau_c}d_q(t,e) + \frac{1}{\tau_c}\int f\left(e',e\right)d_q\left(t,e'\right)\,\mathrm{d}e'. \tag{7.13}$$

Here $\hbar = 1$ and H is the Hamiltonian of the system

$$H(e) = H_{\mathrm{vib}} + \frac{J^2}{2I} + V(e),$$

vibrational–rotational interaction being neglected. In an isotropic liquid all directions of the orienting field are equiprobable. Therefore, the density of its angular distribution $\varphi(e)$, and the initial condition for (7.13) is

$$d_q(t=0,e) = \varphi(e)\,d_q(0) = \frac{1}{4\pi}\,d_q\,. \tag{7.14}$$

The isotropy of the liquid rigidly restricts the conditional probability density $f(e',e)$ for the cell axis to change as a result of an elementary jump. It depends only on the angle between successive directions of the field

$$f\left(e',e\right) = f\left(e'\cdot e\right). \tag{7.15}$$

It is normalized

$$\int f\left(e',e\right)\,\mathrm{d}e = 1 \tag{7.16}$$

and conserves the stationary angular distribution

$$\varphi(e) = \int f\left(e',e\right)\varphi\left(e'\right)\,\mathrm{d}e'. \tag{7.17}$$

7.3 General solution

The particular features of the problem, namely the properties of kernel (7.15) of the relaxational part of (7.13) and of Hamiltonian (7.12) in the dynamical part of (7.13), allow one to advance essentially in solving kinetic equation (7.13).

For this purpose, let us use invariance of the matrix product trace under cyclic permutation of factors and represent (7.11) as

$$K(t) = \mathrm{Tr} \sum_q (-1)^q \int \mathrm{d}\boldsymbol{e} \mathrm{D}(\Omega) \rho d_q(\boldsymbol{e}, t)\, d_{-q} D^+(\Omega). \tag{7.18}$$

Here $D(\Omega) = D(\alpha, \beta, \gamma)$, Euler angles α, β and γ being chosen so that the first two coincide with the spherical angles determining orientation $\boldsymbol{e} = \boldsymbol{e}(\beta, \alpha)$. Using the theorem about transformation of irreducible tensor operators during rotation [23], we find

$$D(\Omega) d_{-q} D^+(\Omega) = \sum_{q}^{\prime} D^1_{q'\,-q}(\Omega) d_{q'} . \tag{7.19}$$

It is noteworthy that $d_q(\boldsymbol{e}, t)$ does not satisfy this relation, as equality $[J_\mu, d_q] = \sqrt{2}\, C^{1\,q+\mu}_{1q1\mu} d_{q+\mu}$ (the definition of an irreducible tensor operator) does not hold for it [23]. Integration in (7.18), performed over the spherical angles of vector $\boldsymbol{e}$, may be completed up to an integral over the full rotational group due to the axial symmetry of the Hamiltonian relative to the field. This, together with (7.19), yields

$$\begin{aligned} K(t) &= \sum_{qvjmq'v'ln} (-1)^q \langle vjm|\rho \int \frac{\mathrm{d}\Omega}{2\pi} D(\Omega) d_q(\Omega, t) D^+(\Omega) |v'\ell n\rangle \\ &\times\ D_{q'-q}(\Omega) \langle v'\ell n| d_{q'} |vjm\rangle = \sum_{q'vjmv'ln} \langle vjm|\rho \tilde{d}_{q'}(t)|v'\ell n\rangle \\ &\times\ (-1)^{j-m} C^{1q'}_{\ell n j-m} \frac{\langle v'\ell || \boldsymbol{d} || vj\rangle}{\sqrt{3}} . \end{aligned} \tag{7.20}$$

In the latter expression the matrix element of operator $d_{q'}$ is transformed according to the Wigner–Eckart theorem and the definition used is

$$\tilde{d}_{q'}(t) = \sum_q (-1)^q \int \frac{\mathrm{d}\Omega}{2\pi} D(\Omega) d_q(\Omega, t) D^+(\Omega) D^1_{q'-q}(\Omega) , \tag{7.21}$$

where $\langle v'l||\boldsymbol{d}||vj\rangle$ is a reduced matrix element of the dipole moment operator.

It is remarkable that integro-differential Eq. (7.13) can be reduced to a differential one, closed with respect to $\langle vjm|\tilde{d}_{q'}(t)|v'ln\rangle$. The final set of equations obtained is formally identified with the impact theory master equation. For this, axial symmetry of Hamiltonian (7.12) and of the kernel of Eq. (7.15) relative to the field is sufficient. In order to prove that, let us substitute in (7.13) integration over Ω' for integration over $\boldsymbol{e}'$. After this, multiplying its left-hand side by $(-1)^q D(\Omega)$, and the right-hand side by $(2\pi)^{-1} D^1_{q-q'}(\Omega) D(-\Omega)$, we integrate it over Ω and sum over q, like in (7.21). As a result we get in Liouville space

$$\frac{\partial}{\partial t}\tilde{d}_{q'}(t) = \mathrm{i}H^{\times}\tilde{d}_{q'}(t) - \tau_c^{-1}$$
$$\times \left[\tilde{d}_{q'}(t) - \sum_q (-1)^q \int \frac{\mathrm{d}\Omega}{2\pi} \int \frac{\mathrm{d}\Omega'}{2\pi} D(\Omega) d_q(\Omega', t) D^+(\Omega) D^1_{q'-q}(\Omega) f(\boldsymbol{e}\cdot\boldsymbol{e})\right], \tag{7.22}$$

where $H^{\times}\tilde{d}_{q'} = [H\tilde{d}_{q'}]$ and

$$H = H_{\text{vib}} + \frac{J^2}{2I} - Ed_0 . \tag{7.23}$$

Representing the integral term of (7.22) as

$$\sum_{qq''\tilde{q}} (-1)^q \int \frac{\mathrm{d}\Omega}{2\pi} \int \frac{\mathrm{d}\Omega'}{2\pi} D(\Omega) D^+(\Omega') D(\Omega') d_q(\Omega', t) D^+(\Omega') D(\Omega') D^+(\Omega)$$
$$\times D^1_{q''-q}(\Omega') D^{1+}_{\tilde{q}-q''}(\Omega') D_{q'\tilde{q}}(\Omega) f\left(\boldsymbol{e}'\cdot\boldsymbol{e}\right) \tag{7.24}$$

one can make sure [23] that product $D(\Omega)D(-\Omega') = D(\tilde{\Omega})$ causes such rotation that $\cos\tilde{\beta} = \boldsymbol{e}'\cdot\boldsymbol{e}$, where β is one of the Euler angles, corresponding to $\tilde{\Omega} = \{\tilde{\alpha}, \tilde{\beta}, \tilde{\gamma}\}$. According to the theorem on invariant integration over a full rotational group, we can proceed from integration over Ω to integration over $\tilde{\Omega}$. Owing to this expression (7.24) transforms into

$$\sum_{q''} \int \frac{\mathrm{d}\tilde{\Omega}}{2\pi} D(\tilde{\Omega}) \left[\sum_q (-1)^q \int \frac{\mathrm{d}\Omega'}{2\pi} \mathrm{D}(\Omega') d_q(\Omega', t) D^+(\Omega') D^1_{q''-q}(\Omega')\right]$$
$$\times D^+(\tilde{\Omega}) D^1_{q'q''}(\tilde{\Omega}) f(\tilde{\Omega}) . \tag{7.25}$$

In the latter expression variable (7.21) is identified by square brackets. After this Eq. (7.22) can be easily closed relative to it. In basis $|vjm\rangle$ it has the following form

$$\langle vjm|\dot{\tilde{d}}_{q'}(t)|v'\ell n\rangle = \mathrm{i}\sum_{j'm'\ell'n'} \langle vjm, v'\ell n|H^{\times}|vj'm', v\ell' n\rangle \times \langle vj'm'|\tilde{d}_{q'}(t)|v'\ell' n'\rangle - \frac{1}{\tau_c}\Bigg[\langle vjm|\tilde{d}_{q'}(t)|v'\ell n\rangle - \sum_{q''m'n'} \langle vjm, v'\ell n|\hat{P}_{q'q''}|vjm', v'\ell n'\rangle\langle vjm'|\tilde{d}_{q''}(t)|v'\ell n'\rangle\Bigg], \quad (7.26)$$

where

$$\langle vjm, v'\ell n|\hat{P}_{q'q''}|vjm', v'\ell n'\rangle = \int \mathrm{d}\tilde{\Omega}\, f(\tilde{\Omega}) D^{j}_{mm'}(\tilde{\Omega}) D^{\ell +}_{n'n}(\tilde{\Omega}) D^{1}_{q'q''}(\tilde{\Omega}) \frac{1}{2\pi}. \quad (7.27)$$

Using (7.14), (7.21), and taking into account (7.19), we find the initial condition for (7.26)

$$\tilde{d}_{q'} = \sum_{q\tilde{q}} (-1)^q \int \mathrm{d}\Omega\, D^1_{q'-q}(\Omega) D^1_{\tilde{q}q'}(\Omega)\, \frac{1}{8\pi^2} d_{\tilde{q}} = (-1)^{q'} d_{-q'}. \quad (7.28)$$

Thus the kinetic equation may be derived for operator (7.21), though it does not exist for an average dipole moment. Formally, the equation is quite identical to the homogeneous differential equation of the impact theory with the 'collisional operator' (7.27). It is of importance that this equation holds for 'collisions' of arbitrary strength, i.e. at any angle of the field reorientation. From Eq. (7.10) and Eq. (7.20) it is clear that the shape of the IR spectrum

$$F(\omega) = \frac{1}{\pi}\mathrm{Re} \sum_{q'vjmv'\ell n} \langle vjm|\rho\tilde{d}_{q'}(\omega)|v'\ell n\rangle (-1)^{j-m} C^{1q'}_{\ell nj-m} \frac{\langle v'\ell||d||vj\rangle}{\sqrt{3}} \quad (7.29)$$

may be expressed via the Fourier transform of the solution of Eq. (7.26)

$$\tilde{d}_{q'}(\omega) = \int_0^\infty \tilde{d}_{q'}(t)\, \exp\,(\mathrm{i}\omega t)\, \mathrm{d}t.$$

On this basis let us consider the simplest stochastic models of orienting field fluctuations.

7.4 Anticorrelated perturbation

Let us consider first the same motion of the orienting field as in [18]. Every jump changes its direction to the opposite one. This model corresponds to the kernel of the integral part of operator (7.27)

$$f(\Omega) = \frac{1}{2\pi}\delta(\cos\beta + 1). \tag{7.30}$$

Using (7.30), one can easily perform integration, required by (7.27). Substituting the result into (7.26), taking account of initial condition (7.28) we have the following matrix equation relative to the Fourier-transformed variables $\tilde{d}_{q'}(\omega)$

$$\begin{aligned} -\mathrm{i}\omega\langle 0jm|\tilde{d}_{q'}(\omega)|1\ell n\rangle &= (-1)^{\ell-m} C^{1q'}_{\ell n j-m}\frac{\langle j||\boldsymbol{n}||\ell\rangle d_{0\ell}}{\sqrt{3}} \\ &+ \mathrm{i}\sum_{j'm'\ell'n'}\langle 0jm, 1\ell n|H^{\times}|0j'm', 1\ell'n'\rangle \\ &\langle 0j'm'|\tilde{d}_{q'}(\omega)|1\ell'n'\rangle - \tau_c^{-1}\left[\langle 0jm|\tilde{d}_{q'}(\omega)|1\ell n\rangle\right. \\ &+ \left.(-1)^{j+\ell}\delta_{q',n-m}\langle 0j-m|\tilde{d}_{-q'}(\omega)|1\ell-n\rangle\right]. \end{aligned} \tag{7.31}$$

Here the Liouvillian of system $H^{\times}$ is determined by relation (7.23). This, in line with condition (7.5), yields

$$\begin{aligned} &\langle 0jm, 1\ell n|H^{\times}|0j'm', 1\ell'n'\rangle = \left(\omega_{01} - \omega_{j\ell}\right)\,\delta_{jj'}\,\delta_{\ell\ell'} \\ &-V\left[(-1)^{j'-m'}\,C^{10}_{jmj'-m'}\langle j||\boldsymbol{n}||j'\rangle\,\delta_{\ell\ell'}\,\delta_{nn'}\right. \\ &-(-1)^{\ell-n}\,C^{10}_{\ell'n'\ell-n}\,\rangle\ell'||\boldsymbol{n}||\ell\langle\,\delta_{jj'}\,\delta_{mm'}\,]\,, \end{aligned} \tag{7.32}$$

where

$$\boldsymbol{n} = \boldsymbol{d}/|\boldsymbol{d}|, \quad V = -E|\boldsymbol{d}|, \quad \langle j||\boldsymbol{n}||\ell\rangle = (2\ell+1)^{1/2}\,C^{j0}_{\ell 010}\,.$$

Equation (7.31) can be easily reduced to the form

$$\begin{aligned} \langle 0jm|\boldsymbol{d}_{q'}(\omega)|1\ell n\rangle \;&=\; \sum_{j'm'\ell'n'}\langle 0jm, 1\ell n|\hat{A}|0j'm', 1\ell'n'\rangle \\ &\times\left[(-1)^{\ell'-m'}C^{1q'}_{\ell'n'j'-m'}\left(\frac{2\ell'+1}{3}\right)^{1/2}C^{j'0}_{\ell'010}\,d_{01}\right. \\ &+\left.\tau_c^{-1}(-1)^{j'+\ell'+1}\delta_{q',n'-m'}\,\langle 0j'-m'|\tilde{d}_{-q'}\,(\omega)|1\ell'-n'\rangle\right], \end{aligned} \tag{7.33}$$

where

$$\hat{A} = \left[-\mathrm{i}\left(\omega\hat{1} + H^{\times}\right) + \tau_c^{-1}\hat{1}\right]^{-1}. \tag{7.34}$$

Let us substitute in (7.33) indices $-m, -n, -q'$ for m, n, q', correspondingly. Taking into account the properties of the Clebsch–Gordan coefficients [23]:

$$C^{1q'}_{jn\ell n} = (-1)^{j+\ell+1} C^{1-q'}_{j-m\ell-n}; \quad C^{10}_{j0\ell 0} = 0 \quad \text{if } j+\ell+1 \text{ is odd} \tag{7.35}$$

we find

$$\langle 0j-m|\tilde{d}_{-q'}(\omega)|1\ell-n\rangle = \sum_{j'm'\ell'n'} \langle 0jm, 1\ell n|\hat{A}|0j'm', 1\ell'n'\rangle$$

$$\times \left[(-1)^{\ell'-m'} C^{1q'}_{\ell'n'j'-m'} \left(\frac{2\ell'+1}{2}\right)^{1/2} C^{j'0}_{\ell'010} d_{01}\right.$$

$$\left. + \quad \tau_c^{-1}(-1)^{j'+\ell'_1}\delta_{q',n'-m'}\langle 0j'm'|\tilde{d}_{q'}(\omega)|1\ell'n\rangle\right]. \tag{7.36}$$

The system of matrix equations (7.33), (7.36) may be solved in general form relative to the matrix element:

$$\langle 0jm|\tilde{d}_{q'}(\omega))|1\ell n\rangle = \sum_{j'm'\ell'n'} \langle 0jm, 1\ell n|\hat{G}_A^{-1}|0j'm', 1\ell'n'\rangle$$

$$\times(-1)^{\ell'-m'} C^{1q'}_{\ell'n'j'-m'} C^{j'0}_{\ell'010} \left(\frac{2\ell'+1}{3}\right)^{1/2} d_{01}. \tag{7.37}$$

Here

$$\langle 0jm, 1\ell n|\hat{G}_A|0j'm', 1\ell'n'\rangle = \Big[\langle 0jm, 1\ell n| - i(\omega\hat{1} - H^{\times})|0j'm', 1l'n'\rangle$$

$$+\tau_c^{-1}\left(1+(-1)^{j+\ell}\right)\delta_{jj'}\delta_{\ell\ell'}\Big]\delta_{mm'}\delta_{nn'}. \tag{7.38}$$

As is clear from (7.38) and (7.32), the matrix of operator $\hat{G}_A$, which is of principal importance, is diagonal over the m-projections of the rotational moment. Owing to this, (7.38) separates into parts, which are not connected with each other. Each of them contains transitions with the same m-components of the initial rotational level and the same n-components of the final level. Substituting (7.37) into (7.29), one gets the following expression for the spectral contour:

$$F(\omega) = \frac{|d_{10}|^2}{3\pi} \text{Re}$$

$$\sum_{jm\ell nj'm'\ell'n'\tilde{j}q'} \left[(2\tilde{j}+1)(2j'+1)\right]^{1/2} C^{1q'}_{\ell n\tilde{j}-m} C^{\ell 0}_{\tilde{j}010} C^{1q'}_{\ell'n'j'-m'} C^{\ell'0}_{j'010}$$

$$\times(-1)^{\tilde{j}+j}\langle 0jm, 1\ell n|\hat{G}_A^{-1}|0j'm', 1\ell'n'\rangle\rho_{0\tilde{j}mjm}. \tag{7.39}$$

So, the calculation of the shape of an IR spectrum in the case of anticorrelated jumps of the orienting field in a complete vibrational–rotational basis reduces to inversion of matrix (7.38). This may be done with routine numerical methods, but it is impossible to carry out this procedure analytically. To elucidate qualitatively the nature of this phenomenon, one should consider a simplified energy scheme, containing only the states with $j = 0, 1$. In [18] this scheme had four levels, because the authors neglected degeneracy of states with $j = 1$. Solution (7.39) [275] is free of this drawback and allows one to get a complete notion of the spectrum of such a system.

In addition to the block, which consists of four vertical transitions, corresponding to the four-level model ($j, l \leq 1$; $m = n = 0$), there are also four pairs of transitions, which correspond to the two spectral doublets P–Q ($j = 0, 1$; $m = 0$; $l = 1$; $|n| = 1$) and Q–R ($j = 1$, $|m| = 1$; $l = 0, 1$; $n = 0$). So, each of these doublets is doubly degenerate and transforms with increase of τ_c^{-1} independently of the other and of the spectral triplet of the four-level problem. Transitions between levels $j = l = 1$, $|m| = |n| = 1$ are forbidden optically and they are not connected by relaxation with the other ones. Therefore they do not appear in the spectrum even if interaction of the rotator with the orienting field is taken into account; thus they may be excluded from further consideration.

The following part of matrix (7.38) corresponds to transitions between $j, l \leq 1$, $m = n = 0$:

$$\begin{array}{cc||cccc||}
 & \ell' j' & 10 & 01 & 00 & 11 \\
\hline
 & 10 & -i(\Delta\omega - \omega_0) & 0 & i\tilde{V} & -i\tilde{V} \\
\ell j & 01 & 0 & -i(\Delta\omega + \omega_0) & -i\tilde{V} & i\tilde{V} \\
 & 00 & i\tilde{V} & -i\tilde{V} & -i\,\Delta\omega + 2\tau_c^{-1} & 0 \\
 & 11 & -i\tilde{V} & i\tilde{V} & 0 & -i\,\Delta\omega + \tau_c^{-1}
\end{array} . \tag{7.40}$$

Here $\Delta\omega = \omega - \omega_{10}$, $\tilde{V} = V/\sqrt{3}$, $\omega_0 = 2B$, $B = 1/(2I)$. Inverting (7.40) and substituting the result into (7.39), one can easily obtain the following expression for the triplet spectrum:

$$F_{\text{triplet}}(\nu) \sim \tau_c y^2 \left\{ \nu^2 \left[x^2(\nu^2 - 1) - y^2 \right]^2 + 4x^2(\nu^2 - 1) \right\}^{-1} , \tag{7.41}$$

where $\nu = \Delta\omega/\omega_0$. Spectrum (7.41) coincides with the spectrum obtained in [18]. For the sake of convenience of analysis, dimensionless combinations of parameters are introduced:

$$x = \omega_0 \tau_c, \quad y = 2\tilde{V}\tau_c . \tag{7.42}$$

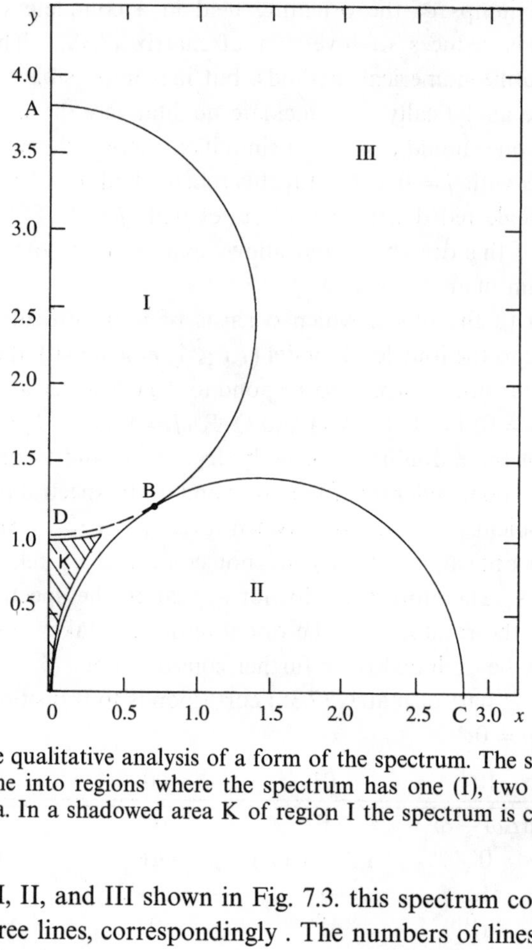

Fig. 7.3. The qualitative analysis of a form of the spectrum. The solid curves cut the x–y plane into regions where the spectrum has one (I), two (II) and three (III) maxima. In a shadowed area K of region I the spectrum is collapsed.

In regions I, II, and III shown in Fig. 7.3. this spectrum consists of one, two and three lines, correspondingly . The numbers of lines, like in [18], is supposed to be equal to the number of local maxima. In this reference system the regions are separated by two semicircles

$$\left(y - \sqrt{6}\right)^2 + x^2 = 2, \qquad y^2 + \left(x - \sqrt{2}\right)^2 = 2 \tag{7.43}$$

which touch each other at point B. This is not the case in the analysis in [18], which was not precise enough. Owing to this, region OBD turns out to be a subregion of III, while in reality it belongs to I. Therefore, as will be demonstrated, in this region a singlet spectrum is observed rather than a triplet. For the strong interaction the edge lines are forbidden and their widths are about τ_c^{-1}, while the central spectral component is permitted and its width is about $\left(\omega_0/V\right)^2 \tau_c^{-1}$. If $x > y$, then, on the contrary, the intense side lines of width about $(V/\omega_0)^2\tau_c^{-1}$ correspond to permitted rotational transitions, and a small peak in the centre with intensity about $(V/\omega_0)^2$ is a weakly resolved component of the Q-branch of width about τ_c^{-1}. Beneath the lower semicircle OBC the spectrum is a doublet, whose lines at $x > 1$ still have width about $(V/\omega_0)^2$, and at $x < 1$ width about $(V/\omega_0)\tau_c^{-1}$. Roughly speaking, in region III the spectrum corresponds to a strong or weak Stark effect, and in region II to a quasi-free rotation.

In region I (ABO) the spectral structure either is not resolved, or does not exist. The latter refers to region K (Fig. 7.3), in which the spectrum expands into a sum of three lines, centred at the vibrational frequency. The narrowest and most intense one has width $(\omega_0/V)^2\tau_c^{-1}$. In the remainder of the region these lines are split, but the intensity of the edge components is not large enough for them to manifest in the wings of the central components. In particular, for a strong interaction in region DBA, the satellites have width about τ_c^{-1}, exceeding $\tilde{V}$, and their intensity is equal to $(\omega_0/\tilde{V})^2$. As a result, they can be observed only as shoulders of the central resolved component, whose width is about $(\omega_0/V)^2\tau_c^{-1}$.

Region K is the only one in which the homogeneously broadened spectrum is observed. Outside this region the spectrum is either resolved (III) or partially resolved (II), or inhomogeneously broadened (I). When transitting from these regions into K, the spectrum collapses and broadens with increasing rate of orienting field fluctuation.

In order to have a full notion of the spectrum, it is enough now to complete the analysis by consideration of spectral doublets. The doublet, corresponding to transitions between levels ($j = 1$, $|m| = 1$; $l = 0, 1$; $n = 0$), is described by the following block of matrices (7.38):

$$
\begin{array}{cc|cccc}
 & \ell' n' j' m' & 0011 & 1011 & 001-1 & 101-1 \\
\hline
 & 0011 & -\mathrm{i}(\Delta\omega+\omega_0) & \mathrm{i}\tilde{V} & 0 & 0 \\
\ell n j m & 1011 & \mathrm{i}\tilde{V} & -\mathrm{i}\,\Delta\omega + 2\tau_c^{-1} & 0 & 0 \\
 & 001-1 & 0 & 0 & -\mathrm{i}(\Delta\omega+\omega_0) & \mathrm{i}\tilde{V} \\
 & 101-1 & 0 & 0 & \mathrm{i}\,\tilde{V} & -\mathrm{i}\,\Delta\omega + 2\tau_c^{-1}
\end{array} . \tag{7.44}
$$

Inverting (7.44) and substituting the result into (7.39), one can see that the spectrum consists of a pair of Lorentzian lines

$$F^I_{\text{doublet}}(\nu) \sim \operatorname{Re}\left[A_+/(\nu-\nu_+) + A_-/(\nu-\nu_-)\right] . \tag{7.45}$$

Their position and width are determined by the roots of the determinant of matrix (7.44):

$$\nu_{\pm} = -1/2 + \mathrm{i}/x \pm K \tag{7.46}$$

and the intensity is determined as follows

$$A_{\pm} = \pm\frac{1}{2K}\left(-2 - \mathrm{i}x\,\nu_{\pm}\right), \tag{7.47}$$

where

$$K = \left[(y/x)^2 + (\mathrm{i}/x + 1/2)^2\right]^{1/2}. \tag{7.48}$$

When the interaction is weak $y < 1 < x$, the doublet spectrum consists of an intense line with width $(V/\omega_0)^2\tau_c^{-1}$ and of component $\Delta\omega_{1/2} \sim \tau_c^{-1}$, forbidden by parameter $(V/\omega_0)^2$. If $y, x < 1$, then the width of the intensive component is about $V^2\tau_c$ and its spectral satellite is forbidden by $(V\tau_c)^2$. For strong interaction ($y > x, 1$), the intensity and width of both components become equal, and splitting becomes proportional to the field. So, the doublet spectrum may be either resolved or unresolved, but it does not collapse at any x and y. Absence of a homogeneous broadening region makes spectral transformation of the doublet qualitatively different from that of the triplet and of the doublet considered in [174, 158, 176]. In order to obtain the spectrum of another doublet ($j = 0, 1;\ m = 0;\ l = 1,\ |n| = 1$), it is sufficient to substitute $1/2$ for $-1/2$ in (7.46) and (7.48). Its analysis is quite analogous to the one performed above, and the conclusions concerning the behaviour of the corresponding spectrum coincide qualitatively with the above consideration.

In Fig. 7.4. there is shown the spectrum of a rotator, calculated taking account of m-degeneration of rotational levels in a truncated basis $j \leq 1$. For the sake of comparison, the corresponding spectral contour from [8] is shown by a dashed line. It can be seen that taking account of m-degeneration of rotational levels leads to both qualitative and quantitative correction of the results of [18]. It is also evident that, even in region K (Fig. 7.3), the summary spectrum (Fig. 7.4) is inhomogeneously broadened: it is an envelope of three lines, split at ω_0 with ratio of intensities 2:1:2 and with widths about $V^2\tau_c$, $\sim (\omega_0/V)^2\tau_c^{-1}$ and $V^2\tau_c$, correspondingly.

The analysis performed allows one to judge qualitatively about the processes, which go on in a spectrum when the Stark structure of rotational transitions is averaged by fluctuations of the orienting field. If y decreases, x being fixed, the resolved Stark structure with the intense Q-branch in the centre transforms into the spectrum of a quasi-free rotator. If $x < 1$, the spectrum may be singlet in the intermediate region.

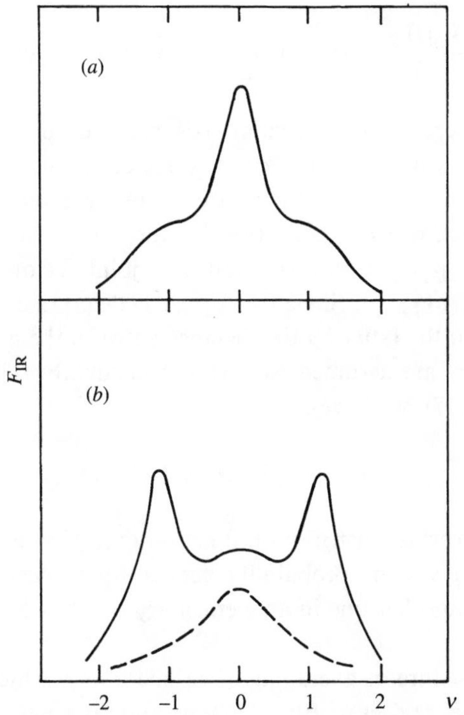

Fig. 7.4. The spectrum of a rotator in the truncated basis ($j \leq 1$): (a) $x = 0.8$, $y = 1.6$, (b) $x = 0.1$, $y = 0.8$.

However, its qualitative difference from the spectrum obtained in the collisional model is that it is inhomogeneously broadened and its structure may be observed in a nonlinear optical experiment.

7.5 Non-correlated perturbation

The model of non-correlated potential fluctuations is of special interest. First, it can be solved analytically, second, the assumption that subsequent values of orienting field are non-correlated is less constrained from the physical point of view. The theory allows for consideration of a rather general orienting field. When the spherical shape of the cell is distorted and its symmetry becomes axial, the anisotropic potential is characterized by the only given axis $\boldsymbol{e}$. However, all the spherical harmonics 'built' on this vector contribute to its expansion, not only the term of lowest order over L as in (7.6)

$$V(t) = \sum_{LM} \frac{4\pi V_L(t)}{2L+1} Y^*_{LM}\,[\boldsymbol{e}(t)]\; Y_{LM}\,(\boldsymbol{n}) = \sum_{L} V_L(t) P_L\,[\cos\alpha(t)]\,. \quad (7.49)$$

Here $\boldsymbol{n}$ is an operator of molecular axis orientation. In the classical description, it is just a unitary vector, directed along the rotator axis. Angle α sets the declination of the rotator from the liquid cage axis. Now a random variable, which is conserved for the fixed form of the cell and varies with its hopping transformation, is a joint set of vectors $\{\boldsymbol{e}, \boldsymbol{V}\}$, where $\boldsymbol{V} = \{V_1, \dots V_L, \dots\}$. Since the former is determined by a break of the symmetry and the latter by the distance between the molecule and its environment, they are assumed to vary independently. This means that in addition to (7.17), we have

$$\psi\,(\boldsymbol{V}) = \int F\left(\boldsymbol{V}', \boldsymbol{V}\right) \psi\left(\boldsymbol{V}'\right)\, \mathrm{d}\boldsymbol{V}', \quad (7.50)$$

where $\psi(\boldsymbol{V})$ is a certain distribution function over $\boldsymbol{V}$ in equilibrium, and kernel $F(\boldsymbol{V}', \boldsymbol{V})$ gives the probability density for the cell to change its configuration to another one in an elementary event of cavity rearrangement.

Using the procedure of averaging described above, which employs the property of kernel $f(\boldsymbol{e}', \boldsymbol{e})$ of Eq. (7.17), we obtain a new master kinetic equation which is still integral with respect to $\boldsymbol{V}$

$$\begin{aligned} \frac{\partial}{\partial t}\tilde{d}_{\ell q'}\,(t, \boldsymbol{V}) &= \mathrm{i}\left(H^{\times}\,(\boldsymbol{V}) - \frac{1}{\tau_c}\right)\tilde{d}_{\ell q'}\,(t, \boldsymbol{V}) \\ &\quad + \sum_{q'} \int F\left(\boldsymbol{V}', \boldsymbol{V}\right) \hat{T}_{q'q}\tilde{d}_{\ell q}\left(t, \boldsymbol{V}'\right) \mathrm{d}\boldsymbol{V}' \quad (7.51) \end{aligned}$$

where the following designations are made:

$$\tilde{d}_{\ell q'}\,(t, \boldsymbol{V}) = \sum_q (-1)^q D(\Omega)\, d_{\ell q}\,(t, \boldsymbol{V}, \boldsymbol{e})\, D(-\Omega) D^{\ell}_{q'-q'}(\Omega) \frac{\mathrm{d}\Omega}{2\pi}\,.$$

Euler angles Ω define the $\boldsymbol{e}$ orientation, and an orientational correlation function of lth order is introduced in the usual way:

$$K_{\theta,\ell}(t) = \mathrm{Tr} \sum_q (-1)^q\, d_{\ell-q} \int \mathrm{d}\boldsymbol{V} \int \mathrm{d}\boldsymbol{e}\; \rho\, d_{\ell q}\,(t, \boldsymbol{V}, \boldsymbol{e})\,. \quad (7.52)$$

After transition into the moving reference system with axis z, which is always directed along the instantaneous symmetry axis of the liquid cage, and integration over variable $\boldsymbol{e}$, only partial integration over $\boldsymbol{V}$ remains in $K_{\theta,\ell}$

$$\begin{aligned} K_{\theta,\ell}(t) &= \mathrm{Tr} \int \mathrm{d}\boldsymbol{V} \int \mathrm{d}\boldsymbol{e}\, D(-\Omega) D(\Omega) \\ & \sum_q (-1)^q d_{\ell-q} D(-\Omega) D(\Omega) \rho d_{\ell q}\,(t, \boldsymbol{V}, \boldsymbol{e}) \\ &= \mathrm{Tr} \int \mathrm{d}\boldsymbol{V} \sum_{q'} \rho \tilde{d}_{\ell q}\,(t, \boldsymbol{V})\, d_{\ell q'}\,. \end{aligned} \tag{7.53}$$

Here it is taken into account that density matrix ρ, being a scalar, commutates with any rotation operator, and $\tilde{d}_{lq}$ defined in Eq. (7.51) is used. After an analogous transformation, in master equation (7.51) there remains the Hamiltonian, which does not depend on $\boldsymbol{e}$:

$$D(\Omega) V\,(\boldsymbol{V}, \boldsymbol{e})\, D(-\Omega) = \sum_L V_L \left(\frac{4\pi}{2L+1}\right)^{\frac{1}{2}} Y_{L0}\,(\boldsymbol{n}) = V\,(\boldsymbol{V})\,. \tag{7.54}$$

The spectrum

$$G_\ell\,(\omega) = \frac{1}{\pi} \,\mathrm{Re} \sum_{q'} \mathrm{Tr} \left\{ \int d\boldsymbol{V}\ \rho\ \tilde{d}_{\ell q'}\,(\omega, \boldsymbol{V})\, d_{\ell q'} \right\} \tag{7.55}$$

is expressed via Fourier transforms of the dipole moment components

$$\tilde{d}_{\ell q'}\,(\omega, \boldsymbol{V}) = \int_0^\infty \tilde{d}_{\ell q'}\,(t, \boldsymbol{V})\, \mathrm{e}^{\mathrm{i}\omega t}\, \mathrm{d}t\,. \tag{7.56}$$

For non-correlated perturbation the kernel of Eq. (7.51) is greatly simplified:

$$F\,(\boldsymbol{V}', \boldsymbol{V}) = \psi\,(\boldsymbol{V})\,. \tag{7.57}$$

As a result its Fourier transform gives the following:

$$\begin{aligned} &\left\{-\mathrm{i}\left[\omega + H^\times(\boldsymbol{V})\right] + \tau_c^{-1}\right\} \tilde{d}_{\ell q'}\,(\omega, \boldsymbol{V}) \\ &= (-1)^{q'} \psi\,(\boldsymbol{V})\, d_{\ell - q'} + \tau_c^{-1} \psi\,(\boldsymbol{V}) \int \mathrm{d}\boldsymbol{V}' \sum_q \hat{T}_{q'q}\, \tilde{d}_{\ell q'}\,(\omega, \boldsymbol{V}')\,. \end{aligned} \tag{7.58}$$

The solution of Eq. (7.58) may be found with extension of the routine method [20] for the case of the matrix operator of the kinetic part of (7.58). However, for the sake of simplicity, let us suppose that the variation of $\boldsymbol{e}$ direction is non-correlated, too. Then

$$f\ \left(\boldsymbol{e}', \boldsymbol{e}\right) = \varphi\,(\boldsymbol{e}) = \mathrm{const}\,. \tag{7.59}$$

As a result, we have the following formula for the $\hat{T}$ operator

$$\begin{aligned}&\langle jm(j'm')^{+}\left|\hat{T}_{q'q}\right| jm_1(j'm'_1)^{+}\rangle\\&=(-1)^{m+m'}(2\ell+1)^{-1}C^{\ell q}_{j'm'j-m}C^{\ell q}_{j'm'_1j-m_1}\,.\end{aligned}\tag{7.60}$$

Taking into account (7.60), we write Eq. (7.58) in the following form

$$\begin{aligned}&\langle jm\left|\tilde{d}_{\ell q'}(\omega,V)\right| j'm'\rangle=\\&\sum_{j_1m_1j'_1m'_1}\langle jm(j'm')^{+}\left|\frac{1}{-\mathrm{i}\,[\omega+H^{\times}(V)]+\tau_c^{-1}}\right| j_1m_1\,(j'_1m'_1)^{+}\rangle(-1)^{j'_1-m_1}\\&\times(2\ell+1)^{-1/2}\,C^{\ell q'}_{j'_1m'_1j_1-m_1}\,\psi(V)\\&\times\left[\langle j_1||d_\ell||j'_1\rangle+(-1)^{j_1+j'_1}\tau_c^{-1}\langle j_1||\overline{d_\ell(\omega)}||j'_1\rangle\right].\end{aligned}\tag{7.61}$$

The matrix element of operator $d_{l-q'}$ is written in terms of the Wigner–Eckart theorem, and the integral part is denoted as

$$\begin{aligned}\langle j_1||\overline{d_\ell(\omega)}||j'_1\rangle &= (2\ell+1)^{1/2}\sum_{q'\tilde{m}\tilde{n}}(-1)^{j_1-\tilde{m}}C^{\ell q'}_{j'_1\tilde{n}j_1-\tilde{m}}\\&\qquad\times\int \mathrm{d}V\,\langle j_1\tilde{m}|d_{\ell q'}(\omega,V)|j'_1\tilde{n}\rangle\,.\end{aligned}\tag{7.62}$$

At this stage the problem may be considered to be solved in principle since one can readily calculate the spectrum (7.55) with $d_{lq'}(\omega, V)$ found from Eq. (7.61).

So,

$$\begin{aligned}G_\ell(\omega) &= \frac{1}{\pi}\,\mathrm{Re}\sum_q \mathrm{Tr}\,\rho\int\tilde{d}_{\ell q'}(\omega,V)d_{\ell q'}\,\mathrm{d}V\\&=\frac{1}{\pi}\,\mathrm{Re}\sum_{jj'}\rho_j\langle j||\overline{d_\ell(\omega)}||j'\rangle\langle j'||d_\ell||j\rangle\end{aligned}\tag{7.63}$$

is expressed through (7.62) relative to which Eq. (7.61) may be closed. For this purpose it suffices for the latter to be multiplied by $(-1)^{j-m}(2l+1)^{-1/2}C^{lq'}_{j'm'\,j-m}$, summed over q', m, m', and integrated over V. The result is an algebraic equation with respect to the quantity required, which can be easily solved. Using it in Eq. (7.63), we obtain the final solution in quadratures:

$$G_\ell(\omega)=\frac{1}{\pi}\mathrm{Re}\sum_{jj'j_1j'_1}\rho_j\langle j'||d_\ell||j\rangle\langle j(j')^{+}\left|\frac{1}{\hat{A}^{-1}(\omega)-\tau_c^{-1}}\right| j_1(j'_1)^{+}\rangle\langle j'_1||d_\ell||j_1\rangle\,,\tag{7.64}$$

where

$$\langle\langle j(j')^+;\ \ell|\hat{A}(\omega)|j_1(j_1')^+;\ell\rangle\rangle$$
$$= \sum_{q'} \int \mathrm{d}V \langle\langle j(j')^+;\ell q \left| \frac{\psi(V)}{-\mathrm{i}\,[\omega + H^\times(V)] + \tau_c^{-1}} \right| j_1(j_1')^+;\ell q\rangle\rangle, \quad (7.65)$$

and

$$|j_1(j_1')^+;\ \ell q\rangle\rangle = \sum_{m_1 m_1'} (-1)^{j_1-m_1}\, C^{\ell q}_{j_1' m_1' j_1 - m_1} |j_1 m_1 (j_1' m_1')^+\rangle (2\ell+1)^{-1/2} \quad (7.66)$$

is the Liouville vector written with account taken of the spectral transition rank.

Calculation of spectra with formula (7.66) leads to some numerical complications, as it requires inversion of matrices of very high rank. Therefore it seems to be reasonable to use for the processing of real experimental data the classical version of the theory, described in Appendix 8, rather than the quantum one.

Even without using numerical methods, one can analyse some physically sound limiting cases of the exact solution for the case of strong collisions (7.64). First of all, (7.64) evidently reduces to the results of Robert and Galatry in the quasi-static case, i.e. when $\tau_c \to \infty$. An opposite limiting case of fast fluctuations

$$V\tau_c \ll 1 \quad (7.67)$$

corresponds to so-called stochastic perturbation theory and may be considered outside the frames of a given model representation, e.g. of a stochastic model of the $V(t)$ random variation. As this case is of special significance, it is considered here in two different versions. In the present section we shall obtain the expansion of the exact solution (7.64) of the strong collision model over parameter (7.67). In the next section the solution of a stochastic Liouville equation (SLE) will be averaged on the sole basis of condition (7.67) without any assumption on the stochastic properties of $V(t)$.

Expanding (7.64) in a convergent series, one can sum over m in the zeroth and first order of expansion (which turns out to be zero) and present the result, accurate to second order in $V\tau_c$, as

$$\langle\langle j(j')^+;\ \ell|\hat{A}(\omega)|j_1(j_1')^+;\ \ell\rangle\rangle = \frac{1}{-\mathrm{i}(\omega - \omega_{j'j}) + \tau_c^{-1}}$$

$$\times \left[\delta_{jj_1}\delta_{j'j_1'} - \sum_{LL'q} \overline{V_L V_{L'}} \langle\langle j(j')^+;\ \ell q|C^\times_{L0}\, \frac{1}{-\mathrm{i}(\omega + H_0^\times) + \tau_c^{-1}} \right.$$

$$C^{\times}_{L0}\,\frac{1}{-\mathrm{i}(\omega+H^{\times}_0)+\tau_c^{-1}}|j_1(j_1')^+;\ \ell q\rangle\rangle\Bigg]\ , \tag{7.68}$$

where $\overline{V_L V_{L'}} = \overline{V_L}\ \overline{V_{L'}}$ at $L \neq L'$ and $\overline{V_L^2} = \int \Phi_L(V_L)V_L^2\, \mathrm{d}V_L$ holds not only for rotational, but also vibrational-rotational spectra, and C_{L0} designates the factor for V_L in the sum of (7.54). It is assumed above that the vectors of (7.66) contain also the vibrational indices v and v' of optically coupled levels, and the eigenvalue matrix of the unperturbed Hamiltonian involves corresponding frequencies $\omega_{vv'}$. When an isolated vibrational transition is under study, its frequency can be removed from Eq. (7.68) by making it coincident with the zero of the frequency scale. It is hence only the frequency shift of the rotational spectral components induced by vibrational–rotational interactions that remains in $\omega_{jj'}$.

Inverting the matrix (7.68) and summing it over m and q up to the end, we obtain in the same order of $V\tau_c$

$$\begin{aligned}&\langle\langle j(j')^+;\ \ell\left|\hat{A}^{-1}(\omega)-\tau_c^{-1}\right|j_1(j_1')^+;\ \ell\rangle\rangle\\ &=\left[-\mathrm{i}\left(\omega-\omega^{\ell}_{j'j}\right)+\chi^{\ell}_{j'j}(\omega)\right]\delta_{jj_1}\delta_{j'j_1'}\\ &-\langle\langle j(j')^+;\ \ell\left|\hat{\Lambda}(\omega)\right|j_1(j_1')^+;\ell\rangle\rangle,\end{aligned} \tag{7.69}$$

where

$$\begin{aligned}\chi^{\ell}_{j'j}(\omega)&=\sum_L\frac{\overline{V_L^2}}{2L+1}\\ &\times\left[\sum\tilde{j}\frac{\left(C^{\tilde{j}0}_{j0L0}\right)^2}{-\mathrm{i}\left(\omega-\omega^{\ell}_{j'\tilde{j}}\right)+\tau_c^{-1}}+\sum_{\tilde{j}}\frac{\left(C^{\tilde{j}0}_{j'0L0}\right)^2}{-\mathrm{i}\left(\omega-\omega^{\ell}_{\tilde{j}j}\right)+\tau_c^{-1}}\right.\\ &\left.-\frac{2\left[(2j+1)(2j'+1)\right]^{1/2}}{-\mathrm{i}\left(\omega-\omega^{\ell}_{j'j}\right)+\tau_c^{-1}}C^{j0}_{j0L0}C^{j'0}_{j'0L0}\begin{Bmatrix} j & j' & \ell\\ j' & j & L\end{Bmatrix}\right]\end{aligned} \tag{7.70}$$

$$\begin{aligned}&\langle\langle j(j')^+;\ \ell|\hat{\Lambda}(\omega)|\ j_1(j_1')^+;\ \ell\rangle\rangle=\\ &\sum_L\frac{\overline{V_L^2}}{2L+1}\left[(2j+1)(2j'+1)\right]^{1/2}\ C^{j_10}_{j0L0}C^{j_1'0}_{j'0L0}\\ &\times\begin{Bmatrix} j & j' & \ell\\ j_1' & j_1 & L\end{Bmatrix}\left[\frac{1}{-\mathrm{i}(\omega-\omega^{\ell}_{j'j_1})+\tau_c^{-1}}+\frac{1}{-\mathrm{i}(\omega-\omega^{\ell}_{j_1'j})+\tau_c^{-1}}\right].\end{aligned}$$

The presence of spectral exchange is proved by the non-diagonal character of the matrix (7.69). The peculiar implications of this very fact are analysed in detail in the next section. Here we restrict ourselves to the secular approximation and consider the spectrum at $\Lambda = 0$, as was done in papers by Bonamy [273] and Frenkel [279], which from the very beginning developed the secular theory. After this, the theory describes only a well-resolved spectrum

$$G_\ell(\omega) = \frac{1}{\pi}\mathrm{Re}\sum_{jj'} \rho_j\,|\langle j'||d_\ell||j\rangle|^2 \frac{1}{-\mathrm{i}(\omega-\omega^\ell_{j'j})+\chi^\ell_{j'j}(\omega)}\,. \tag{7.71}$$

To compare this result with that obtained within perturbation theory [273, 279], one must additionally assume the perturbation correlation function to be exponential, as in [273, 279, 280]. In this case, the purely rotational spectrum [273, 279] and that obtained with Eq. (7.71) coincide, if the ω-dependence of the χ operator is neglected and $(\omega-\omega^\ell_{j'j})\tau_c \ll 1$.

As a result, (7.71) consists of Lorentzian lines, centred near the eigenvalues of the free rotator. Their widths can be found as follows:

$$\Gamma^{(\ell)}_{j'j} = \mathrm{Re}\ \chi^{(\ell)}_{j'j}\left(\omega_{j'j}\right)\,. \tag{7.72}$$

Conserving only the first two terms of the anisotropic potential expansion (7.49) and summing as noted in (7.70) over $\tilde{j}$, we get

$$\Gamma^{(\ell)}_{j'j} = \frac{1}{3}\,\overline{V_1^2}\,\tau_c\,\Gamma^{\ell,1}_{j'j} + \frac{1}{5}\,\overline{V_2^2}\,\tau_c\,\Gamma^{\ell,2}_{j'j}\,. \tag{7.73}$$

The first term is additive relative to the widths of rotational states (for the sake of simplicity we consider the case $\ell = 1$)

$$\Gamma^{\mathrm{IR},1}_{j'j} = f^1(j)+f^1(j');\;\; f^1(j) = \frac{1}{2j+1}\left(\frac{j}{1+(jx)^2}+\frac{j+1}{1+(j+1)^2x^2}\right). \tag{7.74}$$

Additivity is broken by the second term of (7.73), which corresponds to the second harmonics of the potential expansion

$$\begin{aligned}
\Gamma^{\mathrm{IR},2}_{j'j} &= f^2(j)+f^2(j')-2\frac{j(j'+1)}{(2j+3)(2j'-1)}\\
f^2(j) &= \frac{1}{2j+1}\left(\frac{j(j+1)(2j+1)}{(2j+3)(2j-1)}+\frac{3}{2(2j+3)}\,\frac{(j+1)(j+2)}{1+(2j+3)^2x^2}\right.\\
&\quad\left.+\frac{3}{2(2j-1)}\,\frac{j(j-1)}{1+(2j-1)^2x^2}\right);\quad x = 2B\tau_c\,.
\end{aligned} \tag{7.75}$$

The latter three expressions demonstrate that in the present case the widths of the absorption spectra and the anisotropic Raman spectra are

Table 7.1. *Liquid-phase systems having spectra with resolved gas-like structure.*

Spectroscopically active molecule	Solvent	Reference
H_2	Ar, SF_6, CCl_4, C_6H_{12} C_7H_{16}, C_8H_{18}, CF_2Cl_2	[281–284]
D_2	Ar, SF_6, CCl_4	[281, 282, 284]
HF	SF_6	[285, 286]
DF	SF_6	[285]
HCl	Ar, Kr, Xe, SF_6	[16, 287, 285, 288–290]
DCl	SF_6	[288]
HBr	C_2F_6	[288]

determined by both adiabatic (independent of parameter x) and non-adiabatic modulations of the interaction of the rotator with the liquid cage orienting field.

The spectrum calculated in the secular non-adiabatic approximation reproduces some special peculiarities of the spectra observed. In following papers (see Table 7.1) for diatomic molecules the dependence of the resolved spectra components on the rotational quantum number was described. As an example, the experimental dependence $\chi(j) = \Gamma^1_{j+1,j}/\Gamma^1_{2,1}$ is shown in Fig. 7.5.

The values of the half-widths of the components of the rotational absorption spectrum of HCl, dissolved in various noble gases, are borrowed from [291]. In order to make this example obvious, a continuous curve is drawn through the calculated points. Comparison between experimental data and calculated results demonstrates, in line with the qualitative agreement, a good numerical coincidence of the observed j-dependence of the half-widths of the rotational lines with the theoretical one in the case of HCl dissolved in Kr and Xe. This allows one to estimate the model parameters for these systems: dispersion of the potential

$$\frac{1}{3}\,\overline{V_1^2} = \begin{cases} 950\ \text{cm}^{-2} & \text{HCl/Kr} \\ 1050\ \text{cm}^{-2} & \text{HCl/Xe} \end{cases} \tag{7.76}$$

and the frequency of its fluctuations

$$\tau_c^{-1} = \begin{cases} 50\ \text{cm}^{-1} & \text{HCl/Kr} \\ 42\ \text{cm}^{-1} & \text{HCl/Xe}. \end{cases} \tag{7.77}$$

For the system HCl/Ar the values of $\chi(j)$ quantitatively do not agree

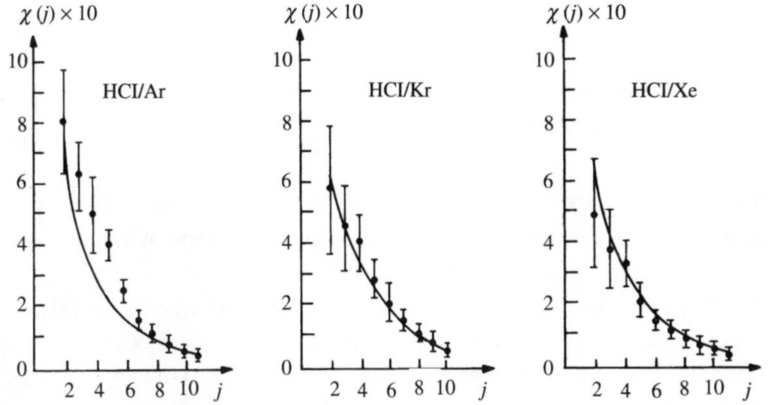

Fig. 7.5. Fitting of the j-dependence of the widths for the rotational absorption spectrum of HCl in liquid noble gases. Experimental data are shown by dots, theoretical calculations by solid curve.

with experiment. The possible explanation is that in this case the next terms of the expansion (7.49) contribute essentially. Let us also note that the extracted parameters of (7.76) and (7.77) hardly satisfy the criterion of perturbation theory. An analogous problem arose also in [273], as the assumption that fluctuation process $E(t)$ is Gaussian does not allow use of expression (7.71) outside the limits of stochastic perturbation theory.

Paper [16] reported unusual behaviour of the widths of rotational components of the P–R doublet of HCl dissolved in SF_6. They decreased with increasing temperature. The widths of spectral lines, obtained with (7.73), really must decrease with increasing temperature, because τ_c decreases due to intensification of thermal motion, and $\overline{V_L^2}$ due to thermal expansion.

Let us note that sometimes spectra are so well resolved that even some asymmetry of lines is registered [283]. This means that the wings of individual components of a spectral band become observable. Hence, non-adiabatic secularization becomes too crude an approximation. The experiment [283] was interpreted by Kouzov [280] taking into account the ω-dependent diagonal part of the relaxational operator.

Let us analyse the consequences of account being taken of the ω-dependence of $\chi_{j,j'}$ in the central part of the liquid phase P–R doublet in the IR spectrum of a linear rotator. For this purpose, we consider the

j-component of the P-branch near the centre of the band

$$\omega < 2B\,. \tag{7.78}$$

Its partial contribution to the spectral density

$$G_1^j \simeq \mathrm{Re}\; j\; \left[-\mathrm{i}(\omega + j)\,x + \chi_{j-1,j}(\omega)\right]\,, \tag{7.79}$$

where for simplicity of analysis it was assumed that $L = 1$. When $x < 1$, due to (7.78) at the same time $\omega x < 1$. Neglecting frequency-dependence of $\chi_{j'j}$, we find that the partial contribution of the j-component is determined by the Lorentzian line wing, centred at frequency $2Bj$, and the complete spectrum (which is a sum of symmetrical components of both branches) has in this case a gap in the centre. However, if $x > 1$, then the centre of the band corresponds to the partial component wing, and (7.79) yields

$$G_{IR}^j(\omega) \sim \frac{4j\overline{V_1^2}}{3(4j^2-1)\,(2B)^2}\;\frac{\tau_c^{-1}}{\left[\omega - (2\overline{V_1^2}/3B)\,(j/4j^2-1)\right]^2 + \tau_c^{-2}}\,. \tag{7.80}$$

Eq. (7.80) describes a Lorentzian line centred near the zero of frequencies, of width τ_c^{-1} and integral intensity $\overline{V_1^2/B}$. Owing to this, a component appears in the complete spectrum, which is forbidden for an unperturbed linear rotator. Its maximum intensity

$$G_{\mathrm{IR}}(0) \approx \frac{\overline{V_1^2}}{B^2}\,\tau_c \tag{7.81}$$

is small in comparison with the j-components of P and R branches. The complexes that are formed seem to be responsible for appearance of an essentially intense Q-branch, observed in [16].

In conclusion we should point out once more that appearance of the P–R doublet in liquids is a very interesting phenomenon, interpretation of which is still a challenging problem. Here it is good to recall an idea put forward by Bulanin and Orlova in the late fifties [292]. In accord with it, the rotation of a part of a molecule with small kinetic energy is stopped by anisotropic interaction, and molecules with energy exceeding the barrier height are hindered just slightly. The IR spectrum of a solution of this type has to have a weakly perturbed high-frequency P–R doublet for large-frequency detuning from the centre and a suppressed rotational structure near the vibrational frequency. The model suggests estimation of the height of the barrier relative to the energy of the last (i.e. that with largest number j) rotational component, which is not observed in the central part of the spectrum. A comparative analysis of the voluminous experimental data shows (see Table 7.1) that the model oversimplifies the physical picture: the deviation between the estimations of energy of interaction of a rotator with an anisotropic environment from IR and Raman spectra of the same solution was too large.

The liquid phase cage model accounts for appearance in the spectrum of resolved rotational components by effective 'isotropization' of the rapidly fluctuating interaction. This interpretation of the gas-like spectral manifestations seems to be more adequate to the nature of the liquid phase, than the impact description or the hypothesis of over-barrier rotation. Whether it is possible to obtain in the liquid cage model triplet IR spectra of linear rotators with sufficiently intense Q-branch and gas-like smoothed P–R structure has not yet been investigated. This investigation requires numerical calculations for spectra at an arbitrary value of parameter $V\tau_V$.

7.6 The Hubbard relation in the frames of the liquid cage model

In Chapter 2 the Hubbard relation was derived. Its derivation is described as in the original version [30]: the variation of orientation with time is governed by the Liouville equation with angular momentum as a stochastic variable, fluctuations of the latter being considered as fast enough. From this point of view, the liquid cage model is the next logical step in development of the method. Really, now time variation of both orientation and angular momentum is determined by the stochastic Liouville equation (SLE), in which input noise is an anisotropic part of the intermolecular potential. In this case the situation of fast fluctuation of the orienting potential $\langle V^2\rangle\tau_V^2 \ll 1$ contains two limiting regions with respect to the relation between molecular rotational constant B and effective rotational damping $\langle V^2\rangle\tau_V$. If the following inequality holds:

$$\langle V^2\rangle\, \tau_V < 2B\ell < \bar{\omega}\,\ell \tag{7.82}$$

then the angular momentum relaxes very slowly and the width of the resolved rotational components $\langle V^2\rangle\tau_V$ tends to be around τ_J^{-1}. So, the change of inequality sign

$$\langle V^2\rangle\, \tau_V > \bar{\omega}\,\ell \tag{7.83}$$

resembles the condition $\bar{\omega}\ell\tau_J < 1$, which leads to the Hubbard relation (see Chapter 2). It is quite realistic to check this relation by NMR methods. This was done for a number of systems. It was demonstrated in Chapter 6 that the spectral equivalent of the Hubbard relation is proportionality between rotational relaxation time and the width of the Lorentzian line, which is formed by collapsed rotational branches (by P–R doublet in IR, or by O–Q–S triplet in Raman spectra). Spectral verification of the Hubbard relation is straightforward, and therefore more reliable. However, spectral recording as such allows direct verification of isothermal narrowing of the collapsed line only in the gas phase. For liquids it is impossible, until the temperature- and density-dependence of τ_J are known independently. In this respect the combined NMR–optical measurements are the most hopeful. The relaxation time, measured with NMR, is of single-particle nature, as cross-correlational contributions are strongly suppressed due to a short-range decrease of intermolecular magnetic coupling. So, if the Hubbard relation is proved experimentally for a given system, this fact implies that diffusional evolution of orientation is built by frequently interrupted inertial rotations, when angular momentum is a well-defined dynamical variable. This assertion is not trivial because the numerical factor in the Hubbard relation corresponds to the moment of inertia of a single molecule, consequently, angular momentum, mentioned above, is a one-particle dynamical variable.

This brief analysis explains why it is very important to know whether the Hubbard relation is reproduced in the liquid cage model. The existence of the Hubbard limit means that orientational relaxation is insensitive to the precise details of the interaction. Below, it is shown that this is the case.

Separating the Hamiltonian into parts describing a free rotator and the orienting effect of a liquid cage, we have in a coordinate representation

$$H = H_0 + V(t),$$

where

$$H_0 = B\hat{J}^2, \tag{7.84a}$$

$$V(t) = \sum_{LM} \frac{4\pi V_L}{2L+1} Y^*_{LM}(\boldsymbol{N}(t))\, Y_{LM}(\boldsymbol{n}). \tag{7.84b}$$

The vector $\boldsymbol{n}$ is directed along the molecular axis and $\boldsymbol{N}$ is, in our approximation, the only vector describing the orientation of the directing field for every order L of the expansion of V.

Along the trajectory within the ensemble of $\boldsymbol{N}(t)$ realizations, any molecular operator X of a target molecule satisfies a SLE

$$\frac{\partial}{\partial t} X = \mathrm{i}\, H^{\times}\{\boldsymbol{N}(t)\}\, X, \tag{7.85}$$

where $\hbar = 1$. To find the correlation function of x, it is necessary to average the general solution of Eq. (7.85):

$$K_X(t) = \sum_{jm} \langle jm|\rho X^+ \overline{X(t)}|jm\rangle \,, \tag{7.86}$$

where averaging over random noise $N(t)$ is designated by the overbar, X^+ is the Hermitian conjugate, ρ is the density matrix and $|jm\rangle$ is the usual set of eigenfunctions of $\hat{J}$. For the first non-vanishing term of a stochastic perturbative expansion in the G-ordering representation we have

$$\frac{\partial}{\partial t}\bar{X}(t) = \mathrm{i}H_0^\times \bar{X}(t) - \int_0^t \mathrm{d}t' \; \Gamma\left(t-t'\right) \; \bar{X}(t'); \;\; \bar{X}(t=0) = X \,, \tag{7.87}$$

where

$$\Gamma(\tau) = \overline{V^\times(\tau) \; \exp\;\left(\mathrm{i}H_0^\times \tau\right) \; V^\times(0)} \,. \tag{7.88}$$

Owing to liquid isotropy, the averaged matrix elements of $\bar{X}$ are expressed in the Wigner–Eckart form (omitting further on the overbar denoting averaging)

$$\langle jm|X_{\ell q}|j'm'\rangle = \frac{1}{(2j+1)^{\frac{1}{2}}} \; C^{jm}_{j'm'\ell q} \langle j||X_\ell||j'\rangle \,, \tag{7.89}$$

where ℓ, as earlier, is the tensor rank of the operator X, q is the component of the tensor irreducible representation coefficients. Solving (7.87) with the help of the Laplace transform and substituting the solution into (7.86), we obtain after summation over magnetic quantum numbers

$$\begin{aligned} \tilde{K}_{X_\ell}(p) \;\; &= \;\; \int_0^\infty \mathrm{e}^{-pt} K_{X_\ell}(t)\,\mathrm{d}t = \sum_{j'j} \rho_j \langle j||X_\ell||j'\rangle \; \langle j'||\tilde{X}_\ell(p)||j\rangle \\ &= \;\; \sum_{\{j\}q} \frac{1}{2\ell+1} \; \rho_j \langle j||X_\ell||j'\rangle \; \langle\langle T(j'j)_{\ell q} \left| \frac{1}{p - \mathrm{i}H_0^\times + \tilde{\Gamma}(p)} \right| \\ &\quad \times T\left(j_1'j_1\right)_{\ell q}\rangle\rangle \; \langle j_1'||X_\ell||j_1\rangle \,. \end{aligned} \tag{7.90}$$

In (7.90) a slightly modified notation is introduced for convenience for the 'bra' and 'ket' vectors in the Liouville space for the resolvent superoperator

$$|T\;\left(j'j\right)_{\ell q}\rangle\rangle = \sum_{mm'} (-1)^{j-m} \; C^{\ell q}_{j'm'j-m} \;\; |j'm\rangle \; \langle jm| \,. \tag{7.91}$$

The corresponding explicit representation of the memory operator is the four-indexed one:

$$\tilde{\Gamma}^{\ell}_{j'jj'_1j_1}(p) = \langle\langle T(j'j)_{\ell q}\,|\tilde{\Gamma}(p)|\,T\,(j'_1j_1)_{\ell q}\rangle\rangle$$
$$= \sum_L \frac{2V_L^2\tau_L}{2L+1} \times \left[\delta_{jj_1}\delta_{j'j'_1} - (-1)^{j'-j+\ell}[(2j+1)(2j'+1)]^{\frac{1}{2}} C^{j'_1 0}_{j'0L0} C^{j_1 0}_{j0L0} \right.$$
$$\left. \begin{Bmatrix} j'_1 & j' & L \\ j & j_1 & \ell \end{Bmatrix}\right] . \tag{7.92}$$

Two simplifications were made in (7.92) without significant loss of generality: for the correlation functions of the directing field

$$\tilde{K}^L_V(p) = \int_0^\infty e^{-pt}\overline{V_L(t)\,V_L(0)}\,dt \tag{7.93}$$

the relaxation times τ_L are small in comparison with the proper rotational frequencies, $\omega_{jj_1}\tau_L \ll 1$, and with the corresponding amplitudes, $V_L\tau_L \ll 1$, in line with the aforementioned $\langle V^2\rangle\tau_V^2 < 1$. Note that the symmetry properties of the $6j$ symbols forbid expression of spectral line amplitudes $\langle j\|X_\ell\|j'\rangle$ with different ℓ through each other (e.g. Raman scattering cannot be mixed with IR absorption).

Let us analyse the properties of a few $\tilde{\Gamma}^\ell$ operators for $\ell = 0, 1$. By direct summation it is easy to verify that

$$\tilde{\Gamma}^0|1\rangle\rangle \;=\; \sum_{j_1} \tilde{\Gamma}^0_{jjj_1j_1}\,(2j_1+1)^{\frac{1}{2}} = 0\,, \tag{7.94a}$$

$$\tilde{\Gamma}^1|J\rangle\rangle \;=\; \sum_{j_1} \tilde{\Gamma}^1_{jjj_1j_1}\,[j_1(j_1+1)(2j_1+1)] = 0\,. \tag{7.94b}$$

The physical meaning of $\tilde{\Gamma}^0_{jjj_1j_1}$ and $\tilde{\Gamma}^1_{jjj_1j_1}$ is obvious: they govern the relaxation of rotational energy and angular momentum, respectively. The former is also an operator of the spectral exchange between the components of the isotropic Raman Q-branch. So, equality (7.94a) holds, as the probability conservation law. In contrast, the second one, Eq. (7.94b), is wrong, because, after substitution into the definition of the angular momentum correlation time

$$\tau_J = \frac{\tilde{K}_J(p=0)}{K_J(t=0)} = \frac{1}{K_J(t=0)}$$
$$\times \sum_{jj_1} \rho_j[jj_1(j+1)(j_1+1)(2j+1)(2j_1+1)]^{\frac{1}{2}}\langle j\left|\frac{1}{\tilde{\Gamma}^1(p=0)}\right|j_1\rangle \tag{7.95}$$

it gives an unreasonable result. Moreover, both operators Γ^1 and Γ^0 do not satisfy the detailed balance principle. This 'bad' behaviour has its origin in our main approximation (7.85), where parametric modulation of the Liouville equation was introduced; in other words, neglect of the system back reaction on the bath must be rejected. An attempt to correct Eq. (7.85) would result in the impossibility of a simple transition by averaging from the stochastic equation (7.85) to the kinetic master equation (7.87). So, an alternative procedure, though not rigorous, is preferable. We choose a phenomenological way of changing directly the operator Γ. (For the classical version of correction introducing dissipative friction, see Appendix 8.) In Chapter 5 we dealt with two-indexed governing operators, whose non-diagonal elements have the meaning of transition probabilities. This simplification disappears in this case, but we will show that a proper quasi-classical approximation permits us to reduce the four-indexed operator Γ effectively to the two-indexed ones for any given X_ℓ and to propose a new correction procedure [293] based on correspondence between results of quasi-classical and classical formalisms of Chapter 7.

First of all, we perform an important reordering, introducing the new discrete variables

$$\mu = j' - j; \quad \nu = j_1' - j_1 \tag{7.96}$$

and rewriting

$$|T\left(j_1' \, j_1\right)_{\ell q}\rangle\rangle \equiv |\nu j_1\rangle \quad |0 j_1\rangle \equiv |j_1\rangle\,. \tag{7.97}$$

Secondly, due to the smallness of the rotational temperature for the majority of molecules (only hydrogen and some of its derivatives being out of consideration), under temperatures higher than, say, 100 K, we replace further on the corresponding summation over rotational quantum numbers by an integration. We also exploit the asymptotic expansion for the Clebsch–Gordan coefficients and $6j$ symbol [23] ($j, j_1, j_2,\ L \gg \mu, \nu, \ell$)

$$C^{j0}_{j_1 0 j_2 0} \approx (-1)^{(j_1+j_2-j)/2}\left(\frac{2(2j+1)}{\pi}\right)^{\frac{1}{2}}$$

$$\times \frac{\delta_{j+j_1+j_2,2n}}{\left[2\left(j^2 j_1^2 + j^2 j_2^2 + j_1^2 j_2^2\right) - \left(j^4 + j_1^4 + j_2^4\right)\right]^{1/4}}, \tag{7.98a}$$

$$\begin{Bmatrix} j_1 & j & L \\ j+\mu & j_1+\nu & \ell \end{Bmatrix} \approx \frac{(-1)^{j+j_1+L+\ell+\mu}}{[(2j+1)(2j_1+1)]^{\frac{1}{2}}}\, d^\ell_{\mu\nu}(\alpha), \tag{7.98b}$$

where n is an integer, j_1, j_2 and j satisfy the triangle rule, and the Wigner function $d^\ell_{\mu\nu}(\alpha)$ depends on the angle α between quasi-classical momenta

$\boldsymbol{J}_1$ and $\boldsymbol{J}$ obeying the vector sum $\boldsymbol{J}_1 + \boldsymbol{L} = \boldsymbol{J}$:

$$\cos\alpha \approx \frac{(j+1/2)^2 + (j_1+1/2)^2 - (L+1/2)^2}{2\,(j+1/2)\,(j_1+1/2)}.$$

Limiting ourselves to derivation of the Hubbard relation in the simplest case $\ell = 1$ (for $\ell = 2$ see Appendix 9), we write out the definition of orientational relaxation time

$$\tau_{\theta,1} = \sum_{j_1 j \mu\nu} \rho_j\, [(2j+1)(2j_1+1)]^{\frac{1}{2}}\, C^{j+\mu,0}_{j010}\, C^{j_1+\nu,0}_{j_1 010} \left\langle \mu j \left| \frac{1}{-\mathrm{i}H_0^{\times} + \tilde{\Gamma}^1(p=0)} \right| \nu j_1 \right\rangle. \tag{7.99}$$

Again, direct substitution of $\tilde{\Gamma}^1$ from (7.92) into (7.99) is impossible. Using the asymptotic expressions (7.98), the operator in the denominator of (7.99) may be represented as

$\mu \backslash \nu$	-1	1	0
-1	$-2\mathrm{i}Bj\delta_{jj_1} + \frac{1}{2}\left(\tilde{\Gamma}^J_{jj_1} + \tilde{\Gamma}^Q_{jj_1}\right)$	$\frac{1}{2}\left(\tilde{\Gamma}^J_{jj_1} - \tilde{\Gamma}^Q_{jj_1}\right)$	0
1	$\frac{1}{2}\left(\tilde{\Gamma}^J_{jj_1} - \tilde{\Gamma}^Q_{jj_1}\right)$	$-2\mathrm{i}Bj\delta_{jj_1} + \frac{1}{2}\left(\tilde{\Gamma}^J_{jj_1} + \tilde{\Gamma}^Q_{jj_1}\right)$	0
0	0	0	$-\mathrm{i}\omega_{jj}\delta_{jj_1} + \tilde{\Gamma}^J_{jj_1}$

(7.100)

where the initial four-indexed operator is transformed into a supermatrix of rank 3, each element of this matrix being a two-indexed operator. For these operators convenient notations are introduced

$$\tilde{\Gamma}^Q_{jj_1} = \tilde{\Gamma}^0_{jjj_1j_1}, \quad \tilde{\Gamma}^J_{jj_1} = \tilde{\Gamma}^1_{jjj_1j_1} \tag{7.101}$$

and terms of order $1/j \ll 1$ are neglected. Note that frequencies ω_{jj} differ from 0 if rotation–vibration interaction is taken into account. As follows from (7.100), the Q-branch is separated from the P–R branches, which are connected by the Gordon P–R exchange.

Now we are ready to formulate the basic idea of the correction algorithm: in order to correct the four-indexed operator $\tilde{\Gamma}^1$, it is enough to correct the two-indexed operators $\tilde{\Gamma}^Q$ and $\tilde{\Gamma}^J$ in the supermatrix representation (7.100). The real advantage of this proposal is its compatibility with any definite way of $\tilde{\Gamma}^Q$ and $\tilde{\Gamma}^J$ correction [61, 294]. The matrix inversion demanded in (7.99) is divided into two stages. In the μ, ν subspace it is possible to find the inverse matrix analytically with the help of the Frobenius formula that is well known in matrix algebra [295]. The existence of an inverse operator for the 'right' $\tilde{\Gamma}^J$ is sufficient to obtain

$$\sum_{\mu\nu} C_{j010}^{j+\mu,0} C_{j_1010}^{j_1+\nu,0} \left\langle \mu j \left| \frac{1}{-\mathrm{i}H_0^\times + \tilde{\Gamma}^1} \right| \nu j_1 \right\rangle = \left\langle j \left| \frac{1}{\tilde{\Gamma}^\varrho + \Omega \frac{1}{\tilde{\Gamma}^J} \Omega} \right| j_1 \right\rangle, \quad (7.102)$$

where $\Omega_{jj_1} = -2jB\delta_{jj_1}$. Replacing in (7.99) summation over j by an integration, we have

$$\tau_{\theta,1} = \int\int \mathrm{d}j\, \mathrm{d}j_1\, 2\rho_j\, (jj_1)^{\frac{1}{2}} \left(\frac{1}{\tilde{\Gamma}^\varrho + \Omega \frac{1}{\tilde{\Gamma}^J} \Omega} \right)_{jj_1} . \quad (7.103)$$

The last stage of inversion is performed within perturbation theory approximation, the second operator in the denominator being small according to (7.83). The 'right' $\tilde{\Gamma}^\varrho$ has zero eigenvalue and ρ_j is a corresponding eigenfunction. So it is necessary to find the first non-vanishing correction to this zero eigenvalue

$$\tau_{\theta,1} = \left(\int\int \mathrm{d}j\, \mathrm{d}j_1\, 2\rho_j (jj_1)^{\frac{1}{2}} \left\langle j \left| \Omega \frac{1}{\tilde{\Gamma}^J} \Omega \right| j_1 \right\rangle \right)^{-1} = \frac{I}{2kT\tau_J}, \quad (7.104)$$

where (7.95) is used.

Appendix 1

The analytical expressions corresponding to (2.37) have the form

$$= (-\mathrm{i})^2 \sum_{\alpha_1 \alpha_2} L_{\alpha_1} L_{\alpha_2} \langle \omega_{\alpha_2}(t)\, \omega_{\alpha_2}(t') \rangle \eta\,(t-t')\,, \qquad \text{(A1.1a)}$$

$$= (-\mathrm{i})^4 \sum_{\{\alpha_i\}} L_{\alpha_1} L_{\alpha_2} L_{\alpha_3} L_{\alpha_4} \langle \omega_{\alpha_1}(t)\, \omega_{\alpha_4}(t') \rangle_c$$

$$\times \int_{t'}^{t} \mathrm{d}t_2 \int_{t'}^{t_2} \mathrm{d}t_3 \, \langle \omega_{\alpha_2}(t_2)\, \omega_{\alpha_3}(t_3) \rangle_c \, \eta\,(t-t'), \qquad \text{(A1.1b)}$$

$$= (-\mathrm{i})^4 \sum_{\{\alpha_i\}} L_{\alpha_1} L_{\alpha_2} L_{\alpha_3} L_{\alpha_4} \int_{t'}^{t} \mathrm{d}t_2 \int_{t'}^{t_2} \mathrm{d}t_3$$

$$\times \langle \omega_{\alpha_1}(t)\, \omega_{\alpha_3}(t_3) \rangle_c \, \langle \omega_{\alpha_2}(t_2)\, \omega_{\alpha_4}(t') \rangle_c \, \eta\,(t-t'), \qquad \text{(A1.1c)}$$

$$= (-\mathrm{i})^4 \sum_{\{\alpha_i\}} L_{\alpha_1} L_{\alpha_2} L_{\alpha_3} L_{\alpha_4} \int_{t'}^{t} \mathrm{d}t_2 \int_{t'}^{t_2} \mathrm{d}t_3$$

$$\times \langle \omega_{\alpha_1}(t) \omega_{\alpha_2}(t_2) \omega_{\alpha_3}(t_3) \omega_{\alpha_4}(t') \rangle_c \eta(t-t'). \qquad \text{(A1.1d)}$$

The averaging denoted in (A1.1) by the angular brackets with index c (connected), according to the definition of a cumulant [38] differs from the usual averaging, which is fulfilled when searching for a correlator of the same order. All the correlated moments of the quantities included in the cumulant are eliminated. For example:

$$\begin{aligned}
\langle \omega_{\alpha_i}(t)\, \omega_{\alpha_j}(t') \rangle_c &= \langle \omega_{\alpha_i}(t)\, \omega_{\alpha_j}(t') \rangle - \langle \omega_{\alpha_i}(t) \rangle \, \langle \omega_{\alpha_j}(t') \rangle \\
&= \delta_{ij} \langle \omega_{\alpha_i}(t) \omega_{\alpha_i}(t') \rangle \qquad \text{(A1.2a)} \\
\langle \omega_{\alpha_1}(t)\, \omega_{\alpha_2}(t_2)\, \omega_{\alpha_3}(t_3)\, \omega_{\alpha_4}(t') \rangle_c &= \langle \omega_{\alpha_1}(t)\, \omega_{\alpha_2}(t_2)\, \omega_{\alpha_3}(t_3)\, \omega_{\alpha_4}(t') \rangle \\
&\quad - \langle \omega_{\alpha_1}(t)\, \omega_{\alpha_2}(t_2) \rangle \, \langle \omega_{\alpha_3}(t_3)\, \omega_{\alpha_4}(t') \rangle \\
&\quad - \langle \omega_{\alpha_1}(t)\, \omega_{\alpha_3}(t_3) \rangle \, \langle \omega_{\alpha_2}(t_2)\, \omega_{\alpha_4}(t') \rangle \\
&\quad - \langle \omega_{\alpha_1}(t) \omega_{\alpha_4}(t') \rangle \, \langle \omega_{\alpha_2}(t_2)\, \omega_{\alpha_3}(t_3) \rangle. \qquad \text{(A1.2b)}
\end{aligned}$$

It is taken into account that relaxation of different Cartesian components of ω_{α_i} proceeds independently, and $\langle \omega_{\alpha_i} \rangle = 0$. One can easily see that every cumulant in (A1.1b) when integrated yields the corresponding power of τ_{α_i}, i.e., of rotational relaxation time of the α_ith component. Therefore,

$$\mathscr{H}_0 = \omega_\alpha^2 \, \tau_\alpha^2$$

is still a natural parameter of the series.

Appendix 2

For linear and spherical tops (2.86) yields the following equation

$$\frac{\partial}{\partial t} K_\ell(t) = \int_0^t \mu(t-t')\, K_\ell(t')\, \mathrm{d}t', \tag{A2.1}$$

where $\mu = M_{00}$. Three-fold differentiation of this equation leads to

$$I_4 = K_\ell^{(4)}(0) = [\mu(0)]^2 + \ddot{\mu}(0). \tag{A2.2}$$

It is readily seen that second-order terms of (A1.1b) do not contribute to $\mu(0)$, correcting only the $\ddot{\mu}(0)$ value. In order to calculate this correction, one has to substitute (A1.2) into (A1.1b) and to reduce the similar terms in (2.82). This leads to the following expression for μ:

$$\begin{aligned} \mu(t-t') \;=\; & -\sum_{\alpha_1\alpha_2} L_{\alpha_1} L_{\alpha_2} \langle \omega_{\alpha_1}(t)\, \omega_{\alpha_2}(t')\rangle + \\ & + \sum_{\{\alpha_i\}} \int_{t'}^{t} \mathrm{d}t_2 \int_{t'}^{t_2} \mathrm{d}t_3\; L_{\alpha_1} L_{\alpha_2} L_{\alpha_3} L_{\alpha_4} \\ & \times \left[\langle \omega_{\alpha_1}(t)\, \omega_{\alpha_2}(t_2)\, \omega_{\alpha_3}(t_3)\, \omega_{\alpha_4}(t')\rangle \right. \\ & \left. - \langle \omega_{\alpha_1}(t) \omega_{\alpha_2}(t_2)\rangle \langle \omega_{\alpha_3}(t_3)\, \omega_{\alpha_4}(t')\rangle \right]. \end{aligned} \tag{A2.3}$$

Calculating the second derivative of this expression, one can see that only the result of successive differentiation over the upper limit does not vanish at $t = 0$:

$$(-1)^4 \sum_{\{\alpha_i\}} L_{\alpha_1} L_{\alpha_2} L_{\alpha_3} L_{\alpha_4} \left[\langle \omega_{\alpha_1} \omega_{\alpha_2} \omega_{\alpha_3} \omega_{\alpha_4} \rangle - \langle \omega_{\alpha_1} \omega_{\alpha_2} \rangle \langle \omega_{\alpha_3} \omega_{\alpha_4} \rangle \right]. \tag{A2.4}$$

Calculation of these correlator and matrix elements of L_{α_i} provides the following addition to the fourth moment compared with (2.65b):

$$\Delta I_4 = \begin{cases} \left[2\ell^2\,(\ell+1)^2 - \ell\,(\ell+1)\right]\, (kT/I)^2 & \text{spherical tops} \\ 2\left[2\ell^2\,(\ell+1)^2 - \ell\,(\ell+1)\right]\, (kT/I)^2 & \text{linear tops.} \end{cases}$$

Appendix 3

Reduction of integral equations to their differential form is equivalent to proceeding from the Smoluchovski equation to the Fokker–Planck equation. This procedure is similar for the isotropic and anisotropic scattering or for the IR absorption. In the further consideration only the corresponding index ℓ differs. It is easy to see that direct substitution of $\gamma = 1$ into Eq. (6.14) is impossible. For a correct procedure for the limiting transition $\gamma \to 1$, $\tau_0 \to 0$, $\tau_0(1-\gamma)^{-1} = \tau_J$ let us first expand the integral expression (6.14) into a Taylor series over difference $(J' - J)$. In the expansion, we restrict ourselves to the first three terms. As is clear from the above consideration, the other terms vanish during the limiting transitions. We have

$$d_0^\ell\left(t, J'\right) \approx d_0^\ell(t, J) + (J' - J)\,\frac{\partial}{\partial J}\,d_0^\ell(t, J) + \frac{1}{2}(J' - J)^2\,\frac{\partial^2}{\partial J^2}\,d_0^\ell(t, J)\,. \quad \text{(A3.1)}$$

After substitution of (A3.1) into (6.14), several integrals of the same type must be calculated. These integrals can be expressed via the degenerate hypergeometrical function $\Phi(\cdots, \cdots, \cdots)$ and gamma-function $\Gamma(\cdots)$:

$$\begin{aligned} J_m^{q''} &= \int \mathrm{d}J' \frac{J}{d\left(1-\gamma^2\right)}\,\exp\left(-\frac{1}{2d\left(1-\gamma^2\right)}\,\left(J^2 + \gamma^2 J'^2\right)\right)\,I_{q''}(z)(J')^m \\ &= \frac{J}{2d\left(1-\gamma^2\right)}\,\exp\left(-\frac{J^2}{2d\left(1-\gamma^2\right)}\right)\frac{\Gamma\left(\left(m+q''+1\right)/2\right)}{\Gamma\left(q''+1\right)} \\ &\quad \frac{B^{q''}}{p^{(m+q''+1)/2}}\,\Phi\left(\frac{m+q''+1}{2}, m+1, \frac{B^2}{p}\right), \end{aligned} \quad \text{(A3.2)}$$

where $z = \gamma J' J/[d(1-\gamma^2)]$, $B = \gamma J/[2d(1-\gamma^2)]$ and $p = J^2/[2d(1-\gamma^2)]$. At $\gamma \to 1$ argument $B^2/p \to \infty$, and asymptotic expansion Φ may be employed. After that, to first order with respect to small parameter p/B^2 we get at $\gamma \to 1$

$$J_m^{q''} \approx \gamma^{-1}\left(\frac{J}{\gamma}\right)^m \left\{1 + \frac{2d\left(1-\gamma^2\right)}{J^2}\,\left[(1-m)^2 - \left(q''\right)^2\right]\right\}. \quad \text{(A3.3)}$$

Reducing similar terms in (6.14) taking account of (A3.1) and (A3.3), we obtain the following expressions containing $d_0^\ell(t, J)$, $(\partial/\partial J)\,d_0^\ell(t, J)$ and

$(\partial^2/\partial J^2)\, d_0^\ell(t,J)$:

$$\frac{1-\gamma}{\gamma\tau_0}\, d_0^\ell(t,J) + \frac{(1-\gamma^2)\,d}{2\gamma\, J^2\,\tau_0}\, d_0^\ell(t,J)\, \left[1 - \left(\hat{L}_x^2\right)_{00}\right], \tag{A3.4a}$$

$$\frac{1}{\tau_0}\left[\frac{J(1-\gamma)}{\gamma^2} - \frac{(1-\gamma^2)\,d}{2\gamma J}\right]\frac{\partial}{\partial J}\, d_0^\ell(t,J), \tag{A3.4b}$$

$$\frac{(1-\gamma)d}{\tau_0}\,\frac{\partial^2}{\partial J^2}\, \mathrm{d}_0^\ell(t,J). \tag{A3.4c}$$

Finally the following differential equation is obtained:

$$\begin{aligned} \frac{\partial}{\partial t}\, d_0^\ell(t,J) &= \mathrm{i}\alpha J^2 d_0^\ell(t,J) + \frac{1}{\tau_J}\left\{1 + \frac{d}{J^2}\left[1 - \left(\hat{L}_x^2\right)_{00}\right]\right\} d_0^\ell(t,J) \\ &+ \frac{J}{\tau_J}\left(1 - \frac{d}{J^2}\right)\frac{\partial}{\partial J} d_0^\ell(t,J) + \frac{d}{\tau_J}\,\frac{\partial^2}{\partial J^2} d_0^\ell(t,J). \end{aligned} \tag{A3.5}$$

In order to obtain from (A3.5) Eq. (3.39) for isotropic scattering, one should take $\ell = 0$ and introduce $K = d_0^0(t,J)$ that is rotationally invariant.

Appendix 4

Equation

$$\frac{\partial}{\partial t}d_0^\ell(t,J) = \mathrm{i}\alpha J^2 d_0^\ell(t,J) + \frac{1}{\tau_J}\left(1+\frac{a_\ell d}{J^2}\right)d_0^\ell(t,J)$$

$$+\frac{J}{\tau_J}\left(1-\frac{d}{J^2}\right)\frac{\partial}{\partial J}d_0^\ell(t,J) + \frac{d}{\tau_J}\frac{\partial^2}{\partial J^2}d_0^\ell(t,J) \quad \text{(A4.1)}$$

can be solved exactly with the method of separation of variables. Let us expand $d_0^\ell(t,J)$ over eigenfunctions ψ_λ of eigenvalue λ of the operator on the right-hand side of (A4.1),

$$d_0^\ell(t,j) = \sum_\lambda \mathrm{e}^{\lambda t}\psi_\lambda(J). \quad \text{(A4.2)}$$

Substitution of (A4.2) into (A4.1) leads to a second-order differential equation

$$\psi_\lambda'' + \frac{J}{d}\left(1-\frac{d}{J^2}\right)\psi_\lambda' + \left(\frac{1}{d}+\frac{a_\ell}{J^2}+\frac{\mathrm{i}\alpha\tau_J J^2}{d}-\frac{\lambda\tau_J}{d}\right)\psi_\lambda = 0. \quad \text{(A4.3)}$$

Proceeding to a new function $\psi = u\ \exp\left[-J^2/(4d)\right]\sqrt{J^3}$, (A5.3) becomes the Schrödinger equation for a particle moving in a potential $U(r) = A_1 r^{-2} + A_2 r^2$

$$u'' + \frac{2}{J}u' + \left(\frac{1}{d} - \frac{\lambda\tau_J}{d} + \frac{a_\ell}{J^2} - \frac{3}{4J^2} + \mathrm{i}\alpha\,\tau_J\frac{J^2}{d} - \frac{J^2}{4d^2}\right)u = 0. \quad \text{(A4.4)}$$

After this, the change of variables $x = \sqrt{B}\,J^2/(2d)$, where $B = 1 - 2\mathrm{i}/\Gamma$, yields

$$xu'' + \frac{3}{2}u' + \left(\frac{1-\lambda\tau_J}{2\sqrt{B}} + \frac{4a_\ell - 3}{16x} - \frac{x}{4}\right)u = 0. \quad \text{(A4.5)}$$

Now it is possible to employ the eigenfunctions given in [121], which are proportional to the generalized Laguerre polynomials. The corresponding eigenvalues are equidistant:

$$\lambda_k = \frac{1}{\tau_J}\left[1 - 2\sqrt{B}\left(k + S_\ell + \frac{3}{4}\right)\right]. \quad \text{(A4.6)}$$

Parameter S_ℓ in (A4.6) is a root of the square equation

$$S_\ell\left(S_\ell + 1/2\right) = (3 - 4a_\ell)/16 \quad \text{(A4.7)}$$

This parameter

$$S_\ell = \left[-1 \pm 2(1-a_\ell)^{\frac{1}{2}}\right]/4$$

determines the order of the corresponding Laguerre polynomial

$$u_k \sim e^{-x/2} x^{S_\ell} L_k^{2S_\ell+1/2}. \qquad (A4.8)$$

For isotropic and anisotropic scattering, correspondingly, we have

$$S_0 = -\frac{1}{4}; \quad S_2 = \left(2\sqrt{3}-1\right)/4. \qquad (A4.9)$$

The final form of the equation will demonstrate that the choice of the latter root provides the convergency at $t \to \infty$.

With (A4.8) and (A4.6) we represent (A4.2) as

$$d_0^\ell(t,J) = \sum_{k=0}^{\infty} d_k^\ell \exp\left\{\frac{t}{\tau_J}\left[1-2\sqrt{B}\left(k+S_\ell+\frac{3}{4}\right)\right]\right\}\sqrt{J^3}$$

$$\times \exp\left[-\frac{(1+\sqrt{B})}{4d}J^2\right]\left(\frac{\sqrt{B}J^2}{2d}\right)^{S_\ell} L_k^{2S_\ell+1/2}\left(\sqrt{B}\frac{J^2}{2d}\right). \qquad (A4.10)$$

Undetermined coefficients d_k^ℓ can be easily found using the initial condition (6.3) and orthogonality of the Laguerre polynomials [37]

$$d_k^\ell = 2\,d_0^\ell(0)\,(B)^{1/8}\,(2d)^{-5/4}\,\frac{\Gamma\left(S_\ell+5/4\right)}{\Gamma\left(2S_\ell+3/2\right)}\left(\frac{2\sqrt{B}}{1+\sqrt{B}}\right)^{S_\ell+5/4}$$

$$\times {}_2F_1\left(-k,\, S_\ell+\frac{5}{4},\, 2S_\ell+\frac{3}{2},\, \frac{2\sqrt{B}}{1+\sqrt{B}}\right)\frac{\Gamma\left(k+2S_\ell+3/2\right)}{k\,!} \qquad (A4.11)$$

via the hypergeometrical function ${}_2F_1$ and the gamma-function $\Gamma(...)$. To complete calculation of $d_0^\ell(t)$ it is necessary to integrate over J. This may be done, if the corresponding integral is represented in the form of the Laplace transform of a generalized Laguerre polynomial [37]. As a result, one has

$$d_0^\ell(t) = d_0^\ell(0)\,\frac{\Gamma^2\left(S_\ell+5/4\right)}{\Gamma\left(2S_\ell+3/2\right)}\,\frac{1}{\sqrt{B}}\left(\frac{2\sqrt{B}}{1+\sqrt{B}}\right)^{2S_\ell+5/2}$$

$$\times \sum_{k=0}^{\infty} \exp\left\{\frac{t}{\tau_J}\left[1-2\sqrt{B}\left(k+S_\ell+\frac{3}{4}\right)\right]\right\}\frac{\left(2S_\ell+3/2\right)_k}{k!}$$

$$\times \left[{}_2F_1\left(-k,\, S_\ell+\frac{5}{4},\, 2S_\ell+\frac{3}{2},\, \frac{2\sqrt{B}}{1+\sqrt{B}}\right)\right]^2. \qquad (A4.12)$$

Appendix 5

Derivation of the isotropic Q-branch spectra for the case of linear molecules is analogous to the case for spherical molecules. The integral part of the kinetic equation determines the set of eigenfunctions of the collisional operator

$$\int_0^\infty f(J', J)\Psi_n(J')\,\mathrm{d}J' = \lambda_n \Psi_n(J) \qquad \text{(A5.1)}$$

with two-dimensional kernel (3.28)

$$f(J', J) = \frac{J}{d(1-\gamma^2)} I_0\left(\frac{\gamma J J'}{d(1-\gamma^2)}\right) \exp\left(-\frac{J^2 + \gamma^2 J'^2}{2d(1-\gamma^2)}\right), \qquad \text{(A5.2)}$$

where $d = IkT$ and I_0 is a modified Bessel function of zeroth order. Instead of the generalized Laguerre polynomials for three-dimensional $\boldsymbol{J}$-space, we have for two-dimensional angular momentum a set of ordinary Laguerre polynomials

$$\int_0^\infty f(J'J)\varphi_B(J')L_n(J'^2/2d)\,\mathrm{d}J' = (\gamma^2)^n \varphi_B(J) L_n(J^2/2d). \qquad \text{(A5.3)}$$

After substitution of the solution

$$K(t, J) = \varphi_B(J) \sum_{n=0}^{\infty} a_n(t) L_n\left(\frac{J^2}{2d}\right) \qquad \text{(A5.4)}$$

into kinetic equation (3.26) and performing the Fourier transform, we obtain the continuous fraction representation

$$\begin{aligned} a_0(\omega) &= \frac{1}{\mathrm{i}(\omega - \omega_Q) + \mathrm{i}\omega_Q / b_{\omega_Q}(\omega)}, \\ b_n(\omega) &= \frac{a_n(\omega)}{a_{n+1}(\omega)} = 2 + \frac{1}{n+1} \\ &\times \left(1 - \frac{\omega}{\omega_Q} + \frac{\mathrm{i}}{\tau_E \omega_Q} \sum_{m=0}^{n} \gamma^{2m}\right) - \frac{[1 + (1/n+1)]}{b_{n+1}(\omega)}. \end{aligned} \qquad \text{(A5.5)}$$

Here it is possible to have τ_E and γ^2 variables everywhere, as time τ_0 is included in the formulae only in the combination $\tau_0/(1-\gamma^2) = \tau_E$.

The algorithm proposed above may be also realized in the presence of vibrational dephasing, exactly in the same way as is done in Chapter 3 for spherical tops. For this the following substitution should be made:

$$\omega \longrightarrow \omega - \frac{\mathrm{i}}{\tau_{\mathrm{dp}}}; \quad \frac{1}{\tau_E} \longrightarrow \frac{1}{\tau_E} - \frac{1-\gamma^2}{\tau_{\mathrm{dp}}}. \qquad \text{(A5.6)}$$

In a general case parameters τ_E, τ_{dp} and γ must be determined by self-consistent two-parameter fitting. Owing to the property of orthogonality of Laguerre polynomials, one has for the spectral band shapes

$$I_{\mathrm{IS}}^{\mathrm{CARS}}(\omega) \propto |a_0(\omega)|^2 ,$$
$$I_{\mathrm{IS}}^{\mathrm{RAMAN}}(\omega) = \frac{1}{\pi}\ \mathrm{Re}\ a_0(\omega) . \tag{A5.7}$$

The following truncation improves the convergency of the numerical calculation of the continuous fraction:

$$b_N(\omega) \approx 1 + \left[\frac{1}{N} \left(\frac{\mathrm{i}}{\omega_Q \tau_E} \sum_{m=0}^{N} \gamma^{2m} - \frac{\omega}{\omega_Q} \right) \right]^{1/2} , \tag{A5.8}$$

where N is the number of the stage of truncation.

Appendix 6

Let us solve the problem for eigenvalues

$$\int_0^\infty f\left(J',\, J\right)\psi\left(J'\right)\, \mathrm{d}J' = \lambda\psi(J) \tag{A6.1}$$

with integral kernel (3.52) [163]. Having proceeded to the new variable $x = J^2$, we perform a two-fold integration of (A6.1) over x and multiply it by x

$$\lambda x \frac{\mathrm{d}^2}{\mathrm{d}x^2}\psi = \frac{\lambda}{2}\frac{\mathrm{d}}{\mathrm{d}x}\psi - \frac{\lambda}{2x}\psi - \lambda\beta x\left(\lambda\frac{\mathrm{d}}{\mathrm{d}x}\psi - \frac{\lambda}{2x}\psi + \beta\lambda\psi\right) + L$$

$$L = \left(\frac{\beta}{\pi}\right)^{\frac{1}{2}} \int_0^\infty \mathrm{e}^{-\beta(x+\gamma^2 x')}\,\gamma\beta^2\,(xx')^{\frac{1}{2}}\,\sinh\left[2\beta\gamma\left(xx'\right)^{\frac{1}{2}}\right]\,\psi\left(x'\right)\frac{\mathrm{d}x'}{2\sqrt{x'}}, \tag{A6.2}$$

where $\beta = 1/[2d(1-\gamma^2)]$. One can easily see that the result of differentiation of (A6.1) over γ can be expressed through L. Owing to this, (A6.2) may be closed relative to ψ

$$x\frac{\mathrm{d}^2}{\mathrm{d}x^2}\psi = \left(\frac{1}{2} - \beta x\,\left(1-\gamma^2\right)\right)\frac{\mathrm{d}}{\mathrm{d}x}\psi - \frac{1}{2}\left(\frac{1}{x} + \beta\,\left(1-\gamma^2\right)\,(2n+1)\right)\psi, \tag{A6.3}$$

and by a trivial substitution $\xi = x/2d$, $\psi = \xi \mathrm{e}^{-\xi}\varphi$, it reduces to

$$\xi\,\frac{\mathrm{d}^2}{\mathrm{d}\xi^2}\,\varphi(\xi) + \left(\frac{3}{2} - \xi\right)\frac{\mathrm{d}}{\mathrm{d}\xi}\,\varphi(\xi) + n\varphi(\xi) = 0. \tag{A6.4}$$

In (A6.3) and (A6.4) the notation $2n = \mathrm{d}\left(\ln|\lambda|\right)/\mathrm{d}\left(\ln|\gamma|\right)$ is introduced. For integer and non-negative n the solutions of this equation form a complete set of orthogonal functions – the Laguerre generalized polynomials $L_n^{1/2}(\xi)$ [37]. So, the integral operator in (A6.1) is proved to have the discrete non-degenerate spectrum of eigenvalues

$$\lambda_n = \gamma^{2n}; \quad n = 0,\, 1,\, 2\ldots \tag{A6.5}$$

with corresponding eigenfunctions

$$\psi_n(J) = \varphi_B(J)\; L_n^{1/2}\left(J^2/d\right)\,. \tag{A6.6}$$

Appendix 7

Let us consider the quasi-classical formulation of impact theory. A rotational spectrum of ℓth order at every value of ω is a sum of spectral densities at a given frequency of all J-components of all branches

$$G_\ell(\omega) = \int_0^\infty \mathrm{d}J \sum_q G_q^\ell(\omega, J)\varphi_B(J). \tag{A7.1}$$

Here the relative intensities of the components of each branch are determined by the Boltzmann factor $\varphi_B(J)$, and the contour of a separate J-component in the qth branch is defined taking account of the interference of all lines. Correlation function $K_q^\ell(t, J)$, corresponding to $G_q^\ell(\omega, J)$, is obviously the correlation function of a transition matrix element in Heisenberg representation

$$K_q^\ell(t, J) = \langle A_q^\ell(t, j) \cdot \{A_q^\ell(0, j)\}^+\rangle = \langle A_q^\ell(t, j)\, A_{-q}^\ell(0, j)\, (-1)^q\rangle \tag{A7.2}$$

where $J = \hbar j$ and $A_q^\ell(t, j)|_{t=0} = A_q^\ell(0, j)$. The integral intensity of each component

$$\int \mathrm{d}\omega\, G_q^\ell(\omega, J) = K_q^\ell(0, J) = \langle |A_q^\ell(0, j)|^2\rangle \tag{A7.3}$$

is determined by dependence of the intrinsic matrix element $A_q^\ell(0, j) = \langle j|d^\ell|j+q\rangle$ of the tensor operator $\mathbf{d}^\ell$ on the number of the branch (q) and of the component (j).

We are interested in a quasi-classical expression in the limit of large j, which is the same for all j-components of one branch:

$$\begin{aligned} A_q^\ell(0) &= \lim_{j\to\infty} \langle j|d^\ell|j+q\rangle = (-1)^{(\ell+q)/2} \\ &\quad \left(\frac{(\ell+q-1)!!(\ell-q-1)!!}{(\ell+q)!!(\ell-q)!!}\right)^{\frac{1}{2}} \delta_{\ell+q,2k}, \end{aligned} \tag{A7.4}$$

where $k = 0, 1, ..., \ell$. The dependence of quasi-classical $A_q^\ell(t, J)$ on angular momentum J is now only subject to the different precession frequency $\Omega_q(J)$ during the free path with a given J:

$$\frac{\mathrm{d}}{\mathrm{d}t} A_q^\ell(t, J) = \mathrm{i}\Omega_q(J) A_q^\ell(t, J). \tag{A7.5}$$

The correlation function $K_\ell(t)$ defined by Eq. (6.1) and (6.2) may be represented in a form analogous to (A7.1) as

$$K_\ell(t) = \int_0^\infty \mathrm{d}J \sum_q K_q^\ell(t, J)\varphi_B(J) . \tag{A7.6}$$

Substituting (A7.2) into this definition we get

$$\begin{aligned} K_\ell(t) &= \left\langle \sum_q \int \mathrm{d}J\ A_q^\ell(t, J) A_{-q}^\ell(0)(-1)^q \varphi_B(J) \right\rangle \\ &= \sum_q \int \mathrm{d}J\ d_q^\ell(t, J) d_q^\ell\ (0)(-1)^q . \end{aligned} \tag{A7.7}$$

Averaging over all time realizations of the random process should be extended only to the time-dependent component

$$d_q^\ell(t, J) = \langle A_q^\ell(t, J) \rangle \varphi_B(J) , \tag{A7.8}$$

satisfying the initial condition

$$d_q^\ell(0, J) = d_q^\ell(0)\varphi_B(J) , \tag{A7.9}$$

where

$$d_q^\ell(0) = A_q^\ell(0) . \tag{A7.10}$$

Let us proceed now to calculation of (A7.7). As every time realization reduces to the resulting spatial turn, and the scalar product in (A7.7) is invariant relative to rotation, it can be calculated in any reference system. It is more convenient to use the moving system of coordinates (all quantities set in it are primed). In this system the initial condition $[d_q^\ell(0)]' = d_q^\ell(t, J)$ depends effectively on time, and $[d_q^\ell(t, J)]' = d_q^\ell(0)$, on the contrary, is fixed. Let us consider that the moving system is chosen so that vector $\boldsymbol{u}$ is directed along axis x. Orientation of the angular momentum vector in this coordinate system depends on the choice of the Gordon frame (GF) or molecular frame (MF). In the GF it is always directed along the z axis, and in the MF it can take an arbitrary position in plane y–z. In this latter case, a somewhat more general partial component should be introduced, which depends on the whole vector $\boldsymbol{J}$ rather than on its modulus:

$$d_q^\ell(t) = \int \mathrm{d}\boldsymbol{J}\ d_q^\ell\,(t, \boldsymbol{J}) = \int d_q^\ell\,(t, J_y, J_z)\ \mathrm{d}J_y\ \mathrm{d}J_z . \tag{A7.11}$$

As a result, $\varphi_B(J)$ must be substituted by $\varphi_B(\boldsymbol{J})$.

A closed kinetic equation may be derived not for correlation function $K_\ell(t)$ directly, but for its partial (marginal) value

$$d_q^\ell(t, \boldsymbol{J}) = U(t, \boldsymbol{J})\; d_q^\ell(0), \tag{A7.12}$$

where

$$U(t, \boldsymbol{J}) = \sum_n \int U_n(t, \boldsymbol{J})\, \phi_n \prod_{i=1}^{n} \mathrm{d}t_i\, \mathrm{d}\boldsymbol{J}_i\, \mathrm{d}g_i \tag{A7.13}$$

is the evolution operator averaged over all paths that bring the molecule at time t to the state with angular momentum $\boldsymbol{J}$. An operator $U_n(t, \boldsymbol{J})$ carries out a rotation resulting from a realization that had n collisions during time period $(0, t)$. The probability density to find the n-collision path with given properties

$$\phi_n = (\tau_0)^{-n} \exp\left(-\frac{t}{\tau_0}\right) \varphi_B(\boldsymbol{J}_1) \prod_{i=1}^{n} F(\boldsymbol{J}_i, \boldsymbol{J}_{i+1}, g_i) \tag{A7.14}$$

depends on all intermediate values of angular momentum $\boldsymbol{J}_i$ and every angular displacement g_i during collision. Here $\boldsymbol{J}_{n+1} = \boldsymbol{J}$ and $F(\boldsymbol{J}_i, \boldsymbol{J}_{i+1}, g_i)$ is the conditional probability density of changing at the ith collision the angular momentum $\boldsymbol{J}_i$ to $\boldsymbol{J}_{i+1}$ and the molecule orientation by g_i.

Following all reorientations of the molecule, a particular evolution operator U_n determines the successive transformation of the reference system, and finally transfers the initial value $d_q^\ell(0, J)$ to a system that coincides with MF at moment t. Each new rotation is determined relative to the preceding orientation of the reference system, so that the rotation operator corresponding to the later moment of time is placed to the right of the earlier one [23]. Therefore $U_n(t, \boldsymbol{J})$ can be represented as a multiplication of elementary rotations, dependent only on the relative change of orientation. During a single free path (t_{i-1}, t_i) the reference system turns relative to its position at $t = t_{i-1}$ by angle $(J/I)(t_i - t_{i-1})$ around vector $\boldsymbol{J}$, so that

$$U_0(t_i - t_{i-1}, \boldsymbol{J}) = \exp\left(-\mathrm{i}\,(t_i - t_{i-1})\, \frac{\boldsymbol{J} \cdot \hat{\boldsymbol{L}}}{I}\right). \tag{A7.15}$$

Then the ith collision occurs and the orientation changes by g_i. The form of the corresponding collisional operator $\hat{T}(g_i)$ is to be determined below. Finally

$$U_n(t, \boldsymbol{J}) \;=\; \exp\left(-\mathrm{i}t_1\, \frac{\boldsymbol{J}_1 \cdot \hat{\boldsymbol{L}}}{I}\right) \hat{T}(g_1)$$

$$\times \exp\left(-\mathrm{i}(t_i - t_{i-1})\,\frac{\boldsymbol{J}_i\cdot\hat{\boldsymbol{L}}}{I}\right)\,\hat{T}\,(g_i)$$

$$\times\hat{T}\,(g_n)\,\exp\left(-i\,(t - t_n)\,\frac{\boldsymbol{J}\cdot\hat{\boldsymbol{L}}}{I}\right). \qquad \text{(A7.16)}$$

It must be stressed that every $U_n\,(t, \boldsymbol{J})$ brings $d_q^\ell(0)$ to a different coordinate system. Consequently, the averaged operator (A7.13) is actually a weighted sum of the quantities in differently oriented reference systems. It can nevertheless be used to find the scalar product (A7.7), that is the orientational correlation function.

Without essential limitation of generality it may be assumed that the orientation of the molecule and its angular momentum are changed by collision independently, therefore $F\,(\boldsymbol{J}_i, \boldsymbol{J}_{i+1}, g_i) = f\,(\boldsymbol{J}_i, \boldsymbol{J}_{i+1})\,\psi\,(g_i)$. At the same time the functions $f\,(\boldsymbol{J}_i, \boldsymbol{J}_{i+1})$ and $\psi(g_i)$ have common variables. There are two reasons for this. First, it may be due to the fact that the angle between $\boldsymbol{J}$ and $\boldsymbol{u}$ must be conserved for linear rotators for any transformation. Second, a transformation $\hat{T}$ includes rotation of the reference system by an angle sufficient to combine axis z with vector $\boldsymbol{J}$. After substitution of (A7.16) and (A7.14) into (A7.13), one has to integrate over those variables from the set g_i, which are not common with the arguments of the function $f\,(\boldsymbol{J}_i, \boldsymbol{J}_{i+1})$. As a result, in the MF operator $\hat{T}$ becomes the same for all i and depends on the moments of ψ as parameters.

Insofar as the averaged evolution operator is written in the form (A7.13), one can use the procedure, developed in [20], to sum the realizations. It proves that Eq. (A7.13) is the solution of the integral equation

$$\frac{\partial}{\partial t}U\,(t, \boldsymbol{J}) = \mathrm{i}U\,(t, \boldsymbol{J})\,\frac{\boldsymbol{J}\cdot\hat{\boldsymbol{L}}}{I} - \frac{1}{\tau_0}U\,(t, \boldsymbol{J}) + \frac{1}{\tau_0}\,\tilde{U}\,(t, \boldsymbol{J})\,, \qquad \text{(A7.17)}$$

where $U\,(0, \boldsymbol{J}) = \mathbf{I}\varphi_B\,(\boldsymbol{J})$, and

$$\tilde{U}\,(t, \boldsymbol{J}) = \int f\,(\boldsymbol{J}', \boldsymbol{J})\,\psi(g)U\,(t, \boldsymbol{J}')\,T(g)\,\mathrm{d}\boldsymbol{J}'\,\mathrm{d}g\,. \qquad \text{(A7.18)}$$

Equation (A7.17) can be rewritten in a convenient equivalent

$$\frac{\partial}{\partial t}\,U\,(t, \boldsymbol{J}) = \mathrm{i}\,\frac{\boldsymbol{J}\cdot\hat{\boldsymbol{L}}}{I}\,U\,(t, \boldsymbol{J}) - \frac{1}{\tau_0}\,U\,(t, \boldsymbol{J}) + \frac{1}{\tau_0}\,\tilde{U}\,(t, \boldsymbol{J}) \qquad \text{(A7.19)}$$

making use of the fact that scalar operator $\boldsymbol{J}\cdot\hat{\boldsymbol{L}}$ commutes with any rotational operator $U(t, \boldsymbol{J})$ [120]. The first term on the right-hand side of (A7.19) corresponds to the dynamical equation (A7.5). It means that the molecule, rotating with frequency J/I, absorbs or scatters light at a frequency divisible by the rotational one. The other two terms in (A7.19) are analogous to the collisional term in the Boltzmann kinetic equation. The 'income' of the subensemble of given orientation and given $\boldsymbol{J}$ is formed by particles with different orientation and $\boldsymbol{J}'$. During collision they undergo reorientation on $g = g(t_{coll} + 0) - g(t_{coll} - 0)$ and simultaneously $\boldsymbol{J}'$ is changed to $\boldsymbol{J} = \boldsymbol{J}' + (\Delta \boldsymbol{J})_{coll}$. Equation (A7.17) has an impact–Markovian form [20], and integration over g in (A7.18) is analogous to averaging over impact parameters (see Chapter 4).

The equation for $d_q^\ell(t, \boldsymbol{J})$ can be easily derived from (A7.19):

$$\frac{\partial}{\partial t} d_q^\ell\,(t, \boldsymbol{J}) = \mathrm{i} \sum_{q'} \left(\frac{\boldsymbol{J} \cdot \hat{\boldsymbol{L}}}{I} \right)_{q'q} d_{q'}^\ell\,(t, \boldsymbol{J}) - \frac{1}{\tau_0}\, d_q^\ell\,(t, \boldsymbol{J})$$
$$+ \frac{1}{\tau_0} \sum_{q'=-\ell}^{\ell} \int \mathrm{d}\boldsymbol{J}\, \mathrm{d}g\; T_{q'q}(g) f\left(\boldsymbol{J}', \boldsymbol{J}\right) \psi(g) d_{q'}^\ell\left(t, \boldsymbol{J}'\right). \quad \text{(A7.20)}$$

Generally speaking, neither dynamical, nor relaxation parts of (A7.20) are diagonal over index q. It is simpler to diagonalize the dynamical term, proceeding to the GF, where scalar product $\left(\boldsymbol{J} \cdot \hat{\boldsymbol{L}}\right)_{q'q} = \left(J_z \hat{L}_0\right)_{q'q} = \delta_{q'q} J_z q$. The partial index of $d_q^\ell(t, \boldsymbol{J})$ is, instead of vector $\boldsymbol{J}$, its modulus J, because the orientation of $\boldsymbol{J}$ is always fixed in the GF. As a result we have in this reference system an equation

$$\frac{\partial}{\partial t} d_q^\ell(t, J) = \mathrm{i} q \frac{J}{I} d_q^\ell(t, J) - \frac{1}{\tau_0} d_q^\ell(t, J)$$
$$+ \frac{1}{\tau_0} \sum_{q'} \int \mathrm{d}J\, \mathrm{d}\alpha\; f\left(J', J, \alpha\right) T_{q'q}(\alpha) d_{q'}^\ell(t, J'). \quad \text{(A7.21)}$$

In the integral part of (A7.22) the only angle remaining from the whole set of g is $\alpha = \arccos\left(\boldsymbol{J}' \cdot \boldsymbol{J}/J'J\right)$.

In the MF equation (A7.20) can be simplified essentially only in the particular case of non-adiabatic collisions, which do not change the molecular orientation: $\psi(g) = \delta(g)$. In this case operator $\hat{T} = \hat{I}$, and the relaxation part of (A7.20) can be diagonalized over index q:

$$\frac{\partial}{\partial t} d_q^\ell\,(t, \boldsymbol{J}) = \mathrm{i} \sum_{q'} \left(\frac{\boldsymbol{J} \cdot \hat{\boldsymbol{L}}}{I} \right)_{q'q} d_{q'}^\ell\;(t, \boldsymbol{J}) \quad \text{(A7.22)}$$
$$- \frac{1}{\tau_0} d_q^\ell\,(t, \boldsymbol{J}) + \frac{1}{\tau_0} \int \mathrm{d}\boldsymbol{J}'\; f\left(\boldsymbol{J}'\boldsymbol{J}\right) d_q^\ell\left(t, \boldsymbol{J}'\right).$$

In the general case of collisions, changing the molecular reorientation, (A7.23) contains a fully averaged $\langle \hat{T} \rangle = \int \hat{T}(g)\psi(g)\,\mathrm{d}g$, where the integral operator should satisfy the only obvious condition

$$\boldsymbol{u}' \cdot \boldsymbol{J}' = \boldsymbol{u} \cdot \boldsymbol{J} = 0\,. \tag{A7.23}$$

Here $\boldsymbol{u}'$ is the orientation of vector $\boldsymbol{u}$ before the collision. This only has physical sense for angle δ = arccos $(\boldsymbol{u}'\boldsymbol{u})$, which is an even argument of the function ψ. This means that, as a result the collision vector $\boldsymbol{u}$ may be oriented inside a cone, whose axis coincides with the position of $\boldsymbol{u}'$. A characteristic value of conical angle δ depends on the width of function ψ.

Appendix 8

The present appendix represents a detailed derivation of the kinetic equations of the fluctuating liquid cage model in the classical formalism. A natural generalization is done for the case of partially ordered media, e.g. nematic liquid crystals. One of the simplest ways to take into account the back reaction is demonstrated, namely to introduce friction.

For simplicity, let us restrict ourselves to the case when fluctuations of the anisotropic potential are connected only with variation of its direction, determined by the Euler angles Σ. The energy of interaction of a rotator with the directing field depends only on the difference angle

$$U(t) = U\,[\Sigma(t) - \Omega(t)]\,, \tag{A8.1}$$

where the angles Ω determine orientation of the molecular axis $\boldsymbol{n}$. Along every random trajectory of the $\Sigma(t)$ variation the distribution function over orientations $\delta[\Omega - \Omega(t)]$ satisfies the stochastic equation

$$\frac{\partial}{\partial t}\delta\,[\Omega - \Omega(t)] = -\mathrm{i}\boldsymbol{\omega} \cdot \hat{\boldsymbol{L}}_{\Omega}\,\delta\,[\Omega - \Omega(t)]\,. \tag{A8.2}$$

Unlike (2.19), in the present case the angular velocity vector $\boldsymbol{\omega}(t)$ is not an input random noise, but, in its turn, satisfies the stochastic equation

$$\frac{\partial}{\partial t}\boldsymbol{\omega}(t) = \frac{1}{I}\boldsymbol{M}\,(t)\,, \tag{A8.3}$$

which coincides with (1.60). In its turn, the torque may be expressed through the field and molecular orientation at a given moment of time t

$$\boldsymbol{M} = \mathrm{i}\,\hat{\boldsymbol{L}}_{\Omega(t)}U\,[\Sigma(t) - \Omega(t)]\,. \tag{A8.4}$$

System (A8.2)–(A8.4) defines completely the time variation of orientation and angular velocity for every path $\Sigma(t)$. One can easily see that (A8.2)–(A8.4) describe the system with parametrical modulation, as the $\Sigma(t)$ variation is an input noise and does not depend on behaviour of the solution of $\{\Omega(t),\ \boldsymbol{\omega}(t)\}$. In other words, the 'back reaction' of the rotator to the collective motion of the closest neighbourhood is neglected. Since the spectrum of fluctuations $\Sigma(t)$ does not possess a carrying frequency, in principle, for the rotator the conditions of parametrical resonance and excitation (unrestricted heating of rotational degrees of freedom) are always fulfilled. In reality the thermal equilibrium is provided by dissipation of rotational energy from the rotator to the environment and in the simplest case may be introduced as friction in (A8.3)

$$\dot{\omega}(t) = \frac{1}{I}\boldsymbol{M}(t) - \xi\omega\,. \tag{A8.5}$$

Introducing the distribution function $\delta\,[\omega - \omega(t)]$, analogous to (A8.2), we use the properties of δ-functions

$$\frac{\partial}{\partial t}\delta\,[\omega - \omega(t)] = -\frac{\partial}{\partial\omega}\left[\delta\,[\omega - \omega(t)]\,\left(\frac{1}{I}\mathrm{i}\hat{\boldsymbol{L}}_{\Omega(t)}\,U\,[\Sigma(t) - \Omega(t)] - \xi\omega\right)\right]. \tag{A8.6}$$

In order to obtain the instantaneous joint distribution $\mathscr{P} = \delta[\Omega - \Omega(t)]\delta[\omega - \omega(t)]$, let us multiply, correspondingly, (A8.2) by $\delta(\omega - \omega(t))$ and (A8.6) by $\delta(\boldsymbol{\Omega} - \boldsymbol{\Omega}(t))$. After summation we get

$$\frac{\partial}{\partial t}\mathscr{P} = -\mathrm{i}\omega\hat{\boldsymbol{L}}_{\Omega}\mathscr{P} - \frac{\partial}{\partial\omega}\left\{\mathscr{P}\left[\frac{1}{I}\mathrm{i}\hat{\boldsymbol{L}}_{\Omega}\,\mathrm{U}\,[\Sigma(t) - \Omega] - \xi\omega\right]\right\}. \tag{A8.7}$$

Stochastic equation (A8.7) is linear over $\mathscr{P}$ and contains the operators $\hat{\boldsymbol{L}}_{\Omega}$ and ∇_{ω} of differentiation over time-independent variables Ω and ω. Therefore, if we assume that the time fluctuations of the liquid cage axis orientation $\Sigma(t)$ are Markovian, then the method used in Chapter 7 yields a kinetic equation for the partially averaged distribution function $P(\Omega, \omega, t, \Sigma)$. The latter allows us to calculate the searched averaged distribution function

$$P\,(\Omega, \omega, t) = \int \mathrm{d}\Sigma\; P\,(\Omega,\; \omega,\; t,\; \Sigma) = \langle\mathscr{P}\rangle_{\Sigma}\,, \tag{A8.8}$$

where angular brackets denote averaging over the whole ensemble of $\Sigma(t)$ realizations. As a result, for $P(\Omega, \omega, t, \Sigma)$ we have a classical analogue of the kinetic equation (7.51)

$$\frac{\partial P}{\partial t} = -\mathrm{i}\omega\cdot\hat{\boldsymbol{L}}_{\Omega}\,P - \frac{\partial}{\partial\omega}\left\{P\,\left[\frac{1}{I}\,\mathrm{i}\,\hat{\boldsymbol{L}}_{\Omega}\,U\,(\Sigma - \Omega) - \xi\omega\right]\right\} + L_{\Sigma}\,P\,, \tag{A8.9}$$

where L_{Σ} is the operator governing the Markovian process $\Sigma(t)$. It is of interest that in this form equation (A8.9) describes isotropic liquids, as well as nematic liquid crystals, if Σ is considered as a local orientation of a director [296] or an orientation of a liquid crystal cage in the model of two-stage orientational relaxation in partially ordered media [297]. In the latter case the operator L_{Σ} has the form

$$L_{\Sigma}P\,(\Sigma, t) = -\frac{1}{\tau_0}P\,(\Sigma, t) + \frac{1}{\tau_0}\int f\,(\Sigma', \Sigma)\,P\,(\Sigma', t)\;\mathrm{d}\,\Sigma'\,. \tag{A8.10}$$

For partially ordered media the stationary solution (A8.10) is an eigenfunction of the integral operator in (A8.10), belonging to the eigenvalue 1:

$$\int f\left(\Sigma', \Sigma\right) P_0\left(\Sigma'\right)\, \mathrm{d}\Sigma' = P_0\left(\Sigma\right);$$

$$P_0\left(\Sigma\right) = \exp\left(\frac{-U_0(\Sigma)}{kT}\right)\left[\int\left(\frac{-U_0(\Sigma)}{kT}\right)\, \mathrm{d}\Sigma\right]^{-1}. \quad \text{(A8.11)}$$

where $U_0(\Sigma)$, e.g., is the Maier–Saupe potential. The other eigenfunctions must be found proceeding from a given form of the kernel $f(\Sigma', \Sigma)$. For isotropic liquid, P_0=const., the kernel becomes a function of a difference argument

$$f\left(\Sigma', \Sigma\right) = f\left(\Sigma - \Sigma'\right),$$

and, like in the Valiev–Ivanov model, the eigenfunctions of the operator are Wigner functions.

In conclusion, we would like to note that methods of correction for detailed balance in kinetic equations are somewhat different in quantum and classical formalisms of the liquid cage model. As is demonstrated in Chapter 7, this correction may be performed in the region of fast fluctuations $V\tau_V < 1$. However, the four-index corrected operator of spectral exchange does not exist in the equations determining evolution of partially averaged density matrix $\rho(t, \Sigma)$. Therefore, for values $V\tau_V > 1$ the 'right' relaxation operator is unknown. In contrast, developing stochastic perturbation theory ($V\tau_V < 1$) in the classical formalism, one can see that the fluctuation–dissipative theorem leads in this case to the connection between the rotational friction coefficient and the potential characteristics. In principle, this eliminates the additional phenomenological parameter ξ in (A8.9). After this, this system of equations may be solved for any value of $V\tau_V$. In other words, in the classical formalism the simplified method of correction for detailed balance may hold outside the applicability limits of perturbation theory.

Appendix 9

The Γ-operator may be obtained from the general formulae (7.90)–(7.92) for $\ell = 2$

$$\begin{aligned}\Gamma^2_{\mu\nu}(jj_1) &= \sum_N R_N \Bigg\{ \delta_{jj_1}\,\delta_{\mu\nu} \sum_k \Big[K_N(\omega_{jk}+\omega)\,\left(C^{k0}_{j+\mu 0N0}\right)^2 \\ &\quad + K^*_N(\omega_{j+\mu k}-\omega)\,\left(C^{k0}_{j0N0}\right)^2\Big] \\ &\quad - [(2j+1)(2j+2\mu+1)]^{\frac{1}{2}}\,(-1)^{\mu} C^{j_1 0}_{j0N0} C^{j_1+\nu 0}_{j+\mu 0N0} \begin{Bmatrix} j_1+\nu & j_1 & 2 \\ j & j+\mu & N \end{Bmatrix} \\ &\quad \times [K^*_N(\omega_{j+\mu j_1}-\omega) + K_N(\omega_{jj_1+\nu}+\omega)] \Bigg\}. \end{aligned} \tag{A9.1}$$

The conditions determined by the selection rules for Clebsch–Gordan coefficients and $6j$ symbols provide the following block-diagonal structure of the operator

$$\Gamma^2_{\mu\nu}(jj_1) = \begin{array}{c} \mu\setminus\nu \\ -2 \\ 0 \\ 2 \\ -1 \\ 1 \end{array} \begin{array}{c} \begin{array}{ccccc} -2 & 0 & 2 & -1 & 1 \end{array} \\ \left\| \begin{array}{ccccc} \Gamma^2_{-2-2} & \Gamma^2_{-20} & \Gamma^2_{-22} & 0 & 0 \\ \Gamma^2_{0-2} & \Gamma^2_{00} & \Gamma^2_{02} & 0 & 0 \\ \Gamma^2_{2-2} & \Gamma^2_{20} & \Gamma^2_{22} & 0 & 0 \\ 0 & 0 & 0 & \Gamma^2_{-1-1} & \Gamma^2_{-11} \\ 0 & 0 & 0 & \Gamma^2_{1-1} & \Gamma^2_{11} \end{array} \right\| \end{array} \tag{A9.2}$$

We are interested in the upper left-hand block, whose diagonal and off-diagonal elements govern, respectively, exchange inside and between the O, Q and S branches of the Raman spectrum. Therefore, further on we may consider only 3 × 3 matrices, meaning that the left-hand upper block does not couple with the right-hand lower one. The quasi-classical expression for the relaxation operator

$$\Gamma^2_{\mu\nu}(jj_1) = \sum_N R_N$$

$$\left[\delta_{\mu\nu}\delta(j-j_1) f^j_N - (-1)^{(\mu-\nu)/2} \left(\frac{j}{j_1}\right)^{\frac{1}{2}} f^{jj_1}_N d^2_{\mu\nu}(\alpha)\,\delta_{\mu+\nu,2n} \right] \tag{A9.3}$$

allows one to obtain a supermatrix representation for $\Gamma^2_{\mu\nu}$ [298]

$$\Gamma^2_{\mu\nu} = \frac{1}{6}\left\|\begin{matrix} \Gamma^2_{jj_1} + 3\Gamma^1_{jj_1} + 2\Gamma^0_{jj_1} & \sqrt{6}\left(\Gamma^2_{jj_1} - \Gamma^0_{jj_1}\right) & \Gamma^2_{jj_1} - 3\Gamma^1_{jj_1} + 2\Gamma^0_{jj_1} \\ \sqrt{6}\left(\Gamma^2_{jj_1} - \Gamma^0_{jj_1}\right) & \Gamma^2_{jj_1} & \sqrt{6}\left(\Gamma^2_{jj_1} - \Gamma^0_{jj_1}\right) \\ \Gamma^2_{jj_1} - 3\Gamma^1_{jj_1} + 2\Gamma^0_{jj_1} & \sqrt{6}\left(\Gamma^2_{jj_1} - \Gamma^0_{jj_1}\right) & \Gamma^2_{jj_1} + 3\Gamma^1_{jj_1} + 2\Gamma^0_{jj_1} \end{matrix}\right\| \tag{A9.4}$$

Here a new operator $\Gamma^2_{jj_1}$ appears, which governs exchange within the Q-branch of the Raman spectrum. To diagonalize (A9.4), let us use the eigenvector matrix

$$\langle\mu\,|\,T\,|\,\nu\rangle = \begin{matrix} \mu\setminus\nu \\ -2 \\ 0 \\ 2 \end{matrix}\quad \begin{matrix} -2 & 0 & 2 \\ -\frac{1}{\sqrt{2}} & -\left(\frac{3}{8}\right)^{\frac{1}{2}} & \frac{1}{2\sqrt{2}} \\ \frac{1}{2} & \frac{1}{2} & \frac{\sqrt{3}}{2} \\ \frac{1}{\sqrt{2}} & -\left(\frac{3}{8}\right)^{\frac{1}{2}} & \frac{1}{2\sqrt{2}} \end{matrix} \tag{A9.5}$$

Diagonalization results in

$$\langle\mu j|\hat{\mathrm{T}}^{-1}\left(-\mathrm{i}\,\mathrm{H}_0^{\times} + \hat{\Gamma}^2\right)\hat{\mathrm{T}}|\nu j_1\rangle = \left\|\begin{matrix} \Gamma^1_{jj_1} & -\sqrt{3}\mathrm{i}\Omega_{jj_1} & \mathrm{i}\,\Omega_{jj_1} \\ -\sqrt{3}\mathrm{i}\Omega_{jj_1} & \Gamma^0_{jj_1} & 0 \\ \mathrm{i}\Omega_{jj_1} & 0 & (4/3)\Gamma^2_{jj_1} - (1/3)\Gamma^0_{jj_1} \end{matrix}\right\| \tag{A9.6}$$

The asymptotic expressions for the corresponding Clebsch–Gordan coefficients take the form

$$C^{j+\mu 0}_{j020} \approx \quad \begin{matrix} \mu & -2 & 0 & 2 \\ & \left(\frac{3}{8}\right)^{\frac{1}{2}} & -\frac{1}{2} & \left(\frac{3}{8}\right)^{\frac{1}{2}} \end{matrix} \tag{A9.7}$$

and after this, summation over $\mu\nu$ leads to a simple expression for $\tau_{\theta,2}$

$$\tau_{\theta,2} = \int_0^\infty \mathrm{d}j\,\mathrm{d}j_1\, 2(jj_1)^{\frac{1}{2}}\ \rho_j\ a_{00}\,(j, j_1)\,, \tag{A9.8}$$

where

$$a_{00}\,(j, j_1) = a_{\mu\nu}\,(j, j_1)\ |_{\mu=\nu=0} = \langle\mu j|\,\frac{1}{T^{-1}\left(-\mathrm{i}H_0^{\times} + \hat{\Gamma}^2\right)T}\,|\nu\, j_1\rangle|_{\mu,\nu=0} \tag{A9.9}$$

is found by two-fold application of the Frobenius formula [295]. As a result

$$\tau_{\theta,2} = \int_0^\infty \mathrm{d}j\,\mathrm{d}j_1\ 2\,(jj_1)^{\frac{1}{2}}\ \rho_j\left(\frac{1}{\hat{\Gamma}^0 + 3\hat{\Omega}(1/\Gamma^1)\,\hat{\Omega}}\right)_{jj_1}. \tag{A9.10}$$

Here $\Omega_{jj_1} = -2Bj\delta_{jj_1}$. Let us write out again a quasi-classical expression for the rotational time

$$\tau_J = \frac{1}{K_J(t=0)} \int_0^\infty \mathrm{d}j\,\mathrm{d}j_1\; 2(jj_1)^{\frac{1}{2}}\; jj_1\,\rho_j \left(\frac{1}{\hat{\Gamma}^1}\right)_{jj_1} . \tag{A9.11}$$

Rewriting (A9.10) and (A9.11) in the compact form as scalar product

$$\tau_J = \frac{1}{K_J(0)} \langle\langle J|\frac{1}{\hat{\Gamma}^1}|J\rangle\rangle\,, \tag{A9.12}$$

$$\tau_{\theta,2} = \langle\langle 1 \left| \frac{1}{\hat{\Gamma}^0 + 3\hat{\Omega}(1/\hat{\Gamma}^1)\hat{\Omega}} \right| 1\rangle\rangle\,, \tag{A9.13}$$

and taking into account that $|1\rangle\rangle$ is the eigenvector of Γ^0, belonging to the zero eigenvalue, we find in the first non-vanishing order

$$\begin{aligned}\tau_{\theta,2} &= \frac{1}{\langle\langle 1|\,3\hat{\Omega}\,(1/\hat{\Gamma}^1)\hat{\Omega}|1\rangle\rangle} = \frac{1}{12B^2\,\langle\langle 1|J\,(1/\hat{\Gamma}^1)J|1\rangle\rangle} \\ &= \frac{1}{12B^2\tau_J K_J(0)} = \frac{I}{6kT\tau_J}\,,\end{aligned} \tag{A9.14}$$

i.e. the famous Hubbard relation (in the last equality the definition $B = \hbar/(2I)$ was used). Generalization for arbitrary ℓ does not evoke any difficulties.

References

[1] Debye P. *Polar Molecules* (Dover, New York) (1929).

[2] Gordon R. G. Semiclassical theory of spectra and relaxation in molecular gases, *J. Chem. Phys.* **45**, 1649–55 (1966).

[3] Neilson W. B., Gordon R. G. On a semiclassical study of molecular collisions. 1. General method, *J. Chem. Phys.* **58**, 4143–8 (1973).

[4] Burshtein A. I., Strekalov M. L., Temkin S. I. Spectral exchange in collisional broadening of rotational structure. *Sov. Phys. JETP.* **39**, 433–9 (1974). [ZhETP **66**, 894–906, (1974)].

[5] De Santis A., Sampoli M., Morales P., Signorelli G. Density evolution of Rayleigh and Raman depolarized scattering in fluid N_2. *Mol. Phys.* **35**, 1125–40 (1978).

[6] Marabella L. J. Molecular motion and band shapes in liquids, *Appl. Spectr. Rev.*, **7**, 313–55 (1974).

[7] Bloom M., Bridges F., Hardy W. N. Nuclear spin relaxation in gaseous methane and its deuterated modifications, *Can. J. Phys.* **45**, 3533–54 (1967).

[8] Gordon R. G. Correlation functions for molecular motion, *Adv. Magn. Res.* **3**, 1–42 (1968).

[9] Pople J. A., Schneider W. G., Bernstein H. J. *High-resolution Nuclear Magnetic Resonance* (McGraw-Hill, New York) (1959). [Russian translation IIL (1962)].

[10] Keilson J., Storer J. E. On a Brownian motion, Boltzmann equation and the Fokker–Planck equation, *Quart. Appl. Math.*, **10**, 243–53 (1952).

[11] Valiev K. A., Ivanov E. N. Rotational Brownian motion, *Sov. Phys. Usp.* **16**, 1–16 (1973). [*Uspehi Phys. Nauk* **109**, 31–64 (1973)].

[12] Hubbard P. S. Rotational Brownian motion, *Phys. Rev.* **A6**, 2421–33 (1972).

[13] Rautian S. G., Sobelman I. I. The effect of collisions on the Doppler broadening of spectral lines, *Sov. Phys. Usp.* **9**, 701–16 (1967). [*Uspehi Phys. Nauk* **90**, 209–36 (1966)].

[14] Bouché Th., Drier Th., Lange B., Wolfrum J., Franck E. U., Schilling W. Collisional narrowing and spectral shift in coherent anti-Stokes Raman spectra of molecular nitrogen up to 2500 bar and 700 K, *Appl. Phys.* **B50**, 527–33 (1990).

[15] Vu H. Perturbation des bandes de rotation-vibration de quelques molécules diatomiques polaires comprimées et 'paires orbitantes' à rotation bloquée, *J. des Recherches du CNRS* **53**, 313–64 (1960).

[16] Orlova N. D., Pozdniakova L. A. Profiles of infrared absorption bands and rotational motion of molecules in liquids. Quantum rotation of hydrogen-chloride molecules, *Opt. & Spectr.* **35**, 624–7 (1973). [*Optika i Spectr.* **35**, 1074–7 (1973)].

[17] Robert R., Galatry L. Infrared absorption of diatomic molecules in liquid solutions, *J. Chem. Phys.*, **55**, 2347–59 (1971).

[18] Talin B., Galatry L., Klein L. Relaxation processes and spectra in liquids and dense gases. *J. Chem. Phys.* **66**, 2789–800 (1977).

[19] Serebrennikov Y. A., Temkin S. I., Burshtein A. I. Infrared and Raman spectra of a linear rotator in the fluctuating liquid cage model, *Chem. Phys.* **81**, 31–40 (1983).

[20] Burshtein A. I. *'Lectures in Quantum Kinetics'* (Novosibirsk State University, Novosibirsk) (1968), [in Russian].

[21] Gnedenko B. V. *The Theory of Probability* (Chelsea, New York) (1962). [Translated from Russian: *Teoria veroyatnostey* Moscow GITTL, (1954)].

[22] Gordon R. G. On the rotational diffusion of molecules, *J. Chem. Phys.*, **44**, 1830–6 (1966).

[23] Varshalovich D. A., Moskalev A. N., Khersonski V. K. *Quantum Theory of Angular Momentum* (World Scientific, Singapore) (1986). [*Kvantovaia teoria uglovogo momenta* Leningrad, Nauka (1975)].

[24] Doktorov A. B., Burshtein A. I. Manifestation of the frequency migration mechanism in decay of echo signals, *Sov. Phys. JETP.* **36**, 411–17 (1973), [*ZhETP* **63**, 784–98 (1972)].

[25] Burshtein A. I., Kofman A. G. Light-induced relaxation in an intense Rayleigh field, *Sov. Phys. JETP.* **43**, 436–41 (1976). [*ZhETP* **70**, 840–50 (1976)].

[26] Sack R. A. Relaxation process and inertial effects. 2. Free rotation in space. *Proc. Phys. Soc.* (*London*) **B70**, 414–26 (1957).

[27] Burshtein A. I., Temkin S. I. Collapse of the rotational structure of Raman-scattering spectra in dense media, *Sov. Phys. JETP.* **44**, 492–499 (1976). [*ZhETP* **71**, 938–51 (1976)].

[28] Shore B. W. *The Theory of Coherent Atomic Excitation,* Vol. 2. (John Wiley, New York) (1990).

[29] Fixman M., Rider K. Angular relaxation of the symmetric top. *J. Chem. Phys.* **51**, 2425–38 (1969).

[30] Hubbard P. S. Theory of nuclear magnetic relaxation by spin-rotational interactions in liquids. *Phys. Rev.* **131**, 1155–65 (1963).

[31] McClung R. E. D. Rotational diffusion of spherical-top molecules in liquids. *J. Chem. Phys.* **51**, 3842–52 (1969).

[32] St. Pierre A. G., Steele W. A. Collisional effects upon rotational correlations of symmetrical top molecules. *J. Chem. Phys.* **57**, 4638–48 (1972).

[33] Lynden-Bell R. M. The effect of molecular reorientation on the line-shapes of degenerate vibrations in infra-red and Raman spectra of liquids, *Mol. Phys.* **31**, 1653–62 (1976).

[34] Chandrasechar S. Stochastic problems in physics and astronomy. *Rev. Mod. Phys.* **15**, 1–80 (1943).

[35] Nikitin E. E. *Theory of Elementary Atomic and Molecular Processes in Gases* (Clarendon Press, Oxford) (1974). [Moscow, Chimia (1970)].
[36] Burshtein A. I. Hopping mechanism of energy transfer. *Sov. Phys. JETP.* **35**, 882–5. (1972). [*ZhETP* **62**, 1695–701, (1972)].
[37] Bateman H., Erdélyi A. *Higher Transcendental Functions* (McGraw-Hill, New York) (1953).
[38] Kubo R. Generalized cumulant expansion method. *J. Phys. Soc. Japan.* **17**, 1100–20 (1962).
[39] Abragam A. I. *The Principles of Nuclear Magnetism.* (Clarendon Press, Oxford) (1961).
[40] Diestler D. J., Wilson R. S. Quantum dynamics of vibrational relaxation in condensed media, *J. Chem. Phys.* **62**, 1572–8 (1975).
[41] Madden P. A., Lynden-Bell R. M. Theory of vibrational line-widths, *Chem. Phys. Lett.*, **38**, 163–5 (1976).
[42] Zusman L. D., Burshtein A. I. Kinetics of intermolecular energy transfer in condensed media, *J. Appl. Spectr.* **15**, 932–6 (1971) [*Zh. Prikl. Spectroscopii* **15**, 124–30 (1971)].
[43] Rytov S. M. *Principles of Statistical Radiophysics Vol. I Elements of Random Process Theory*. Springer, Berlin. (1987) [*Stochastic Processes.* Moscow, Nauka (1976)].
[44] Purcell E. M. Nuclear spin relaxation and nuclear electric dipole molecules. *Phys. Rev.* **117**, 828–31, (1960).
[45] Petrunina E. B., Romanov V. P., Soloviov V. A. The computation of the relaxation times in liquid in bimolecular collisions model, *Acoustic Journal*, **21**, 782–8 (1975) [in Russian].
[46] Hare W. F. J., Welsh H. L. Pressure induced infrared absorption of hydrogen and hydrogen–foreign gas mixtures in the range 1500–5000 atmospheres, *Can. J. Phys.* **36**, 88–103 (1958).
[47] Cunsolo S., Gush H. P. Collision-induced absorption of H_2 and He–Ar mixtures near $\lambda = 1$mm, *Can. J. Phys.* **50**, 2058–9 (1972).
[48] Nikitin E. E. On the interpretation of pressure induced infrared spectra of gases. *Opt. & Spectr.* **8**, 135–6 (1960) [*Optika i Spectr.* **8**, 264–6 (1960)].
[49] Van Kranendonk J. Intercollisional interference effects in pressure induced infrared spectra. *Can. J. Phys.* **46**, 1173–9 (1968).
[50] Lewis J. C. Theory of intercollisional interference effects. II. Induced absorption in a real gas. *Can. J. Phys.* **50**, 2881–901 (1972).
[51] Berne B. J., Boon J. P., Rice S. A. On the calculation of autocorrelation functions of dynamical variables, *J. Chem. Phys.* **45**, 1086–96 (1966).
[52] Mori H. Transport, collective motion and Brownian motion. *Progr. Theor. Phys.* **33**, 423–55 (1965); A continued-fraction representation of the time-correlation functions, *Ibid.* **34**, 399–416 (1965).
[53] Burshtein A. I., Naberukhin Yu. I. Adiabatic theory of the shapes of spectral lines in gases in terms of the hard sphere molecules, *Opt. & Spectr.* **20**, 521–4 (1966) [*Optika i Spectr.* **20**, 936–43 (1966)].
[54] Rahman A. Correlations in the motion of atoms in liquid argon, *Phys. Rev.* **A136**, 405–11 (1964).
[55] Hiwatari Y. The applicability of the soft core model of fluids to dynamical properties of simple liquids, *Progr. Theor. Phys.* **53**, 915–28 (1975).
[56] Singwi K. S., Tosi M. P. On the velocity autocorrelation in a classical fluid, *Phys. Rev.* **157**, 153–5 (1967).

[57] Burshtein A. I., Storozhev A. V. The angular momentum relaxation due to multiparticle collisions of molecules with atoms, *Chem. Phys.* **164**, 47–55 (1992).

[58] Ch'en S., Takeo M. Broadening and shift of spectral lines due to presence of foreign gases, *Rev. Modern Phys.* **29**, 20–73 (1957) [*Usp. Fiz. Nauk* **66**, 391–474 (1958)].

[59] Fisher I. Z., Zatovskii A. V., Malomuzh N. P. Hydrodynamic asymptotic form of the rotational motion correlation function of a molecule in a liquid, *Sov. Phys. JETP.* **38**, 146–50 (1974) [*ZhETP* **19**, 761–4 (1974)].

[60] Hansen J. P., McDonald I. R. *The Theory of Simple Liquids* (Academic Press, London) (1986).

[61] Burshtein A. I, Storozhev A.V., Strekalov M. L. Rotational relaxation in gases and its spectral manifestations, *Chem. Phys.* **131**, 145–56 (1989).

[62] Storozhev A. V., Lynden-Bell R. M. Rotational relaxation in dense gases, *Chem. Phys. Lett.* **183**, 316–20 (1991).

[63] Anderson P. W., Talman J. D. In: *Proceedings of the Conference on Broadening of Spectral Lines, University of Pittsburg, Bell Telephone Syst. Monogr.* p. 3117 (1955).

[64] Fedorenko S. G., Burshtein A. I. Deviations from linear Stern–Volmer law in hopping quenching theory, *J. Chem. Phys.* **97**, 8223–32 (1992).

[65] Barojas J., Levesque D., Quentrec B. Simulation of diatomic homonuclear liquids, *Phys. Rev.* **A7**, 1092–1105 (1973).

[66] O'Reilly D. E., Peterson E. M., Hogenboom D. L., Scheie C. E. Fluorine-19 nuclear magnetic resonance of the liquid and solid state of fluorine, *J. Chem. Phys.* **54**, 4194–9 (1971).

[67] Gillen K. T., Douglas D. S., Malmberg M. S., Maryott A. A. NMR relaxation study of liquid CCl_3F. Reorientational and angular momentum correlation times and rotational diffusion, *J. Chem. Phys.* **57**, 5170–9 (1972).

[68] (Eds Ciccotti G., Frenkel D., McDonald I. R.) *Simulation of Liquids and Solids: Molecular Dynamics and Monte Carlo Methods in Statistical Mechanics* (North-Holland Physics Publishing, Amsterdam) (1987).

[69] Allen M. P., Tildesley D. J. *Computer Simulation of Liquids.* (Clarendon Press, Oxford) (1989).

[70] Steele W. A. Computer simulation of the depolarized Rayleigh scattering in fluid N_2. *J. Mol. Liquids* **48**, 321–3 (1991).

[71] Ishol L. M., Scott T. A., Goldblatt M. Nuclear spin–lattice relaxation in solid and liquid $^{15}N_2$ and $^{14}N_2$, *J. Magn. Res.* **23**, 313–20 (1976).

[72] Chandler D. Translational and rotational diffusion in liquids. I. Translational single-particle correlation functions. *J. Chem. Phys.* **60**, 3500–507, (1974). Translational and rotational diffusion in liquids. II. Orientational single-particle correlation functions. *J. Chem. Phys.* **60**, 3508–12 (1974).

[73] Verlet L., Weis J. -J. Equilibrium theory of simple liquids, *Phys. Rev.* **A5**, 939–52 (1972).

[74] Einwohner T., Alder B. J. Molecular dynamics. VI. Free-path distributions and collision rates for hard-sphere and square-well molecules, *J. Chem. Phys.* **49**, 1458–73 (1968).

[75] Talbot J. A statistical model for the dynamical properties of hard particle fluids, *Mol. Phys.* **75**, 43–58 (1992).

[76] Burshtein A. I., Temkin S. I., Zharikov A. A. Non-Markovian theory of sudden modulation, *Theor. and Math. Phys.* **66**, 166–73, (1986) [*Teor. Mat. Fiz.* **66**, 253–6 (1986)].
[77] Abdrakhmanov B. M., Burshtein A. I., Temkin S. I. Impact description of the Poley absorption, *Chem. Phys.* **143**, 297–304 (1990).
[78] Zatsepin V. M. Time correlation functions of one-dimensional rotational Brownian motion in n-fold periodical potential. *Theor. and Math. Phys.* **33**, 400–408 (1977).
[79] Revokatov O. P., Gangardt M. G., Parfenov S. V. 'Gaslike-character' of rotational motion of SF_6 molecules in a liquid. *JETP Lett.* **19**, 391–2 (1974) [*Pis'ma ZhETP,* **19**, 761–4 (1974)].
[80] Brueck S. R. J. Vibrational two-photon resonance linewidth in liquid media, *Chem. Phys. Lett.* **50**, 516–20 (1977).
[81] Golubev N. S., Burshtein A. I., Temkin S. I. Experimental verification of the theory of J-diffusion in nitrogen gas, *Chem. Phys. Lett.* **91**, 139–42 (1982).
[82] Jameson C. J., Jameson A. K., Smith N. C. ^{15}N spin-relaxation studies of N_2 in buffer gases. Cross-sections for molecular reorientation and rotational energy transfer, *J. Chem. Phys.* **86**, 6833–8 (1987).
[83] Gordon R. G. Kinetic theory of nuclear spin relaxation in gases, *J. Chem. Phys.* **44**, 228–34 (1966).
[84] Frenkel D., Gravestein D. J., van der Elsken J. Non-linear density dependence of rotational line-broadening of HCl in dense argon, *Chem. Phys. Lett.* **40**, 9–13 (1976).
[85] Powles J. G., Rickayzen G. Correlation times for molecular reorientation, *Mol. Phys.* **33**, 1207–27 (1977).
[86] Brown W. F. Dielectrics. *Handbuch der Physik*, Herausgegeben von S. Flügge, Band XVII, S. 1 (Springer, Berlin), (1956) [Moscow IL (Russian translation) (1961)].
[87] Bratos S., Marechal E. Raman study of liquids. I. Theory of Raman spectra of diatomic molecules in inert solutions, *Phys. Rev.* **A4**, 1078–92 (1971).
[88] Breuillard C., Ouillon R. Infrared and Raman band shapes and dynamics of molecular motions for N_2O in solutions: v_3 band in CCl_4 and liquid SF_6. *Mol. Phys.* **33**, 747–57 (1977).
[89] May A. D., Stryland J. C., Varghese G. Collisional narrowing of the vibrational Raman band of nitrogen and carbon monoxide, *Can. J. Phys.* **48**, 2331–5 (1970).
[90] McConnell J. *Rotational Brownian Motion and Dielectric Theory.* (Academic Press, New York) (1980).
[91] Burshtein A. I. External relaxation, *Sov. Phys. Sol. State* **5**, 908–17 (1963) [*Fiz. Tverd. Tela* **5**, 1243–57 (1963)].
[92] Burshtein A. I. Motion-broadened and motion-narrowed spectra, *Chem. Phys. Lett.* 335–40 (1981).
[93] Kondaurov V. A., Melikova S. M., Shchepkin D. N. Rotational broadening of the IR absorption bands of heavy molecules dissolved in liquefied gases, *Opt. & Spectr.* **56**, 626–30 (1984) [*Optika i Spectr.* **56**, 1020–4 (1984)].
[94] Maryott A. A., Farrar T. C., Malmberg M. S. ^{35}Cl and ^{19}F NMR spin–lattice relaxation time measurements and rotational diffusion in liquid ClO_3F. *J. Chem. Phys.* **54**, 64–71 (1971).

[95] Gillen K. T, Douglas D. C., Malmberg M. S, Maryott A. A. NMR relaxation study of liquid CCl_3F. Orientational and angular momentum correlation times and rotational diffusion, *J. Chem. Phys.* **57**, 5170–9 (1972).

[96] Steele W. A. Molecular reorientation in liquids. II. Angular autocorrelation functions, *J. Chem. Phys.* **38**, 2411–18 (1963).

[97] Berne B. J. A self-consistent theory of rotational diffusion, *J. Chem. Phys.* **62**, 1154–60 (1975).

[98] Gross E. P. Inertial effects and dielectric relaxation, *J. Chem. Phys.* **23**, 1415–23 (1955).

[99] Sack R. A. Relaxation processes and inertial effects. I. Free rotation about a fixed axis, *Proc. Phys. Soc. (London)* **B70**, 402–13 (1957).

[100] Poley J. Ph. Microwave dispersion of some polar liquids, *Appl. Sci. Res.* **B4**, 337–87 (1955).

[101] Zwanzig R. Memory effects in irreversible thermodynamics, *Phys. Rev.* **124**, 983–92 (1961).

[102] De Santis A., Nardone M., Sampoli M. Light scattering orientational memory functions for fluid N_2 at moderate densities, *Mol. Phys.* **40**, 1185–96 (1980).

[103] De Santis A., Nardone M., Sampoli M. Correlation functions from depolarized Raman and Rayleigh spectra of N_2 at high density and 150 K, *Mol. Phys.* **41**, 769–77 (1980).

[104] Anderson P. W. A mathematical model for the narrowing of spectral lines by exchange or motion, *J. Phys. Soc. Japan* **9**, 316–39 (1954).

[105] Dokuchaev A. B., Tonkov M. V. Effect of rotational relaxation on the shape of the ν_3 band of C_2. *Opt. & Spectrosc.* **60**, 664–6 (1986) [*Optika i Spectr.* **60**, 1074–7 (1986)].

[106] De Santis A., Moretti E., Sampoli M. Rayleigh spectra of N_2 fluid. Temperature and density behaviour of the second moment, *Mol. Phys.* **46**, 1271–82 (1982).

[107] Nee T. W., Zwanzig R. Theory of dielectric relaxation in polar liquids. *J. Chem. Phys.* **52**, 6353–63 (1970).

[108] Zatsepin V. M. To experimental determination of angular momentum correlation function in liquids, *Ukrainian Phys. J.* **21**, 48–52 (1976).

[109] Zatovski A. V., Salistra G. I. The theory of depolarized light scattering in solutions, *Ukrainian Phys. J.* **18**, 435–9 (1973).

[110] Zatovskaya A. A., Zatovski A. V. Long-time asymptotics of the correlation functions of the molecular rotation in liquids, *Ukrainian Phys. J.* **19**, 1180–4 (1974).

[111] Burshtein A. I., McConnell J. Spectral estimation of finite collision times in liquid solutions, *Physica* **A157**, 933–54 (1989).

[112] Burshtein A. I., McConnell J. The Poley absorption problem, *J. Mol. Liquids* **43**, 21–39 (1989).

[113] Guillot B., Bratos S. Theoretical analysis of dielectric properties of polar liquids in the far-infrared spectral range, *Phys. Rev.* **A 16**, 424–30 (1977).

[114] Desplanques P., Constant E., Fauquembergue R. In: *Molecular Motions in Liquids,* ed. J. Lascombe, (Reidel, Dordrecht), pp. 133–49 (1974).

[115] Anderson J. E., Ullman R. Angular velocity correlation functions and high-frequency dielectric relaxation, *J. Chem. Phys.* **55**, 4406–14, (1971).

[116] Kushick J., Berne B. J. Methods for experimentally determining the angular velocity relaxation in liquids, *J. Chem. Phys.* **59**, 4486–90 (1973).

[117] Dill J. F., Litovitz T. A., Bucaro J. A. Molecular reorientation in liquids by Rayleigh scattering: pressure dependence of rotational correlation functions, *J. Chem. Phys.* **62**, 3839–50 (1975).

[118] Kluk E., Monkos K., Pasterny K., Zerda T. A means to obtain angular velocity correlation functions from angular position correlation functions of molecules in liquid. Part I. General discussion and its application to linear and spherical top molecules, *Acta Physica Polonica* **A 56**, 109–16 (1979).

[119] Lynden-Bell R. M. In: *Molecular Liquids: Dynamics and Interactions*, eds. A. J. Barnes, W. J. Orville-Thomas, J. Yarwood. (Reidel, Dordrecht) (1984).

[120] Berestetski V. B., Lifshitz E. M., Pitaevski L. P. *Relativistic Quantum Theory* (Nauka, Moscow) (1968).

[121] Landau L. D., Lifshitz E. M. *Quantum Mechanics. Nonrelativistic Theory.* (Pergamon Press, Oxford) (1977).

[122] Kometani K., Shimizu H. Study of the dipolar relaxation by a continued fraction representation of the time correlation function, *J. Phys. Soc. Japan* **30**, 1036–48 (1971).

[123] Burshtein A. I., Temkin S. I. On the discrimination between various mechanisms of orientational relaxation in liquids, *XXth Congress AMPERE (Tallinn), Thesis,* p. 459 (1978).

[124] Marsault-Herail F., Marsault J. P., Michond G., Levi G. Raman scattering: Orientational motion and collision frequency in liquid CF_4 from the triple to critical point, *Chem. Phys. Lett.* **31**, 335–9 (1975).

[125] Snider S., McClung R. E. D. Raman studies of molecular reorientation in liquid sulfur hexafluoride, *Can. J. Phys.* **52**, 1209–14 (1974).

[126] Herzberg G. Spectra of diatomic molecules, (Van Nostrand, Princeton) (1965).

[127] Bratos S., Chestier J. P. Infrared and Raman study of liquids. III. Theory of rotation–vibration coupling effects. Diatomic molecules in inert solutions, *Phys. Rev.* **A9**, 2136–50 (1974).

[128] Anderson P. W. Pressure broadening in the microwave and infrared regions, *Phys. Rev.* **76**, 647–61 (1949).

[129] Wittke J. P., Dicke R. H. Redetermination of the hyperfine splitting in the ground state of atomic hydrogen, *Phys. Rev.* **103**, 620–31 (1956).

[130] Fischer S. F., Laubereau A. Dephasing processes of molecular vibrations in liquids, *Chem. Phys. Lett.* **35**, 6–12 (1975).

[131] Diestler D. J., Manz J. Model for librational dephasing of diatomic molecules in liquids. Liquid N_2 and O_2, *Mol. Phys.* **33**, 227–44 (1977).

[132] Knaap E. W. Vibrational dephasing of diatomic molecules in liquids: role of anharmonicity of the diatom, *Chem. Phys. Lett.* **58**, 221–4 (1978).

[133] Temkin S. I., Burshtein A. I. Pressure-induced transformation of the Q-branch of the rotational–vibrational Raman-scattering spectrum, *JETP Lett.* **24**, 86–9 (1976) [*Pis'ma ZhETF* **24**, 99–103 (1976)].

[134] Temkin S. I., Burshtein A. I. Spectral manifestations of various mechanisms of rotational relaxation in dense media. *JETP Lett.* **28**, 538–41 (1978) [*Pis'ma ZhETF* **28**, 583–7 (1976)].

[135] Temkin S. I., Burshtein A. I. On the shape of the Q-branch of Raman scattering spectra in dense media. Theory, *Chem. Phys. Lett.* **66**, 52–6 (1978).

[136] Akhmanov S. A., Gadjiev F. N., Koroteev N. I., Orlov R. Yu., Shumai I. L. Position and width of Q-branch of nitrogen molecules in cryogenic mixtures. *JETP Lett.* **27**, 243–6 (1978). [*Pis'ma ZhETP* **27**, 260–4 (1978)].

[137] Clouter M. J., Kiefte H, Ali N. Anomalous behaviour in the vibrational Raman spectrum of oxygen under near-critical conditions, *Phys. Rev. Lett.* **40**, 1170–3 (1978)

[138] Letokhov V. S., Chebotayev V. P. *Nonlinear Laser Spectroscopy.* (Springer, Berlin) Springer Series in Optical Science (1977) [*Principles of the Nonlinear Spectroscopy.* (Nauka, Moscow) (1975)].

[139] Rautian S. G., Smirnov G. I., Shalagin A. M. *Non-linear Resonance in Spectra of Atoms and Molecules.* (Nauka, Novosibirsk) (1979) [in Russian].

[140] Burshtein A. I., Kofman A. G. Relaxation of a dip in the atomic velocity distribution due to elastic collisions, *Sov. J. Quant. Electron.* **5**, 274–8 (1975) [*Kvantovaya Elektronika* **2**, 1597–608 (1975)].

[141] Kofman A. G., Burshtein A. I. Kinetics of the Doppler spectrum saturation, *Sov. Phys. JETP* **49**, 1019–26 (1989) [*ZhETP* **76**, 2011–25 (1979)].

[142] Allin E. J., May A. D., Stoicheff B. P., Stryland J. C., Welsh H. L. Spectroscopy research at the McLennan physical laboratories in the University of Toronto, *Appl. Opt.* **6**, 1597–608 (1967).

[143] Temkin S. I., Burshtein A. I. On the shape of the Q-branch of Raman scattering spectra in dense media. Comparison with experiment, *Chem. Phys. Lett.* **66**, 57–61 (1979).

[144] Temkin S. I., Suvernev A. A. Evaluation of the CARS spectra of linear molecules in the Keilson–Storer model, In: *Coherent Raman Spectroscopy.* ed. G. Marowsky, V. V. Smirnov, Springer Proceedings in Physics **63**, pp. 49–53 (Springer, Berlin) (1992).

[145] Clouter M. J., Kiefte H. Temperature dependence of the width of the Raman Q-branch in liquid nitrogen and oxygen. *J. Chem. Phys.* **66**, 1736–9 (1977).

[146] Courtney J. A., Armstrong R. L. A nuclear spin relaxation study of the spin–rotation interaction in spherical top molecules. *Can. J. Phys.* **50**, 1252–61 (1972).

[147] Bauer H. -J., Kneser H. O., Sahm K. F. Relaxation der molekularen Freiheitsgrade des NO-Molekuls, *5ème Congr. Intern. d'Acoustique (Liège)*, p. 22 (1965).

[148] Winter T. G., Hill G. L., Raff L. M. The temperature dependence of the rotational relaxation time in gases, *6th Intern. Congr. on Acoustics (Tokyo)*, J-4-2 (1968).

[149] Greenspan M. Rotational relaxation in nitrogen, oxygen and air, *J. Acoust. Soc. Am.* **31**, 155–60 (1959).

[150] Winter T. G., Hill G. L. High-temperature ultrasonic measurements of rotational relaxation in hydrogen, deuterium, nitrogen and oxygen, *J. Acoust. Soc. Am.* **42**, 848–58 (1967).

[151] Golubev N. S., Orlova N. D., Platonova L. A. Evidence of rotational and vibrational relaxation in the isotropic spectra of CO and N_2 *JETP Lett.* **35**, 76–9 (1982) [*Pis'ma ZhETF* **35**, 65–8 (1982)]

[152] Krynicki K., Rahkamaa E. J., Powles J. G. The properties of liquid nitrogen. II. Nuclear spin relaxation. *Mol. Phys.* **29**, 539–48 (1975).

[153] Laubereau A., Fisher S. F., Spanner K., Kaiser W. Vibrational population lifetimes of polyatomic molecules in liquids. *Chem. Phys.* **31**, 335–44 (1978).

[154] Laubereau A. Picosecond phase relaxation of the fundamental vibrational mode of liquid nitrogen. *Chem. Phys. Lett.* **27**, 600–602 (1974).

[155] Calaway W. F., Ewing G. E. Vibrational relaxation in liquid nitrogen, *Chem. Phys. Lett.* **30**, 485–9 (1975).

[156] Burshtein A. I., Naberukhin Yu. I. The role of excluded volume in the theory of impact broadening of the spectral lines in gases. *Sov. Phys. 'Doklady'* **10**, 545–6 (1965) [*Dokl. AN SSSR, Ser. Fiz.* **162**, 1262–4 (1965)].

[157] Burshtein A. I., Naberukhin Yu. I. Applicability of a perturbation method to phase relaxation and line shape problems in gases, *J. Appl. Spectr.* **3**, 342–4 (1965) [*Zh. Prikl. Spektr.* **3**, 461–3 (1965)].

[158] Burshtein A. I., Naberukhin Yu. I. Phase-memory effects in the theory of spectral line broadening in gases. *Sov. Phys. JETP.* **25**, 799–805 (1967) [*ZhETF,* **52**, 1202–11 (1967)].

[159] Rothschild W. G. Vibrational resonance energy transfer and dephasing in liquid nitrogen near its boiling point: molecular computations, *J. Chem. Phys.* **65**, 2958–61 (1976).

[160] Mikhailov G. V. The influence of temperature and pressure on the Raman spectrum of nitrogen, *Sov. Phys. JETP* **36**, 974–8 (1959) [*Zh. Eksp. Teor. Fiz.* **36**, 1368–73 (1959)].

[161] Mikhailov G. V. Effect of pressure on the Raman spectrum of oxygen, *Sov. Phys. JETP* **37**, 1114–17 (1960) [*Zh. Eksp. Teor. Fiz.* **37**, 1570–4 (1959)].

[162] Volkov S. V., Kozlov D. N., Malikov M. R., Smirnov V. V. Effects of vibrational and rotational relaxation on the behavior of the Q-branch profile of ν_1 vibrations of CH_4 and SiH_4 molecules in dense gases, *Sov. Phys. JETP* **59**, 482–7 (1984) [*ZhETP* **86**, 826–34 (1984)].

[163] Temkin S. I., Suvernev A. A., Burshtein A. I. Pressure transformation of the Q-branch of the CARS spectrum of spherical molecules. *Opt. & Spectr.* **66**, 39–43 (1989) [*Opt. i Spektr.* **66**, 69–76 (1989)].

[164] Alekseev V. A., Malyugin A. V. General features of collision narrowing of spectral lines in gases, *Sov. Phys. JETP,* **53**, 456–65 (1981) [*ZhETP* **80**, 897–915 (1981)].

[165] Suvernev A. A., Temkin S. I. Spin-rotational NMR-relaxation of spherical molecules in gas phase, *Chem. Phys. Lett.* **154**, 49–55 (1989).

[166] Beckmann P. A., Bloom M., Ozier I. Proton spin relaxation in dilute methane gas: a symmetrized theory and its experimental verification, *Can. J. Phys.* **54**, 1712–27 (1976).

[167] Beckmann P. A., Burnell E. E. Nuclear spin relaxation and centrifugal distortion effects in dilute silane gas, *Can. J. Phys.* **55**, 1354–5 (1977).

[168] Kistemaker P. G., Hanna M. M., Tom A., De Vries A. E. Rotational relaxation in mixtures of methane with helium, argon and xenon, *Physica* **60**, 459–71 (1972).

[169] Margenau H., Lewis M. Structure of spectral lines from plasmas, *Rev. Mod. Phys.* **31**, 569–616 (1959).

[170] Kolb A. C., Griem H. R. Theory of line broadening in multiple spectra. *Phys. Rev.* **111**, 514–21 (1958).

[171] Griem H. R., Kolb A. C., Shen K. Y. Stark broadening of hydrogen lines in plasma, *Phys. Rev.* **116**, 4–16 (1959).

[172] Griem H. R., Baranger M., Kolb A. C., Oertel G. Stark broadening of neutral helium lines in a plasma, *Phys. Rev.* **125**, 177–95 (1962).

[173] Alekseev V. A., Sobelman I. I. Influence of collisions on stimulated Raman scattering in gases, *Sov. Phys. JETP.* **28**, 991–4 (1969) [*ZhETF,* **55**, 1847–80 (1969)].

[174] Lightman A., Ben-Reuven A. Line mixing by collisions in the far-infrared spectrum of ammonia, *J. Chem. Phys.* **50**, 351–3 (1969); Cross relaxation in the rotational inversion doublets of ammonia in the far infrared, *J. Quant. Spectrosc. Radiat. Transfer* **12**, 449–54 (1972).

[175] Fano U. Pressure broadening as a gas prototype of relaxation, *Phys. Rev.* **131**, 259–68 (1963).

[176] Strekalov M. L., Burshtein A. I. Collapse of shock-broadened multiplets, *Sov. Phys. JETP.* **34**, 53–8 (1972) [*ZhETF* **61**, 101–11 (1971)].

[177] Anderson P. W. Pressure broadening in the microwave and infrared regions, *Phys. Rev.* **76**, 647–61 (1949).

[178] Doktorov A. B. The impact approximation in the theory of bimolecular quasi-resonant process, *Physica* **A 90**, 109–36 (1978).

[179] Kipriyanov A. A., Doktorov A. B., Burshtein A. I. Binary theory of dephasing in liquid solutions. I. The nonmarkovian theory of encounters, *Chem. Phys.* **76**, 149–62 (1983).

[180] Burshtein A. I., Storozhev A. V., Strekalov M. L. Non-Markovian binary theory of interference between spectral lines, *Sov. Phys. JETP.* **62**, 456–62 (1985) [*ZhETF,* **89**, 796–807 (1985)].

[181] Griem H. *Spectral Line Broadening by Plasma.* (Academic Press, New York) (1974) [Russian translation (Mir, Moscow) p. 186 (1978)].

[182] Janke E., Emde F., Lösch F. *Tafeln Höherer Funktionen.* (B. G. Teubner, Stuttgart) (1960) [Russian translation Nauka, Moscow (1977)].

[183] Bulanin M. O., Muradov G., Tokhadze K. G. Spectroscopic demonstration of local interactions between HCl and impurity molecules in cryosystems, *Opt. & Spectrosc.* **51**, 120–1 (1981) [*Opt. i Spectr.* **51**, 216–18 (1981)].

[184] Ben-Reuven A. Resonance broadening of spectral lines. *Phys. Rev.* **A4**, 2115–20 (1971).

[185] Strekalov M. L., Burshtein A. I. Quantum theory of isotropic Raman spectra changes with gas density, *Chem. Phys.* **60**, 133–48 (1981).

[186] Strekalov M. L., Burshtein A. I. Theory of vibrational line width in dense gases, *Chem. Phys.* **82**, 11–24 (1983).

[187] Smirnov V. V., Fabelinskii V. I. Pressure-induced change of the shape and width of the coherent anti-Stokes light scattering spectrum of the ν_2 oscillations of acetylene. *JETP Lett.* **27**, 123–5 (1978) [*Pis'ma ZhETP,* **27**, 131–4 (1978)].

[188] Kubo R. Note on the stochastic theory of resonance absorption, *J. Phys. Soc. Japan* **9**, 935–46 (1954).

[189] Pack R. T. Relation between some exponential approximations in rotationally inelastic molecular collisions, *Chem. Phys. Lett.* **14**, 393–5 (1972).

[190] Snider R. F., Coombe D. A. Kinetic cross sections in the infinite order sudden approximation, *J. Phys. Chem.* **86**, 1164–74 (1982).

[191] Belikov A. E., Burshtein A. I., Dolgushev S. V., Storozhev A. V., Strekalov M. L., Sukhinin G. I., Sharafutdinov R. G. Rate constants and rotational

relaxation times for N_2 in argon: theory and experiment, *Chem. Phys.* **139**, 239–59 (1989).

[192] Goldflam R., Kouri D. J. On accurate quantum mechanical approximations for molecular relaxation phenomena. Averaged j_z-conserving coupled states approximation, *J. Chem. Phys.* **66**, 542–7 (1977).

[193] Fitz D. E., Marcus R. A. Semiclassical theory of molecular spectral line shapes in gases, *J. Chem. Phys.* **59**, 4380–92 (1973).

[194] De Pristo A. E., Augustin S., Ramaswamy R., Rabitz H. Quantum number and energy scaling for nonreactive collisions, *J. Chem. Phys.* **71**, 850–65 (1979).

[195] Storozhev A. V. Nonresonance effects in the binary relaxation theory, *Chem. Phys.* **138**, 81–8 (1989).

[196] Tusa J., Sulkes M., Rice S. A. Very low energy cross sections for collision-induced rotational relaxation of I_2 seeded in a supersonic free jet, *Proc. Natl. Acad. Sci. USA* **77**, 2367–9 (1980).

[197] Goldflam R., Green S., Kouri D. J. Infinite order sudden approximation for rotational energy transfer in gaseous mixtures, *J. Chem. Phys.* **67**, 4149–61, (1977).

[198] Strekalov M. L. Exponential approximation in the theory of rotational excitation of molecules by atoms, *Opt. & Spectr.* **49**, 18–25 (1981) [*Optika i Spectr.* **49** 36–44 (1980)].

[199] Strekalov M. L. Semiclassical theory of rotational energy relaxation in diffusion approximation, *Khim. Fiz.* **7**, 1182–92 (1988).

[200] Dickinson A. S., Richards D. A semiclassical study of the body-fixed approximation for rotational excitation in atom–molecule collisions, *J. Phys.* **B 10**, 323–43 (1977).

[201] Storozhev A. V., Strekalov M. L. Relaxation cross sections for transfer of rotational angular momentum in a semiclassical approximation, *Chem. Phys.* **153**, 99–113 (1991).

[202] Storozhev A. V., Strekalov M. L. Adiabatic and energetic correction of sudden perturbation method for inelastic rotational transitions, *Khim. Fiz.* **9**, 867–76 (1990).

[203] Strekalov M. L., Burshtein A. I. Temperature dependence of dephasing relaxation in dense media, *Chem. Phys. Lett.* **86**, 295–8 (1982).

[204] Parker G. A., Pack R. T. Rotationally and vibrationally inelastic scattering in the rotational IOS approximation. Ultrasimple calculation of total (differential, integral, and transport) cross sections for non-spherical molecules, *J. Chem. Phys.* **68**, 1585–601 (1978).

[205] Connor J. N. L., Sun H., Hutson J. M. Exact and approximate calculations for the effect of potential anisotropy on integral and differential cross-sections: Ar–N_2 rotationally inelastic scattering, *J. Chem. Soc. Faraday Trans.* **86**, 1649–57 (1990).

[206] Pack R. T. Close coupling test of classical cross-sections for rotationally inelastic Ar-N_2 collisions, *J. Chem. Phys.* **62**, 3143–8 (1975).

[207] Gianturco F. A., Bernardi M., Venanzi M. Accuracy of the IOS approximation for highly inelastic RET collisions, *Chem. Phys. Lett.* **165**, 344–50 (1990).

[208] Gianturco F. A., Toennies P., Bernardi M. Approximating relative rotational energy transfer in molecular collisions, *J. Chem. Soc. Faraday Trans.* **87**, 31–6 (1991).

[209] Pattengill M. D., La Budde P. A., Bernstein R. B., Curtis C. F. Molecular collisions. XVI. Comparison of GPS with classical trajectory calculations of rotational inelasticity for the Ar-N_2 system, *J. Chem. Phys.* **55**, 5517–22 (1971).

[210] Choi B. H., Tang K. T. Close coupling studies of rotational excitations of Ar+N_2 and of H^++H_2 collisions, *J. Chem. Phys.* **65**, 5528–31 (1977).

[211] Ticktin A., Whitaker B. J., McCaffery A. J. Computational test of fitting laws of rotational energy transfer, *Chem. Phys. Lett.* **139**, 571–5 (1987).

[212] Smith E. W., Giraud M., Cooper J. A semiclassical theory for spectral line broadening in molecules, *J. Chem. Phys.* **65**, 1256–67 (1976).

[213] Green S., Monchik L., Goldflam R., Kouri D. J. Computational tests of angular momentum decoupling approximations for pressure broadening cross sections, *J. Chem. Phys.* **66**, 1409–12 (1977).

[214] Jammu K. S., St. John G.E., Welsh H. L. Pressure broadening of the rotational Raman lines of some simple gases, *Can. J. Phys.* **44**, 797–814 (1966).

[215] Burshtein A. I., Storozhev A. V. The quantum theory of collapse of the isotropic Raman scattering spectrum, *Chem. Phys.* **135**, 381–9 (1989).

[216] Kistemaker P. G., de Vries A. E. Rotational relaxation times in nitrogen–noble-gas mixtures, *Chem. Phys.* **7**, 371–82 (1975).

[217] Gordon R. G. On the absorption and dispersion of sound in molecular gases, *Physica* **34**, 398–412 (1967).

[218] Gidiotis G., Forst W. Master equation solution in terms of bulk observations, *Chem. Phys. Lett.* **120**, 381–7 (1985).

[219] Hirschfelder J. O., Curtiss C. F., Bird R. B. *Molecular Theory of Gases and Liquids.* (Wiley, New York) (1954).

[220] De Pristo A. E., Rabitz H. Scaling theoretic deconvolution of bulk relaxation data: state-to-state rates from pressure-broadening linewidths, *J. Chem. Phys.* **68**, 1981–7 (1978).

[221] Belikov A. E., Solovjev I. Yu., Sukhinin G. I., Sharafutdinov R. G. Rotational relaxation time of nitrogen in argon, *J. Appl. Mech. and Techn. Phys.* **28**, 533–9 (1987) [*Zh. Prikl. Mekh. i Tekhn. Fiz.* **28**, 131–8 (1987)].

[222] Russel J. D., Bernstein R. B., Curtiss C. F. Transport properties of a gas of diatomic molecules. VI. Classical trajectory calculations of the rotational relaxation time of the Ar-N_2 system, *J. Chem. Phys.* **57**, 3304–7 (1972).

[223] Gelb A., Kapral R. Rotational relaxation in the Ar-N_2 system, *Chem. Phys. Lett.* **17**, 397–400 (1972).

[224] Belikov A. E., Dubrovskii G. V., Zarvin A. E., Karelov N. V., Pavlov V. A., Skovorodko P. A., Sharafutdinov R. G. Rotational relaxation of nitrogen in a free jet of argon, *J. Appl. Mech. and Techn. Phys.* **27**, 643–51 (1986) [*Zh. Prikl. Mekh. i Tekhn. Fiz.* **27**, 19–29 (1986)].

[225] Koura K. Rotational distribution of N_2 in Ar free jet, *Phys. Fluids* **24**, 401–5 (1981).

[226] Gianturco F. A., Serna S., Sanna N. Dynamical decoupling in the quantum calculations of transport coefficients. I. Coupled state results for He–N_2 gaseous mixture, *Mol. Phys.* **74**, 1071–87 (1991).

[227] Kozlov D. N., Pykhov R. L., Smirnov V. V., Vereschagin K. A., Burshtein A. I., Storozhev A. V. Rotational relaxation of nitrogen in argon: collisional broadening of Q-branch components in coherent Raman spectra of cooled gas, *J. Raman Spectr.* **22**, 403–7 (1991).

[228] Farrow R. L., Rahn L. A. Interpreting coherent anti-Stokes Raman spectra measured with multimode Nd:YAG pump lasers, *J. Opt. Soc. Am.* **B2**, 903–7 (1985).

[229] Bowers M. S., Tang K. T., Toennies J. P. The anisotropic potentials of He-N_2, Ne-N_2, and Ar-N_2, *J. Chem. Phys.* **88**, 5465–74 (1988).

[230] Golubev N. S., Orlova N. D., Khamitov R. Contours of isotropic Raman bands and rotational relaxation of CO and N_2 molecules in dense gas mixtures, *Opt. & Spectr.* **62**, 594–7 (1987) [*Optika i Spectr.* **62**, 1005–10 (1987)].

[231] Koszykowski M. L., Farrow R. L., Palmer R. E. Calculation of collisionally narrowed coherent anti-Stokes Raman spectroscopy spectra, *Opt. Lett.* **10**, 478–80 (1985).

[232] Sala J. P., Bonamy J., Robert D., Lavorel B., Millot G., Berger H. A rotational thermalization model for the calculation of collisionally narrowed isotropic Raman scattering spectra – application to the SRS N_2 Q-branch, *Chem. Phys.* **106**, 427–39 (1986).

[233] Rahn L. A., Palmer R. E., Koszykowski M. L., Greenhalgh D. A. Comparison of rotationally inelastic collision models for Q-branch Raman spectra of N_2, *Chem. Phys. Lett.* **133**, 513–6 (1987).

[234] Lavorel B., Millot G., Bonamy J., Robert D. Study of rotational relaxation fitting laws from calculation of SRS N_2 Q-branch, *Chem. Phys.* **115**, 69–78 (1987).

[235] Bonamy L., Bonamy J., Robert D., Lavorel B., Saint-Loup R., Chaux R., Santos J., Berger H. Rotationally inelastic rates for N_2–N_2 system from a scaling theoretical analysis of the stimulated Raman Q-branch, *J. Chem. Phys.* **89**, 5568–77 (1988).

[236] Looney J. P., Rosasco G. J., Rahn L. A., Hurst W. S., Hahn J. W. Comparison of rotational relaxation rate laws to characterize the Raman Q-branch spectrum of CO at 295 K, *Chem. Phys. Lett.* **161**, 232–8 (1989).

[237] Burshtein A. I., Kolomoitsev D. V., Nikitin S. Yu., Storozhev A. V. Manifestation of adiabaticity and strength of rotational inelastic collisions in time domain CARS spectra of nitrogen, *Chem. Phys.* **150**, 231–5 (1991).

[238] Laubereau A., Kaizer W. Vibrational dynamics of liquids and solids investigated by picosecond light pulses, *Rev. Mod. Phys.* **50**, 607–65 (1978).

[239] Kolomoitsev D. V., Nikitin S. Yu. Analysis of experimental data on nonstationary active spectroscopy of molecular nitrogen in the strong-collision approximation, *Opt. & Spectr.* **66**, 165–8 (1989) [*Optika i Spectr.* **66**, 286–93 (1989)].

[240] Rosasco J. L., Lempert W., Hurst W. S., Fein A. Line interference effects in the vibrational Q-branch spectra of N_2 and CO, *Chem. Phys. Lett.* **97**, 435–40 (1983).

[241] Rahn L. A., Palmer R. E. Studies of nitrogen self-broadening at high temperature with inverse Raman spectroscopy, *J. Opt. Soc. Am.* B **3**, 1164–9 (1986).

[242] Prangsma G. L., Alberga A. H., Beenakker J. J. M. Ultrasonic determination of the volume viscosity of N_2, CO_2, CH_4, and CD_4 between 77 and 300 K. *Physica* **64**, 278–88 (1973).

[243] Koszykowski M. L., Rahn L. A., Palmer R. E., Coltrin M. F. Theoretical and experimental studies of high-resolution inverse Raman spectra of N_2 at 1–10 atm, *J. Phys. Chem.* **91**, 41–6 (1987)

[244] Bonamy L., Thuet J. M., Bonamy J., Robert D. Local scaling analysis of state-to-state rotational energy-transfer rates in N_2 from direct measurements, *J. Chem. Phys.* **95**, 3361–70 (1991).

[245] Temkin S. I., Thuet J. M., Bonamy L., Bonamy J., Robert D. Angular momentum and rotational energy relaxation in N_2–N_2 collisions calculated from coherent and stimulated Raman spectroscopy data, *Chem. Phys.* **158**, 89–104 (1991).

[246] Bulgakov Yu. I., Storozhev A. V., Strekalov M. L. Comparison and analysis of rotationally inelastic collision models describing the Q-branch collapse at high density, *Chem. Phys.* **177**, 145–55 (1993).

[247] Sitz G. O., Farrow R. L. Pump-probe measurements of state-to-state rotational energy transfer rates in N_2 ($v = 1$), *J. Chem. Phys.* **93**, 7883–93 (1990).

[248] Turfa A. F., Knaap H. F. P., Thijsse B. J., Beenakker J. J. M. A classical dynamics study of rotational relaxation in nitrogen gas, *Physica,* **A112**, 18–28 (1982).

[249] Miller D. R., Andres R. P. Rotational relaxation of molecular nitrogen, *J. Chem. Phys.* **46**, 3418–23 (1967).

[250] Wang C. H., Wright R. B. Effect of density on the Raman scattering of molecular fluids. I. A detailed study of the scattering polarization, intensity, frequency shift, and spectral shape in gaseous N_2, *J. Chem. Phys.* **59**, 1706–12 (1973).

[251] Lavorel B., Oksengorn B., Fabre D., Saint-Loup R., Berger H. Stimulated Raman spectroscopy of the Q-branch of nitrogen at high pressure: collisional narrowing and shifting in the 150–6800 bar range at room temperature, *Mol. Phys.* **75**, 397–413 (1992).

[252] Kroon R., Baggen M., Lagendijk A. Vibrational dephasing in liquid nitrogen at high densities studied with time-resolved stimulated Raman gain spectroscopy, *J. Chem. Phys.* **91**, 74–8 (1989).

[253] Kroon R., Sprik R., Lagendijk A. Vibrational dephasing in highly compressed liquid nitrogen studied by time-resolved stimulated Raman gain spectroscopy, *Chem. Phys. Lett.* **161**, 137–40 (1989).

[254] Baranger M. Problem of overlapping lines in the theory of pressure broadening, *Phys. Rev.* **111**, 494–504 (1958).

[255] Nikitin E. E., Burshtein A. I. Relaxation and depolarization of atomic states due to collisions, In: *'Gas lasers'* (Nauka, Novosibirsk) pp. 7–58 (1977).

[256] Strekalov M. L., Burshtein A. I. Quasiresonant spectral line broadening in the impact approximation, *Sov. Phys. JETP* **43**, 448–53 (1976) [*ZhETF* **70**, 862–71 (1976)].

[257] Temkin S. I., Burshtein A. I. On the shape of the Q-branch of Raman scattering spectra in dense media. Anisotropic scattering, *Chem. Phys. Lett.* **66**, 62–4 (1979).

[258] Perchard J. P., Murphy W. F., Bernstein H. J. Raman and Rayleigh spectroscopy and molecular motions. III. Self-broadening and broadening by inert gases of hydrogen halide gas spectra, *Mol. Phys.* **23**, 535–45 (1972).

[259] Le Duff Y. Raman band shape of the N_2 molecule dissolved in liquids, *J. Chem. Phys.* **59**, 1984–7 (1973).

[260] Ivanov A. A., Kamalov V. F., Koroteev N. I., Orlov R. Yu. Nonlinear and luminescence spectroscopy of vibrationally and electronically excited

oxygen in liquids, *Bulletin Acad. Sci. USSR. Physical Series.* **50**, 1238–44 (1987) [*Izv. AN ser. Fiz.* **50**, 188–94 (1987)].
[261] Kiefte H. Clouter M. J., Rich N. H., Ahmad S. F. Pure vibrational Raman spectra of some simple molecular crystals: γ-O_2, O_2 in hcp Ar, β-N_2, *Chem. Phys. Lett.* **70**, 425–9 (1980).
[262] McClung R. E. D. Rotational diffusion of symmetric top molecules in liquids, *J. Chem. Phys.* **57**, 5478–91 (1972).
[263] Bliot F., Abbar C., Constant E. Absorption dipolaire présentée par les molécules toupies symétriques en phase gazeuse condensée dans le domaine hertzien et infra-rouge lointain, *Mol. Phys.* **24**, 241–53 (1972).
[264] Eagles T. E., McClung R. E. D. Reorientational correlation functions and memory functions in the *J*-diffusion limit of the extended rotational diffusion model, *Chem. Phys. Lett.* **22**, 414–18 (1973).
[265] Eagles T. E., McClung R. E. D. Rotational diffusion of spherical top molecules in liquids and gases. IV. Semiclassical theory and applications to the ν_3 and ν_4 band shapes of methane in high pressure gas mixtures, *J. Chem. Phys.* **61**, 4070–82 (1974).
[266] Coulon R. Thèse, Paris, L'Université de Paris, p. 22 (1958).
[267] Campbell J. H., Seymour S. J., Jonas J. Reorientational and angular momentum correlation times in gaseous tetrafluoromethane at moderate densities, *J. Chem. Phys.* **59**, 4151–6 (1973).
[268] Kluk E., Powles J. P. A friction model for reorientation of linear molecules in fluids, *Mol. Phys.* **30**, 1109–16 (1975).
[269] Kluk E. Hubbard relation for linear molecules, *Mol. Phys.* **30**, 1723–8 (1975).
[270] Jameson C. J., Jameson A. K., ter Horst M. A. ^{14}N spin relaxation studies of N_2 in buffer gases. Cross sections for molecular reorientation and rotational energy transfer, *J. Chem. Phys.* **95** 5799–808 (1991).
[271] Speight P. A., Armstrong R. L. Nuclear spin relaxation, *Can. J. Phys.* **47**, 1475–83 (1969).
[272] Gillen K. T., Douglass D. C., Griffiths J. E. Rotating frame NMR relaxation study of gaseous ClF, *J. Chem. Phys.* **69**, 461–7 (1978).
[273] Bonamy L., Nguyen Minh Hoang P. Far infrared absorption of diatomic polar molecules in simple liquids and statistical properties of the interactions. I. Spectral theory, *J. Chem. Phys.* **67**, 4423–30 (1977); Nguyen Minh Hoang P., Bonamy L. Far infrared absorption of diatomic polar molecules in simple liquids and statistical properties of the interactions. II. Statistical theory and spectral application, *J. Chem. Phys.* **67**, 4431–40 (1977).
[274] Burshtein A. I., Serebrennikov Yu. A., Temkin S. I. Orientational relaxation of a rotator in a liquid cell, *Khim. Fiz.* **11**, 1456–63 (1982).
[275] Burshtein A. I., Serebrennikov Yu. A., Temkin S. I. The influence of the cell field fluctuations on the shape of the infrared spectrum of linear molecules in solutions, *Khim. Fiz.* **1**, 10–20 (1983).
[276] Burshtein A. I. *Molecular Physics.* Nauka, Novosibirsk (1986) [in Russian].
[277] Burshtein A. I., Veksler L. S., Shokhirev N. V. Origin and the position of the extremal values of the intermolecular interaction in simple liquids, *J. Struct. Chem.* **18**, 384–98 (1977).
[278] Bartoli F. J., Litovitz T. A. Raman scattering: Orientational motion in liquids, *J. Chem. Phys.* **56**, 413–25 (1972).

[279] Frenkel D., van der Elsken J. Anisotropic density fluctuations in argon at different densities: far infrared measurements and molecular dynamic calculations, *J. Chem. Phys.* **67**, 4243–53 (1977).
[280] Kouzov A. P. Nonadiabatic action of kinetic noise on an isolated spectral line, *Opt. & Spectr.* **49**, 554–5 (1980).
[281] Holzer W., Le Duff Y., Ouillon R. Structure rotationelle du spectre Raman de H_2 et D_2 dissous dans SF_6 liquide, *C. R. Acad. Sci., Paris* **B273**, 313–16 (1971).
[282] Le Duff Y., Holzer W. Raman band shapes of small diatomic molecules dissolved in inert liquids, *Chem. Phys. Lett.* **24**, 212–16 (1974).
[283] Orlova N. D., Pozdnyakova L. A., Khodos E. B. Spectrum of Raman scattering of hydrogen in some liquid solutions, *Opt. & Spectr.* **37**, 340–1 (1974).
[284] Orlova N. D., Pozdnyakova L. A. Vibrational and rotational spectra of H_2 and D_2 in solutions in the 90–313 K temperature range, *Opt. & Spectr.* **48**, 594–8 (1980).
[285] Huong P., Couzi M., Perrot M. Molecular motions of hydrogen halides and deuterium halides in liquid sulfur hexafluoride, *Chem. Phys. Lett.* **7**, 189–90 (1970).
[286] Birnbaum G. Quantized rotational motion in liquids: Far infrared rotational spectrum of HF and NH_3 in liquid SF_6, *Mol. Phys.* **25**, 241–5 (1973).
[287] Birnbaum G., Ho W. Pure rotational spectrum of HCl and DCl in liquid SF_6, *Chem. Phys. Lett.* **5**, 334–6 (1970).
[288] Perchard J., Murphy W., Bernstein H. Rotational Raman spectrum of HCl, DCl, HBr in solution, *Chem. Phys. Lett.* **8**, 559–61 (1971).
[289] van Aalst R., van der Elsken J. The far infrared spectra of hydrogen chloride in liquid argon and in liquid krypton, *Chem. Phys. Lett.* **13**, 631–3 (1972).
[290] van Aalst R., van der Elsken J. The far infrared spectrum of hydrogen chloride in liquid xenon, *Chem. Phys. Lett.* **23**, 198–9 (1973).
[291] Menger P., van der Elsken J. Four time density correlations around a dissolved HCl molecule in dense argon, krypton and xenon as determined from linewidth data, *J. Chem. Phys.* **75**, 17–21 (1981).
[292] Bulanin M. O., Orlova N. D. Spectroscopical investigations of molecular rotational motion in condensed media, In: *Spectroscopy of Interacting Molecules,* ed. M. O. Bulanin, Leningrad State University, Leningrad, pp. 55–97 (1970).
[293] Temkin S. I., Abdrakhmanov B. M. Does the Hubbard relation hold in the liquid cage model? *Phys. Lett.* **A155**, 43–8 (1991).
[294] White J. A., Velasco S., Calvo Hernandez A., Luis D. On the quantum time autocorrelation function, *Phys. Lett.* **A130**, 237–9 (1988).
[295] Gantmakher F. R. *The Theory of Matrices.* (Chelsea, New York) (1960).
[296] Freed J. H. Stochastic-molecular theory of spin-relaxation for liquid crystals, *J. Chem. Phys.* **66**, 4183–99 (1977).
[297] Dozov I., Temkin S., Kirov N. Two-stage model for molecular orientational relaxation in liquid crystals, *Liq. Crystals* **8**, 727–38 (1990).
[298] Bonamy L., Bonamy J., Robert D., Temkin S. I. Consequences of angular momenta coupling on generalized spectroscopic relaxation cross-sections: collisional narrowing in isotropic and anisotropic Raman CO_2 branches, *Proc. 13th ICORS.* (Wiley & Sons, New York) (1992).

Index